AF355967

Integrating Science into Aquatic Conservation

Integrating Science into Aquatic Conservation

Editors

Cong Zeng
Deliang Li

Basel • Beijing • Wuhan • Barcelona • Belgrade • Novi Sad • Cluj • Manchester

Editors
Cong Zeng
Shanghai Jiao Tong University
Shanghai
China

Deliang Li
Hunan Agricultural University
Changsha
China

Editorial Office
MDPI
St. Alban-Anlage 66
4052 Basel, Switzerland

This is a reprint of articles from the Special Issue published online in the open access journal *Biology* (ISSN 2079-7737) (available at: https://www.mdpi.com/journal/biology/special_issues/aquatic_conservation).

For citation purposes, cite each article independently as indicated on the article page online and as indicated below:

Lastname, A.A.; Lastname, B.B. Article Title. *Journal Name* **Year**, *Volume Number*, Page Range.

ISBN 978-3-7258-0221-0 (Hbk)
ISBN 978-3-7258-0222-7 (PDF)
doi.org/10.3390/books978-3-7258-0222-7

Contents

About the Editors

Cong Zeng

Cong Zeng is an associate researcher and doctoral supervisor at Shanghai Jiao Tong University and an adjunct researcher at the Polar Research Institute of China. His research focuses on the conservation and sustainable use of aquatic organisms, and he has presided over a number of projects, including the National and Provincial Natural Science Foundation. So far, he has won the Shen Nong China Agricultural Science and Technology Award, published more than 60 papers, and authored 5 patents.

Deliang Li

Deliang Li is a Professor and doctoral supervisor at Hunan Agricultural University and is also an expert of Hunan Province modern aquatic industry technical systems. He has presided over more than 10 national and provincial research projects such as the National Natural Science Foundation, the Science and Technology Innovation Project of the Ministry of Science and Technology, the public welfare industry research project of the Ministry of Agriculture, and the key research and development project of the Science and Technology Department of Hunan Province. He has also published more than 30 research papers and obtained 2 patents.

Preface

As a key support for human survival and development, how to protect aquatic ecosystems is a hotspot of global concern. This Special Issue was built with the hope of providing scientific references for the conservation of aquatic organisms and ecosystems. Papers cover both freshwater and marine ecosystems, focusing on a single species and multiple species or specific ecosystems. For single-species studies mainly used molecular tools and statistic models to focus on the evolutionary relationships among species, the differentiations among populations within species, and the responses of species to the environment changes. The results of these studies shed light on the evolution, spatial dispersal, and future changes of these species and provided recommendations for the management of population dynamics in these study populations. In multi-species research, biophysical modeling, acoustic technology, and ground surveys were mainly used to predict biodiversity in ecosystems, spatial distribution of different species, and population connectivity among communities. The results of these multi-species studies revealed the spatial patterns of biodiversity, community composition, and ecological corridors from different perspectives, which offer a direct reference for the selection and delineation of marine protected areas. In summary, these papers utilized different tools to reveal the changes or threads faced by important components of water ecosystems from the micro to macro level and provide scientific advice for the conservation and management of protected animals and ecosystems on different spatial scales from local to global.

Cong Zeng and Deliang Li
Editors

Article

Genetic Signature of *Pinctada fucata* Inferred from Population Genomics: Source Tracking of the Invasion in Mischief Reef of Nansha Islands

Binbin Shan [1,2,3], Gang Yu [3], Liangming Wang [2,3], Yan Liu [1,2,3], Changping Yang [2,3], Manting Liu [3] and Dianrong Sun [2,3,*]

[1]   Tropical Aquaculture Research and Development Center, South China Sea Fisheries Research Institute, Chinese Academy of Fishery Sciences, Sanya 572000, China
[2]   Key Laboratory of Marine Ranching, Ministry of Agriculture Rural Affairs, Guangzhou 510300, China
[3]   South China Sea Fisheries Research Institute, Chinese Academy of Fisheries Sciences, Guangzhou 510300, China
*   Correspondence: sundianrong@yeah.net; Tel.: +86-020-8910-0850

Simple Summary: In the present study, we employed population genetics analysis to investigate possible origins of the introduced *Pinctada fucata* population in the Mischief Reef of the South China Sea, as well as diversity and structure of the *P. fucata* populations. Population genomics data clearly revealed a closer genetic relationship between the Mischief Reef introduced population and the Lingshui population, indicating that Lingshui may be the potential geographical origin. Furthermore, several selected genomic regions and genes of the introduced population were identified, some of which may play important roles in the adaptation of temperature and salinity tolerance.

Abstract: Among the anthropogenic stresses that marine ecosystems face, biological invasions are one of the major threats. Recently, as a result of increasingly intense anthropogenic disturbance, numerous marine species have been introduced to their non-native ranges. However, many introduced species have uncertain original sources. This prevents the design and establishment of methods for controlling or preventing these introduced species. In the present study, genomic sequencing and population genetic analysis were performed to detect the geographic origin of the introduced *Pinctada fucata* population in the Mischief Reef of the South China Sea. The results of population genetic structure analysis showed a close relationship between the Mischief Reef introduced population and the Lingshui population, indicating that Lingshui may be the potential geographical origin. Furthermore, lower heterozygosity and nucleotide diversity were observed in the introduced population in Mischief Reef, indicating lower genetic diversity than in other native populations. We also identified some selected genomic regions and genes of the introduced population, including genes related to temperature and salinity tolerance. These genes may play important roles in the adaptation of the introduced population. Our study will improve our understanding of the invasion history of the *P. fucata* population. Furthermore, the results of the present study will also facilitate further control and prevention of invasion in Mischief Reef, South China Sea.

Keywords: Mischief Reef; *Pinctada fucata*; introduced population; geographic origin; population genetics

Citation: Shan, B.; Yu, G.; Wang, L.; Liu, Y.; Yang, C.; Liu, M.; Sun, D. Genetic Signature of *Pinctada fucata* Inferred from Population Genomics: Source Tracking of the Invasion in Mischief Reef of Nansha Islands. *Biology* 2023, 12, 97. https://doi.org/10.3390/biology12010097

Academic Editors: Cong Zeng and Deliang Li

Received: 4 December 2022
Revised: 2 January 2023
Accepted: 6 January 2023
Published: 9 January 2023

1. Introduction

Marine ecosystems are threatened by a number of anthropogenic stresses, including overexploitation of living resources, pollution, climate change, biological invasion and others [1]. Numerous studies have documented that, among these pressures, bioinvasion is one of the major threats that may cause significant changes and degradation of marine biodiversity [2,3]. In general, a biological invasion occurs when species are outside their native range, expanding their original ecological spaces and impacting local biodiversity and resources [4]. In recent decades, marine bioinvasions have become increasingly frequent as a result of anthropogenic disturbance, leading to species translocations through aquaculture, hull fouling, ballast water, etc. [5,6]. The introduction of invasive aquatic species into new habitats has been identified as one of the greatest threats to the world's oceans [7]. Given the accelerating global threat to marine ecosystems, researchers and managers are increasingly focused on tracking, preventing and managing invasions.

Pearl oysters belonging to the genus *Pinctada* (Bivalvia: Pteriidae) are widely distributed between the Indo-Pacific and Western Atlantic tropical and subtropical shallow-water areas, most of which are associated with reef environments [8]. *Pinctada fucata* (Gould, 1850), also named *P. martensii*, is an economically important bivalve native to the coastal waters of the Tropic of Cancer and the Tropic of Capricorn in the Pacific Ocean [9]. It is predominantly cultured for pearl production in China, Japan and Korea [10–12].

Mischief Reef is located in the eastern Nansha Islands of China (Figure 1). In recent decades, with increased development in this area (reclamation, aquaculture and fishery), another species of *Pinctada* can be found in this area. Wang [13] and Chen [14] reported that *Pinctada maculata* was a unique species of the genus *Pinctada* in the Nansha Islands sea area. However, during a recent sampling at Mischief Reef, *P. fucata* was found in extreme abundance. According to the local fishermen, the occurrence of *P. fucata* in Mischief first occurred in 2016 (Figure S1). Even though Mischief Reef is within the theoretical distribution range of *P. maculata*, it is still a non native speices. Similar to many other notable invaders, *P. fucata* has not caused significant ecological or economic problems in its native area. However, as a potential invader, it is unknown whether the introduction and potential subsequent spread of *P. fucata* will damage the ecosystem of Mischief Reef or not.

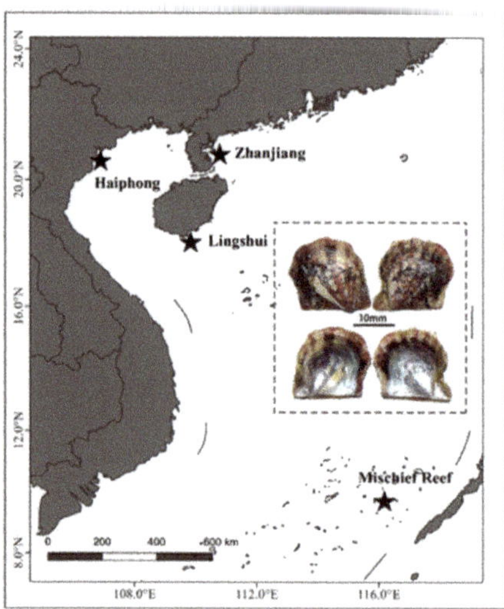

Figure 1. The sample sites in the present study. Stars represent the sampling sites.

In the management of marine ecosystems, reconstructing the complex history of invasions is essential and plays a crucial role in preventing subsequent spread [15,16]. A crucial piece of information in the reconstruction of the invasion history is the investigation of invaders' geographical origin [17]. Furthermore, if the sources and pathways responsible for the introduction could be investigated, then a monitoring and control strategy could be established [3]. However, in practice, such investigation is extremely difficult via ecological surveys or other observational approaches. For bivalve species, their soft tissues and shell morphology are not suitable for identifying source populations or cryptogenic taxa [18].

The development and common application of molecular markers in evolutionary and ecological studies has facilitated the tracking of original sources of these invaders, sometimes with surprising precision [16,19]. Recent years have seen a dramatic rise in the tracing of the biogeographical history of many invasive species in marine systems [20–22]. For instance, combined population genetic and phylogeographic analyses provided evidence of multiple complex invasions of the European green crab *Carcinus maenas* on almost every continent with temperate shores [20]. Another notable case was the invasion of the Pacific oyster *Crassostrea gigas* in NW Europe. Molecular approaches have been effectively utilized when exploring the invasion histories of Pacific oysters [22].

In the present study, we used a population genomics approach utilizing restriction-site associated DNA (RAD) sequencing and comparative genomics to detect the geographic origin of invasive *P. fucata* in Mischief Reef, describing the present invasive population's genomic diversity. Furthermore, we aimed to identify selected genomic regions and genes of the introduced population of *P. fucata* and identify some genes related to environmental adaptation. Our study will improve understanding of the invasion history of *P. fucata* in Mischief Reef.

2. Materials and Methods

2.1. Sampling and Sequencing

From 2017 to 2018, a total of 74 *P. fucata* individuals were obtained from four geographic sites, including the native range (Zhanjiang, $n = 24$; Lingshui, $n = 14$; Haiphong, $n = 18$) and non-native range (Mischief Reef, $n = 18$) (Figure 1). All specimens were frozen at $-80\ °C$ and genomic DNA was extracted from adductor muscle using a TIANGEN TIANamp Marine Animals DNA Kit. Briefly, the RAD-seq library was prepared following the protocol after DNA quality assessment [23]. Then, sequencing was performed on the BGISEQ-500RS (BGI, Shenzhen, China) using a 100-bp paired-end strategy.

2.2. Alignment and Variation Calling

BWA-mem (version: 0.7.17, https://github.com/lh3/bwa, accessed on 24 October 2017) was applied to align the clean reads (default parameters) to the *P. fucata* reference genome sequence [24]. Then, these reads were sorted by the SAMtools "sort" function (version: 1.9, https://github.com/samtools/samtools/releases, accessed on 19 July 2018) [25]. GATK tools version 4.1.2.0 (BaseRecalibrator, HaplotypeCaller and CombineGVCFs, https://github.com/broadinstitute/gatk/releases, accessed on 24 April 2019) was utilized to identify and call high-quality SNPs, with refinement stages similar to those of Todesco et al. [26,27]. Finally, GATK-VariantFiltration was used to filter high-quality SNPs by the following criteria: MQ < 40.0, QD < 2.0, FS > 60.0, SOR > 12.0, MQRankSum < -12.5, ReadPosRankSum < -8.0.

2.3. Population Genetics Analysis

For all *P. fucata* populations, θ_w, θ_π and Tajima's D values were calculated for the assessment of genetic diversity using a sliding window approach and the step and window sizes were set as 5 kb and 10 kb, respectively [28,29]. To measure the differentiation between the four *P. fucata* populations pairwise, the fixation index (F_{ST}) values were also calculated in the same windows [30]. We assessed the genetic relationships between the *P. fucata* population from Mischief Reef and other populations. Principal component analysis was conducted via PLINK (version 2.0, www.cog-genomics.org/plink/2.0/, accessed on 29 January 2022) and GCTA (version 1.93.0, https://yanglab.westlake.edu.cn/software/gcta/#Download, accessed on 9 December 2019) [31]. To estimate the genetic diversity of the *P. fucata* population, PLINK (version 2.0) was also used to calculate the expected heterozygosity (H_e) and observed heterozygosity (H_o). Furthermore, admixture analysis was performed to detect population structure by using ADMIXTURE (version: 1.3.0) [32]. The number of ancestral clusters was assumed to range from 2 to 10.

We also evaluated the genetic connectivity between the four populations by examining the gene flow between them. The analysis was performed based on the directional migration method by estimating the relative migration rates derived from G_{st}, N_m and D [33–35]. The relative migration rates and network were determined by using the function divMigrate from the R package diveRsity, with a bootstrap value of 1000 and a filter threshold of 0.2 [36].

Furthermore, we used VCF2Dis (version 1.47, https://github.com/BGI-shenzhen/VCF2Dis, accessed on 25 July 2022) to calculate the pairwise genetic distances (p distance). Then, PHYLIPNEW (version 3.69, http://evolution.genetics.washington.edu/phylip.html, accessed on 2 October 2019) was utilized to construct an NJ tree (Neighbour-Joining tree). The squared Pearson's correlation coefficient (r^2) was calculated to evaluate the linkage disequilibrium (LD) in four *P. fucata* populations. The analysis was performed by using PopLDdecay version 3.40 (https://github.com/BGI-shenzhen/PopLDdecay/releases, accessed on 16 January 2019) [37]. Then, we measured LD decay based on the distance of each pair of SNPs and r^2 values.

2.4. Potential Selected Regions in the Introduced Population

To identify selected genomic regions and genes of the introduced population of *P. fucata*, we screened the SNPs called across the introduced population and its potential original population (Lingshui population). Then, we estimated pairwise differentiation between the two populations and Tajima's D, F_{ST} and ROD ($1 - \theta_{\pi\text{-}Mischief}/\theta_{\pi\text{-}Lingshui}$) were calculated by using a sliding window approach. The window size was 10 kb and the step size was 5 kb. In our study, an empirical threshold of 5% was set to identify outlier windows of F_{ST} (top 5%), Tajima's D (bottom 5%) and ROD (top 5%) values. The outlier windows were considered candidate regions for selection with the following criteria: the top 5% of F_{ST} values or the top 5% of ROD values with the bottom 5% of negative Tajima's D values.

3. Results

3.1. Variant Calling and Population Genetic Diversity

With the sequencing of 74 samples from four sites (Figure 1), we generated a total of 1207.53 million reads (16.32 million reads per sample) after quality filtering (Supplementary Table S1). Then, all clean reads were mapped against the *Pinctada martensii* reference genome (ftp://parrot.genomics.cn/gigadb/pub/10.5524/100001_101000/100240/, accessed on 18 October 2021). The average mapping rate was 94.89% (range from 93.63% to 95.15%). In the present study, we selected the top 30 samples with the highest sequencing coverage for calibration. SNPs with a missing rate less than 50% and MAF greater than 0.01 were utilized for the downstream analysis. Then, we obtained 3,476,259 SNPs. Of these SNPs, 254,373 SNPs (7.32%) were aligned to exonic regions, 1,017,991 SNPs (29.28%) were aligned to intronic regions and 1,956,307 SNPs (56.28%) were aligned to intergenic regions. Among the 254,373 SNPs in exonic regions, 100,402 were nonsynonymous and 147,749 were synonymous.

The genetic diversity of the four populations was assessed by estimating the nucleotide diversity. As shown in Table 1, the Lingshui and Haiphong populations exhibited the highest genetic diversity, both with θ_w and θ_π values of 0.005. Among the four populations, the θ_w and θ_π of the nonlocal population (Mischief Reef) were lowest, showing a lower gene diversity than in the other three populations (Figure 2A). The results of heterozygosity showed the highest genetic diversity in the Haiphong population and the lowest genetic diversity in the Mischief Reef population.

Table 1. Population genetic parameters of the four populations.

	H_e	H_o	θ_w	θ_π	Tajima's D
Mischief	0.2377	0.0942	0.0037	0.0034	−0.2932
Zhanjiang	0.2403	0.0967	0.0044	0.0045	0.0347
Lingshui	0.2502	0.1233	0.0049	0.0045	−0.3231
Haiphong	0.2622	0.1556	0.0048	0.0051	0.2054

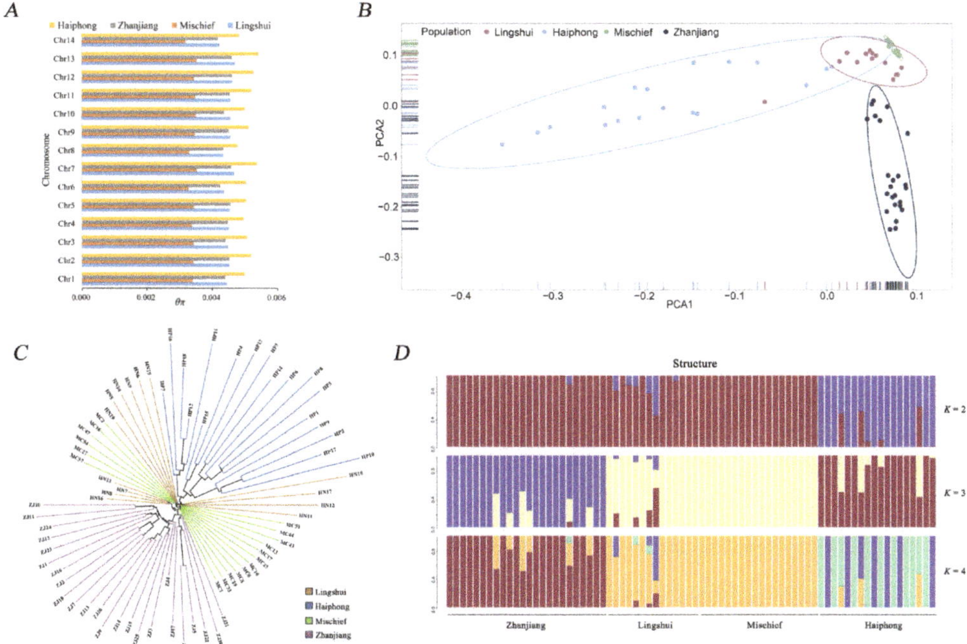

Figure 2. Population genetic diversity and structure of four *P. fucata* populations. (**A**). θ_π values of different chromosomes of the four *P. fucata* populations. (**B**). PCA plot of the first two principal components based on all SNPs. (**C**). The neighbor-joining phylogenetic tree of all samples. (**D**). The structure analysis graph under different *K* values.

3.2. Population Genetic Structure

The results of PCA showed a clear genetic relationship among the four *P. fucata* populations (Figure 2B). The top three principal components (PCs) explained 4.62%, 3.19% and 2.94% of the variation, respectively. In the plot, the populations from Lingshui and Mischief Reef were clustered into a group and those from Zhanjiang and Haiphong were classified into two separate groups. In addition, an NJ phylogenetic tree was constructed (Figure 2C). The topological structure of the tree showed that the Mischief Reef population was more similar to the Lingshui population than the other two populations, indicating a close relationship between the Lingshui population and the introduced population at Mischief Reef.

To further investigate the relationships between the introduced population and other populations, we performed admixture analysis with an assumed number of ancestral clusters from two to six (Figure 2D). The results showed that the cross-validation error (CV error) reached the smallest value when the ancestral cluster number was set as two (Figure S2). However, we considered the estimated cluster number ($K = 2$) unsuitable for two reasons. First, only Haiphong clearly formed an independent group and the introduced population in Mischief Reef was undistinguishable from the Lingshui and Zhanjiang populations. Furthermore, the CV error gap between $K = 2$ and $K = 3$ was not very significant and the second-order rate of change increased after $K = 3$. Thus, the clear genetic relationship when the ancestral cluster number was three provides a more biologically suitable argument than the more conservative number. The fixation index (F_{ST}) was then calculated to measure the genetic difference between the Mischief Reef population and other local populations (Table 2), indicating a closer genetic relationship between the two populations than others.

Table 2. Fixation index (F_{ST}) between different populations.

	Hainan	**Mischief Reef**	**Zhanjiang**
Lingshui			
Mischief	0.048		
Zhanjiang	0.069	0.095	
Haiphong	0.087	0.118	0.102

The relative migration network constructed based on divMigrate analysis indicated gene flow between the four *P. fucata* populations (Figure 3). High and moderate gene flow between the Mischief Reef, Zhanjiang and Lingshui populations were observed. Gene flow between Haiphong and other populations was weak. Among these populations, the Mischief Reef and Lingshui populations showed the strongest migration rates, indicating a closer relationship than between the other populations. However, no strong directionality could be detected.

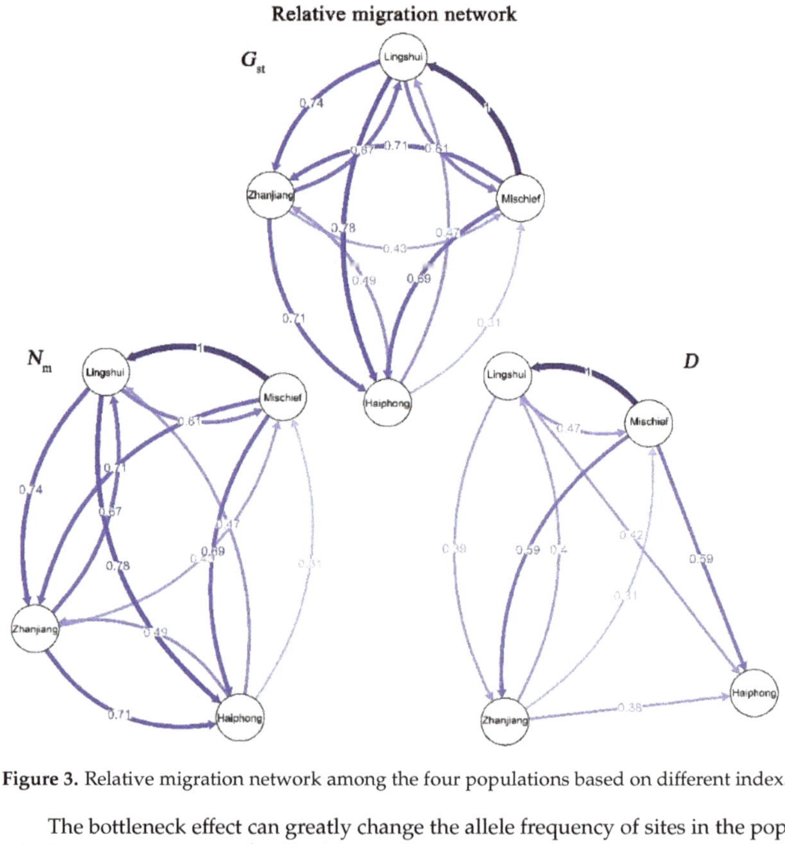

Figure 3. Relative migration network among the four populations based on different index.

The bottleneck effect can greatly change the allele frequency of sites in the population, which is the main reason for the drastic change in Linkage disequilibrium (LD) in a short time [38]. The results of linkage disequilibrium analysis showed that, both over a longer distance ($\geq$20 Kb) and over a short distance ($\leq$20 Kb), the Mischief Reef population had the lowest r^2 value with the fastest LD decay rate among the four populations (Figure 4). Meanwhile, the Haiphong population had the slowest decay rate, indicating potential artificial domestication.

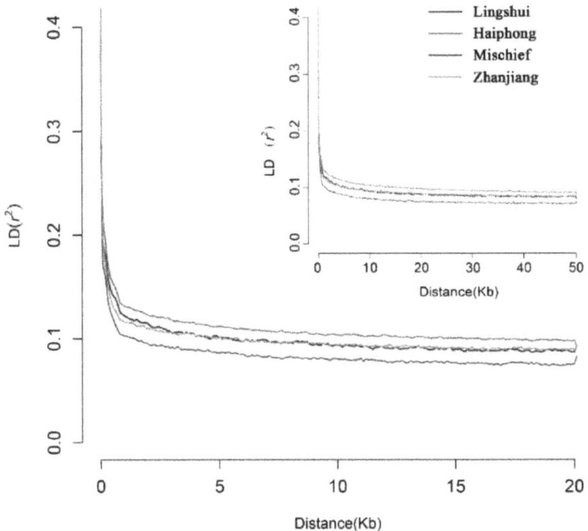

Figure 4. Linkage-disequilibrium decay for different distances of the four populations.

3.3. Population Genetic Structure

In the present study, the values of Tajima's D for populations were significantly different. Among these populations, only the Tajima's D of the Zhanjiang population was close to zero. For the Lingshui population and Mischief Reef population, the values were significantly lower than zero, showing potential selective sweeps.

Furthermore, we identified selected genomic regions and genes of the introduced population of *P. fucata*. As shown in Supplementary Table S2, 448 regions were under selection, containing a total of 481 genes. Of these genes, 82 genes were among the top 5% in terms of F_{ST} values, 436 genes were among the top 5% in terms of ROD values and 37 genes were among the top 5% of both F_{ST} values and ROD values. Among the 481 genes, 416 genes were annotated to the Nr, KEGG or GO database (Supplementary Table S3). Ultimately, we found some notable genes, including the heat shock transcription factor 1 gene, an endoglucanase gene, a sodium channel protein genes and others (Supplementary Table S2).

4. Discussion

In recent decades, numerous invasions of bivalve species have occurred worldwide due to their high fecundity, rapid growth, high filtration rates, early sexual maturation and interactions with human activities [39,40]. The activities of introduced bivalve species, for instance, shell production, bioturbation and filter feeding, may change ecosystem processes and functions and affect biodiversity and the environment [7,41]. Therefore, bivalve invasions have become a serious ecological problem around the world.

The process of invasion can be divided into three distinct phases: initial introduction, establishment and spread in the introduced range [42]. As shown in our investigation, *P. fucata* in Mischief Reef is now in the establishment phase. Although the population does not currently cause any ecological problems in Mischief Reef, *P. fucata* has dispersed beyond its original distribution range. It could be considered a Stage III invasive species in Mischief Reef according to the definition of Collauti and MacIsaac, due to fact that the *P. fucata* population has the potential to establish itself in the Mischief Reef [43].

4.1. Population Genetic Structure and the Potential Origin of the Mischief Reef Population

In the present study, we analysed the population genetic structure of three native *P. fucata* populations and an introduced nonlocal *P. fucata* population. Our work aimed to detect the potential origin of the Mischief Reef population and assess the genetic status of the introduced population. In our study, we selected Lingshui and Zhanjiang populations, because ships or humans from these two sites display more intensive activities. Thus, we considered that the two sites were the potential original source. As shown in Figure 2B, the Mischief Reef population and Lingshui population were clustered into a group, indicating a closer relationship than that of Mischief Reef population with others. Furthermore, both the topology of the NJ phylogenetic tree and the F_{ST} values clearly provided evidence for the close relationship of the Mischief Reef and Lingshui populations. Takeuchi et al. [44] indicated that the western Pacific population could be divided into a southern population (China) and a northern population (Japanese mainland). Overall, the F_{ST} values were low, supporting the results of a previous study [45]. Nevertheless, the F_{ST} values between Lingshui and Mischief Reef were significantly lower than those between other pairs of populations in our study, suggesting that the Lingshui population may be a potential original resource of the introduced population in Mischief Reef. In the structure analysis, $K = 2$ was identified as the most suitable number of clusters. Unfortunately, under the suggested cluster number, only Haiphong could be identified from the mixed population. Then, some clusters coherent with geographical location emerged when the cluster number was increased. The Zhanjiang population first separated from the remaining mixed populations when the cluster number was three. Nevertheless, the Mischief Reef population could not be separated clearly from the Lingshui population when the cluster number was increased continually. Furthermore, the results of gene flow analysis showed a high gene flow between Lingshui and Mischief Reef populations. Overall, all the results of population genetic structure supported the hypothesis that the population in Mischief Reef was potentially introduced from Lingshui.

Previous studies have shown that detecting the precise origin of an introduced population can be extremely challenging [16,45]. In addition to effective approaches, a large area sampling effort is also required to ensure that the real origins are included [16]. Further more, numerous studies have indicated that multiple introductions and multiple origins might be a common phenomenon in most invasions [46]. However, the introduced population in Mischief Reef was less likely to have multiple origins, partly because of political sensitivity. Moreover, in the Mischief Reef population, we did not observe higher genetic variation caused by multiple introductions. Due to the limited sampling size and area, our study has limitations and we could not precisely detect the origin of the Mischief Reef population. However, we reported the presence of the *P. fucata* population in Mischief Reef in the South China Sea for the first time and provided references for further studies reconstructing its introduction history and novel insights into oyster invasions.

4.2. Population Genetic Diversity and LD in P. fucata Populations

In the present study, the nucleotide diversity (θ_π) of the *P. fucata* populations at the whole-genome level was 0.0034–0.0051, indicating significantly higher genetic diversity than in other studies [47]. Huang et al. [48] estimated that cultured *P. fucata* populations in Lingshui and Zhanjiang showed expected heterozygosity values of 0.2636 and 0.2374, respectively. In our study, the expected heterozygosity of the four populations was 0.2377–0.2622. Thus, the genetic diversity of the four populations in our study was moderate. However, the Mischief Reef population showed lower nucleotide diversity than the other populations in the present study. Notably, our data clearly revealed lower genetic diversity in the introduced population than in the potential original population. This result is consistent with previous studies of introduced populations [16]. It is well known that introduced populations experience the loss of genetic variation during the founding event as a result of founder effects and bottlenecks [43,49,50].

As a critical parameter in population genetics, LD could provide deep insight into the complex population history, including selection, hybridization and mutation, of the introduced population during invasion [50]. Unexpectedly, we did not observe a dramatic increase in LD in the introduced population with respect to the native *P. fucata* populations. Furthermore, relatively rapid LD decay was observed in the Mischief Reef population.

Previous studies indicated that a dramatic increase in LD was a common phenomenon in most crop domestication and some invasive organisms [51,52]. We speculated that there were abundant individuals that contributed to the introduction and funding of the Mischief Reef population. Therefore, an increase in LD was not found in the present study. On the other hand, an LD decrease can be attributed to population size and recombination rate [53]. Some researchers have demonstrated that introduced populations may undergo a high level of recombination with mutation as a result of rapid population expansion [50,54]. Hence, we attribute the LD decrease to the rapid expansion of the Mischief Reef population. Furthermore, since the LD decay in the introduced population was rapid, we suspect that the population in Mischief Reef may not have undergone strong natural selection during founding.

4.3. Potential Selected Regions in the Introduced Population

The allele frequency and nucleotide diversity differentiation between populations provide important genetic evidence of selection processes [55]. Meanwhile, Tajima's D represents the frequency of rare variants at a genomic locus. These parameters indicate whether the regions and genes are under selection. In our study, the Tajima's D values of nonnative population and its potential origins were negative and significantly lower than those of other native populations. The results indicated an excess of rare nucleotide site variants compared to the expectation under a neutral model of evolution [56].

Furthermore, we found some genes that were under selection in the introduced population, including the heat shock transcription factor 1 gene, an endoglucanase gene, a serine/threonine-protein phosphatase gene, a sodium channel protein and other genes. However, due to the limited number of selected genes, enrichment pathway analysis could not be performed. According to previous studies, salinity and sea surface temperature are the major factors affecting the distribution of *P. fucata* populations [44,57]. The selected genes are related to temperature and salinity tolerance. Thus, we speculate that these genes play important roles in adaptation to the different environment of Mischief Reef, such as higher salinity and more stable temperature compared with those in Lingshui.

5. Conclusions

In the present study, we analysed the population genetic structure and diversity of three native *P. fucata* populations and an introduced nonlocal *P. fucata* population. Our results revealed a closer relationship between the Mischief Reef introduced population and the Lingshui population, indicating that Lingshui may be the potential geographical origin. Furthermore, our data revealed lower genetic diversity in the introduced population than in the potential original population. In the introduced population, we also identified some selected genomic regions and genes related to temperature and salinity tolerance. Our study will facilitate further aquatic conservation and protection in Mischief Reef, South China Sea.

Supplementary Materials: The following supporting information can be downloaded at: https://www.mdpi.com/article/10.3390/biology12010097/s1, Figure S1: The occurrence of *P. fucata* in Mischief; Figure S2: Cross-validation (CV) error calculated by structure analysis; Table S1: Clean reads statistics; Table S2: Genomic regions and genes under selection of Mischief population; Table S3: The information of potential selected genes; Table S4: The accession information of RAD-seq data in the present study.

Author Contributions: Conceptualization, G.Y., B.S. and Y.L.; methodology, G.Y., B.S. and L.W.; software, B.S. and C.Y.; validation, B.S., Y.L. and C.Y.; formal analysis, B.S. and G.Y. investigation, B.S. and L.W.; resources, G.Y., B.S. and M.L.; data curation, Y.L. and C.Y.; writing—original draft preparation, B.S.; writing—review and editing G.Y., C.Y. and D.S.; visualization, B.S.; supervision, M.L. and D.S.; project administration, G.Y. and D.S.; funding acquisition, D.S. All authors have read and agreed to the published version of the manuscript.

Funding: This study was supported by the Hainan Provincial Natural Science Foundation of China under contract No. 320QN361; Asia Cooperation Fund Project—Modern fishery cooperation between China and neighboring countries around the South China Sea; Biodiversity, germplasm resources bank and information database construction of the South China Sea Project.

Institutional Review Board Statement: Not applicable.

Informed Consent Statement: Not applicable.

Data Availability Statement: All RAD-seq data in the present study were deposited in NCBI. The accession numbers and more details can be found in the Supplementary Table S4.

Conflicts of Interest: The authors declare no conflict of interest.

References

1. Kappel, C.V. Losing pieces of the puzzle: Threats to marine, estuarine, and diadromous species. *Front. Ecol. Environ.* **2005**, *3*, 275–282. [CrossRef]
2. Costello, M.J.; Coll, M.; Danovaro, R.; Halpin, P.; Ojaveer, H.; Miloslavich, P. A Census of Marine Biodiversity Knowledge, Resources, and Future Challenges. *PLoS ONE* **2010**, *5*, e12110. [CrossRef]
3. Ojaveer, H.; Galil, B.S.; Carlton, J.T.; Alleway, H.; Goulletquer, P.; Lehtiniemi, M.; Marchini, A.; Miller, W.; Occhipinti-Ambrogi, A.; Peharda, M.; et al. Historical baselines in marine bioinvasions: Implications for policy and management. *PLoS ONE* **2018**, *13*, e0202383. [CrossRef]
4. Kolar, C.S.; Lodge, D.M. Progress in invasion biology: Predicting invaders. *Trends Ecol. Evol.* **2001**, *16*, 199–204. [CrossRef] [PubMed]
5. Katsanevakis, S.; Moustakas, A. Uncertainty in Marine Invasion Science. *Front. Mar. Sci.* **2018**, *5*, 38. [CrossRef]
6. Seebens, H.; Blackburn, T.M.; Dyer, E.E.; Genovesi, P.; Hulme, P.E.; Jeschke, J.M.; Pagad, S.; Pyšek, P.; Winter, M.; Arianoutsou, M.; et al. No saturation in the accumulation of alien species worldwide. *Nat. Commun.* **2017**, *8*, 14435. [CrossRef]
7. Werschkun, B.; Banerji, S.; Basurko, O.C.; David, M.; Fuhr, F.; Gollasch, S.; Grummt, T.; Haarich, M.; Jha, A.N.; Kacan, S.; et al. Emerging risks from ballast water treatment: The run-up to the International Ballast Water Management Convention. *Chemosphere* **2014**, *112*, 256–266. [CrossRef]
8. Wada, K.T. *The Pearl Oyster Pinctada fucata (Gould) (Family Pteriidae)//Estuarine and Marine Bivalve Mollusk Culture*; CRC Press: Boca Raton, FL, USA, 1991; pp. 245–260.
9. Cunha, R.L.; Blanc, F.; Bonhomme, F.; Arnaud-Haond, S. Evolutionary Patterns in Pearl Oysters of the Genus *Pinctada* (Bivalvia: Pteriidae). *Mar. Biotechnol.* **2010**, *13*, 181–192. [CrossRef]
10. Tomaru, Y.; Udaka, N.; Kawabata, Z.; Nakano, S.-I. Seasonal change of seston size distribution and phytoplankton composition in bivalve pearl oyster *Pinctada fucata martensii* culture farm. *Hydrobiologia* **2002**, *481*, 181–185. [CrossRef]
11. Choi, Y.H.; Chang, Y.J. Gametogenic cycle of the transplanted-cultured pearl oyster, *Pinctada fucata martensii* (Bivalvia: Pteriidae) in Korea. *Aquaculture* **2003**, *220*, 781–790. [CrossRef]
12. Yu, D.; Chu, K.; Jia, X. Preliminary analysis on genetic relationship of common pearl oysters of *Pinctada* in china. *Oceanol. Et Limnol. Sin.* **2006**, *37*, 211.
13. Wang, Z.R.; Chen, R.Q. The species of the Pterioda from the Nansha Islands waters. In *Studies on Marine Species of the Nasha Islands and Neighboring Waters*; Chen, Q.C., Ed.; China Ocean Press: Beijing, China, 1991; pp. 150–151.
14. Chen, Q.C. *Studies on Marine Biodiversity of the Nasha Islands and Neighboring Waters II*; China Ocean Press: Beijing, China, 1996; p. 101.
15. Edelist, D.; Rilov, G.; Golani, D.; Carlton, J.T.; Spanier, E. Restructuring the Sea: Profound shifts in the world's most invaded marine ecosystem. *Divers. Distrib.* **2013**, *19*, 69–77. [CrossRef]
16. Cristescu, M.E. Genetic reconstructions of invasion history. *Mol. Ecol.* **2015**, *24*, 2212–2225. [CrossRef] [PubMed]
17. Handley, L.J.L.; Estoup, A.; Evans, D.M.; Thomas, C.E.; Lombaert, E.; Facon, B.; Aebi, A.; Roy, H.E. Ecological genetics of invasive alien species. *Entomophaga* **2011**, *56*, 409–428. [CrossRef]
18. Wang, H.; Guo, X.; Zhang, G.; Zhang, F. Classification of Jinjiang oysters *Crassostrea rivularis* (Gould, 1861) from China, based on morphology and phylogenetic analysis. *Aquaculture* **2004**, *242*, 137–155. [CrossRef]
19. Muirhead, J.R.; Gray, D.K.; Kelly, D.W.; Ellis, S.M.; Heath, D.D.; Macisaac, H.J. Identifying the source of species invasions: Sampling intensity vs. genetic diversity. *Mol. Ecol.* **2008**, *17*, 1020–1035. [CrossRef]
20. Darling, J.A.; Bagley, M.J.; Roman, J.; Tepolt, C.K.; Geller, J.B. Genetic patterns across multiple introductions of the globally invasive crab genus *Carcinus*. *Mol. Ecol.* **2008**, *17*, 4992–5007. [CrossRef]
21. Rius, M.; Turon, X.; Bernardi, G.; Volckaert, F.A.M.; Viard, F. Marine invasion genetics: From spatio-temporal patterns to evolutionary outcomes. *Biol. Invasions* **2014**, *17*, 869–885. [CrossRef]
22. Troost, K. Causes and effects of a highly successful marine invasion: Case-study of the introduced Pacific oyster *Crassostrea gigas* in continental NW European estuaries. *J. Sea Res.* **2010**, *64*, 145–165. [CrossRef]
23. Baird, N.A.; Etter, P.D.; Atwood, T.S.; Currey, M.C.; Shiver, A.L.; Lewis, Z.A.; Selker, E.U.; Cresko, W.A.; Johnson, E.A. Rapid SNP Discovery and Genetic Mapping Using Sequenced RAD Markers. *PLoS ONE* **2008**, *3*, e3376. [CrossRef]

24. Li, H.; Durbin, R. Fast and accurate short read alignment with Burrows—Wheeler transform. *Bioinformatics* **2009**, *25*, 1754–1760. [CrossRef] [PubMed]
25. Li, H.; Handsaker, B.; Wysoker, A.; Fennell, T.; Ruan, J.; Homer, N.; Marth, G.; Abecasis, G.; Durbin, R. The Sequence Alignment/Map format and SAMtools. *Bioinformatics* **2009**, *25*, 2078–2079. [CrossRef] [PubMed]
26. Todesco, M.; Owens, G.L.; Bercovich, N.; Légaré, J.-S.; Soudi, S.; Burge, D.O.; Huang, K.; Ostevik, K.L.; Drummond, E.B.M.; Imerovski, I.; et al. Massive haplotypes underlie ecotypic differentiation in sunflowers. *Nature* **2020**, *584*, 602–607. [CrossRef] [PubMed]
27. McKenna, A.; Hanna, M.; Banks, E.; Sivachenko, A.; Cibulskis, K.; Kernytsky, A.; Garimella, K.; Altshuler, D.; Gabriel, S.; Daly, M.; et al. The Genome Analysis Toolkit: A MapReduce framework for analyzing next-generation DNA sequencing data. *Genome Res.* **2010**, *20*, 1297–1303. [CrossRef] [PubMed]
28. Watterson, G. On the number of segregating sites in genetical models without recombination. *Theor. Popul. Biol.* **1975**, *7*, 256–276. [CrossRef] [PubMed]
29. Tajima, F. Statistical Method for Testing the Neutral Mutation Hypothesis by DNA Polymorphism. *Genetics* **1989**, *3*, 607–612. [CrossRef] [PubMed]
30. Holsinger, K.E.; Weir, B.S. Genetics in geographically structured populations: Defining, estimating and interpreting FST. *Nat. Rev. Genet.* **2009**, *10*, 639–650. [CrossRef]
31. Yang, J.; Lee, S.H.; Goddard, M.E.; Visscher, P.M. GCTA: A Tool for Genome-wide Complex Trait Analysis. *Am. J. Hum. Genet.* **2011**, *88*, 76–82. [CrossRef]
32. Alexander, D.H.; Novembre, J.; Lange, K. Fast model-based estimation of ancestry in unrelated individuals. *Genome Res.* **2009**, *19*, 1655–1664. [CrossRef]
33. Hedrick, P.W. A standardized genetic differentiation measure. *Evolution* **2005**, *59*, 1633–1638.
34. Jost, L.O.U. GST and its relatives do not measure differentiation. *Mol. Ecol.* **2008**, *17*, 4015–4026. [PubMed]
35. Sundqvist, L.; Keenan, K.; Zackrisson, M.; Prodöhl, P.; Kleinhans, D. Directional genetic differentiation and relative migration. *Ecol. Evol.* **2016**, *6*, 3461–3475. [CrossRef] [PubMed]
36. Keenan, K.; McGinnity, P.; Cross, T.F.; Crozier, W.W.; Prodohl, P.A. diveRsity: An R package for the estimation and exploration of population genetics parameters and their associated errors. *Methods Ecol. Evol.* **2013**, *4*, 782–788. [CrossRef]
37. Zhang, C.; Dong, S.-S.; Xu, J.-Y.; He, W.-M.; Yang, T.-L. PopLDdecay: A fast and effective tool for linkage disequilibrium decay analysis based on variant call format files. *Bioinformatics* **2019**, *35*, 1786–1788. [CrossRef] [PubMed]
38. Hamilton, M.B. *Population Genetics*; John Wiley & Sons: Hoboken, NJ, USA, 2011.
39. McMahon, R.F. Evolutionary and physiological adaptations of aquatic invasive animals: R selection versus resistance. *Can. J. Fish. Aquat. Sci.* **2002**, *59*, 1235–1244. [CrossRef]
40. Gomes, C.; Sousa, R.; Mendes, T.; Borges, R.; Vilares, P.; Vasconcelos, V.; Guilhermino, L.; Antunes, A. Low Genetic Diversity and High Invasion Success of *Corbicula fluminea* (Bivalvia, Corbiculidae) (Müller, 1774) in Portugal. *PLoS ONE* **2016**, *11*, e0158108. [CrossRef]
41. Sousa, R.; Gutiérrez, J.; Aldridge, D. Non-indigenous invasive bivalves as ecosystem engineers. *Biol. Invasions* **2009**, *11*, 2367–2385. [CrossRef]
42. Sakai, A.K.; Allendorf, F.W.; Holt, J.S.; Lodge, D.M.; Molofsky, J.; With, K.A.; Baughman, S.; Cabin, R.J.; Cohen, J.E.; Ellstrand, N.C.; et al. The Population Biology of Invasive Species. *Annu. Rev. Ecol. Syst.* **2001**, *32*, 305–332. [CrossRef]
43. Colautti, R.I.; MacIsaac, H.J. A neutral terminology to define 'invasive' species. *Divers. Distrib.* **2004**, *10*, 135–141. [CrossRef]
44. Takeuchi, T.; Masaoka, T.; Aoki, H.; Koyanagi, R.; Fujie, M.; Satoh, N. Divergent northern and southern populations and demographic history of the pearl oyster in the western Pacific revealed with genomic SNPs. *Evol. Appl.* **2019**, *13*, 837–853. [CrossRef]
45. Ficetola, G.F.; Bonin, A.; Miaud, C. Population genetics reveals origin and number of founders in a biological invasion. *Mol. Ecol.* **2008**, *17*, 773–782. [CrossRef] [PubMed]
46. Fournier, A.; Penone, C.; Pennino, M.G.; Courchamp, F. Predicting future invaders and future invasions. *Proc. Natl. Acad. Sci. USA* **2019**, *116*, 7905–7910. [CrossRef] [PubMed]
47. Gwak, W.S.; Nakayama, K. Genetic variation of hatchery and wild stocks of the pearl oyster *Pinctada fucata martensii* (Dunker, 1872), assessed by mitochondrial DNA analysis. *Aquac. Int.* **2011**, *19*, 585–591. [CrossRef]
48. Huang, J.; Xu, M.; He, M. Discriminant analysis of four cultured populations of the pearl oyster, *Pinctada fucata martensii*, using morphology and single-nucleotide polymorphism markers. *J. World Aquac. Soc.* **2019**, *51*, 1373–1385. [CrossRef]
49. Dlugosch, K.M.; Parker, I.M. Founding events in species invasions: Genetic variation, adaptive evolution, and the role of multiple introductions. *Mol. Ecol.* **2008**, *17*, 431–449. [CrossRef] [PubMed]
50. Flanagan, B.A.; Krueger-Hadfield, S.A.; Murren, C.J.; Nice, C.C.; Strand, A.E.; Sotka, E.E. Founder effects shape linkage disequilibrium and genomic diversity of a partially clonal invader. *Mol. Ecol.* **2021**, *30*, 1962–1978. [CrossRef]
51. Banerjee, A.K.; Hou, Z.; Lin, Y.; Lan, W.; Tan, F.; Xing, F.; Li, G.; Guo, W.; Huang, Y. Going with the flow: Analysis of population structure reveals high gene flow shaping invasion pattern and inducing range expansion of *Mikania micrantha* in Asia. *Ann. Bot.* **2020**, *125*, 1113–1126. [CrossRef]
52. Duk, M.; Kanapin, A.; Rozhmina, T.; Bankin, M.; Surkova, S.; Samsonova, A.; Samsonova, M. The Genetic Landscape of Fiber Flax. *Front. Plant Sci.* **2021**, *12*, 764612. [CrossRef]

53. Hill, W.G. Estimation of effective population size from data on linkage disequilibrium. *Genet. Res.* **1981**, *38*, 209–216. [CrossRef]
54. Bruen, T.C.; Philippe, H.; Bryant, D. A Simple and Robust Statistical Test for Detecting the Presence of Recombination. *Genetics* **2006**, *172*, 2665–2681. [CrossRef]
55. Chen, N.; Juric, I.; Cosgrove, E.J.; Bowman, R.; Fitzpatrick, J.W.; Schoech, S.J.; Clark, A.G.; Coop, G. Allele frequency dynamics in a pedigreed natural population. *Proc. Natl. Acad. Sci. USA* **2018**, *116*, 2158–2164. [CrossRef] [PubMed]
56. de Jong, M.A.; Wahlberg, N.; van Eijk, M.; Brakefield, P.M.; Zwaan, B.J. Mitochondrial DNA Signature for Range-Wide Populations of *Bicyclus anynana* Suggests a Rapid Expansion from Recent Refugia. *PLoS ONE* **2011**, *6*, e21385. [CrossRef] [PubMed]
57. Sun, J.; Chen, M.; Fu, Z.; Yang, J.; Zhou, S.; Yu, G.; Zhou, W.; Ma, Z. A Comparative Study on Low and High Salinity Tolerance of Two Strains of *Pinctada fucata*. *Front. Mar. Sci.* **2021**, *8*, 1039. [CrossRef]

Article

Prediction of Potential Suitable Distribution Areas of *Quasipaa spinosa* in China Based on MaxEnt Optimization Model

Jinliang Hou, Jianguo Xiang *, Deliang Li and Xinhua Liu

College of Animal Science and Technology, Hunan Agricultural University, Changsha 410128, China
* Correspondence: xjg68102000@aliyun.com

Simple Summary: In this study, in order to understand the distribution and optimal living environment of *Quasipaa spinosa* in China, we made predictions about the impacts of different environmental climates on its habitat. Our results show that our model is highly reliable. The mountainous areas in southern China with sufficient water supply are the main suitable areas for this species. The future reduction in emission concentration will be friendly to the ecological environment in the suitable areas of this species, better protecting the reproduction of its natural population.

Abstract: *Quasipaa spinosa* is a large cold-water frog unique to China, with great ecological and economic value. In recent years, due to the impact of human activities on the climate, its habitat has been destroyed, resulting in a sharp decline in natural population resources. Based on the existing distribution records of *Q. spinosa*, this study uses the optimized MaxEnt model and ArcGis 10.2 software to screen out 10 factors such as climate and altitude to predict its future potential distribution area because of climate change. The results show that when the parameters are FC = LQHP and RM = 3, the MaxEnt model is optimal and AUC values are greater than 0.95. The precipitation of the driest month (bio14), temperature seasonality (bio4), elevation (ele), isothermality (bio3), and the minimum temperature of coldest month (bio6) were the main environmental factors affecting the potential range of the *Q. spinosa*. At present, high-suitability areas are mainly in the Hunan, Fujian, Jiangxi, Chongqing, Guizhou, Anhui, and Sichuan provinces of China. In the future, the potential distribution area of *Q. spinosa* may gradually extend to the northwest and north. The low-concentration emissions scenario in the future can increase the area of suitable habitat for *Q. spinosa* and slow down the reduction in the amount of high-suitability areas to a certain extent. In conclusion, the habitat of *Q. spinosa* is mainly distributed in southern China. Because of global climate change, the high-altitude mountainous areas in southern China with abundant water resources may be the main potential habitat area of *Q. spinosa*. Predicting the changes in the distribution patterns of *Q. spinosa* can better help us understand the biogeography of *Q. spinosa* and develop conservation strategies to minimize the impacts of climate change.

Keywords: *Quasipaa spinosa*; MaxEnt; ArcGIS; potential habitat; environment variable

Citation: Hou, J.; Xiang, J.; Li, D.; Liu, X. Prediction of Potential Suitable Distribution Areas of *Quasipaa spinosa* in China Based on MaxEnt Optimization Model. *Biology* **2023**, *12*, 366. https://doi.org/10.3390/biology12030366

Academic Editor: Lucinda Johnson

Received: 2 February 2023
Revised: 17 February 2023
Accepted: 23 February 2023
Published: 25 February 2023

1. Introduction

The geographical distribution of species is the result of adaptation to climate, topography, soil, biology, migration history, and human activities during the long evolutionary process, which reflects the evolutionary history of species, population expansion, and the ability to adapt to the environment [1]. In recent years, habitat destruction due to human disturbance and global climate change has led to the migration, rapid decline, or disappearance of amphibian species [2]. Therefore, it is of great significance to study the suitable areas of species to protect their habitats in situ. The application of niche models to predict and evaluate species habitats is becoming a new hot topic [3,4]. There are more than 10 species distribution models (SDMs) that have been reported, but the MaxEnt model is low cost, simple to operate, short to run, and can simulate the fitness range of species well with a very small number of samples ($n \geq 5$) [5]. At present, it is also widely used in the prediction research of amphibian habitats, such as *Odrrana hainanensis* [2], and five species of *Scutiger* [6], and others such as *Rana Zhenhaiensis* [7], *Buergeria oxycephala* [8], *Nanorana parkeri* [9], *Plethodon* [10], and *Quasipaa boulengeri* [11], and so on.

Quasipaa spinosa, commonly known as stone frog, rock frog, etc., belongs to *Anura*, *Dicroglossidea*, and *Quasipaa* [12]. As a large frog, it breeds in cold water streams and is mainly distributed in the hills of southern China and mountainous areas of northern Vietnam [12–16], which has a special ecological and biological status. Due to overfishing, the wide application of chemical pesticides and environmental changes, the wild population of *Q. spinosa* has decreased dramatically [17]. It has been designated as a vulnerable species by the International Union of the Conservation of Nature (IUCN) Red List and the Red List of Chinese Species [18].

As an important economic frog with high nutritional value and homology of medicine and food, *Q. spinosa* is honored as the "King of Hundred Frogs" [15,19,20]. In the last ten years, the market demand of *Q. spinosa* has been rising, and the price of *Q. spinosa* has increased 20–30 times from 1980s to now [18], which has promoted the rapid development of *Q. spinosa* breeding in China, becoming one of the most important industries for poverty alleviation in mountainous areas. Although artificial breeding has relieved the pressure of hunting for the wild population to some extent, the effect has been limited [21]. At present, research on *Q. spinosa* focuses on gut microbiology [13,14,22], diseases [23,24], genetic transcriptomes [25,26], and captive breeding [16,27–29]. It is very important to better protect and reasonably utilize the existing wild *Q. spinosa* resources to achieve a win–win situation for environmental protection and resource utilization. These have now become a new hot spot in fisheries resources research.

In this paper, we applied a MaxEnt optimization model and ArcGIS technology to comprehensively analyze the key environmental factors affecting the distribution of *Q. spinosa*, with the aim of understanding the current distribution of its habitat suitability, exploring its habitat range, and predicting its potential geographical distribution in order to provide reference for the conservation of *Q. spinosa* diversity, habitat restoration, and biogeography research.

2. Materials and Methods

2.1. Data Collection and Processing

The distribution points information of *Q. spinosa* in this study comes from the Global Biodiversity Information Facility (GBIF; https://www.gbif.org/, accessed on 6 March 2022) [30]. After retrieval, a total of 1779 distribution records of *Q. spinosa* was obtained. Through https://jingweidu.51240.com/, the longitude and latitude information of *Q. spinosa* was determined, and the data without detailed geographic location, duplicate specimens, and location deviations were excluded. Finally, 130 geographical distribution points of *Q spinosa* were screened. We converted the obtained sample point data into .csv format and used ArcGIS 10.2 to draw a map of *Q. spinosa* distribution in China (Figure 1).

Human footprints data were downloaded from the Socioeconomic Data and Applications Center (SEDAC, https://sedac.ciesin.columbia.edu, accessed on 22 April 2022) [31]. Altitude and 19 bioclimatic variables in current (1970–2000), future 2050s (2041–2060), and future 2070s (2061–2080) were downloaded from WorldClim (http://www.worldclim.org/). The future (2050s, 2070s) climatic variables were selected from two of the four greenhouse gases (RCP2.6 and RCP8.5) under the general climate system mode proposed in the fifth assessment report of IPCC—AR5—representing the lowest and highest scenarios of greenhouse gas emissions [32]. The Chinese province boundary map was downloaded from the Institute of Geographic Sciences and Natural Resources Research, Chinese Academy of Sciences (http://www.resdc.cn/). All data have a unified resolution of 2.5 min, the coordinate system was WGS 1984, and was converted to "asc" with ArcGIS 10.2 software (Environmental Systems Research Institute, Inc., America).

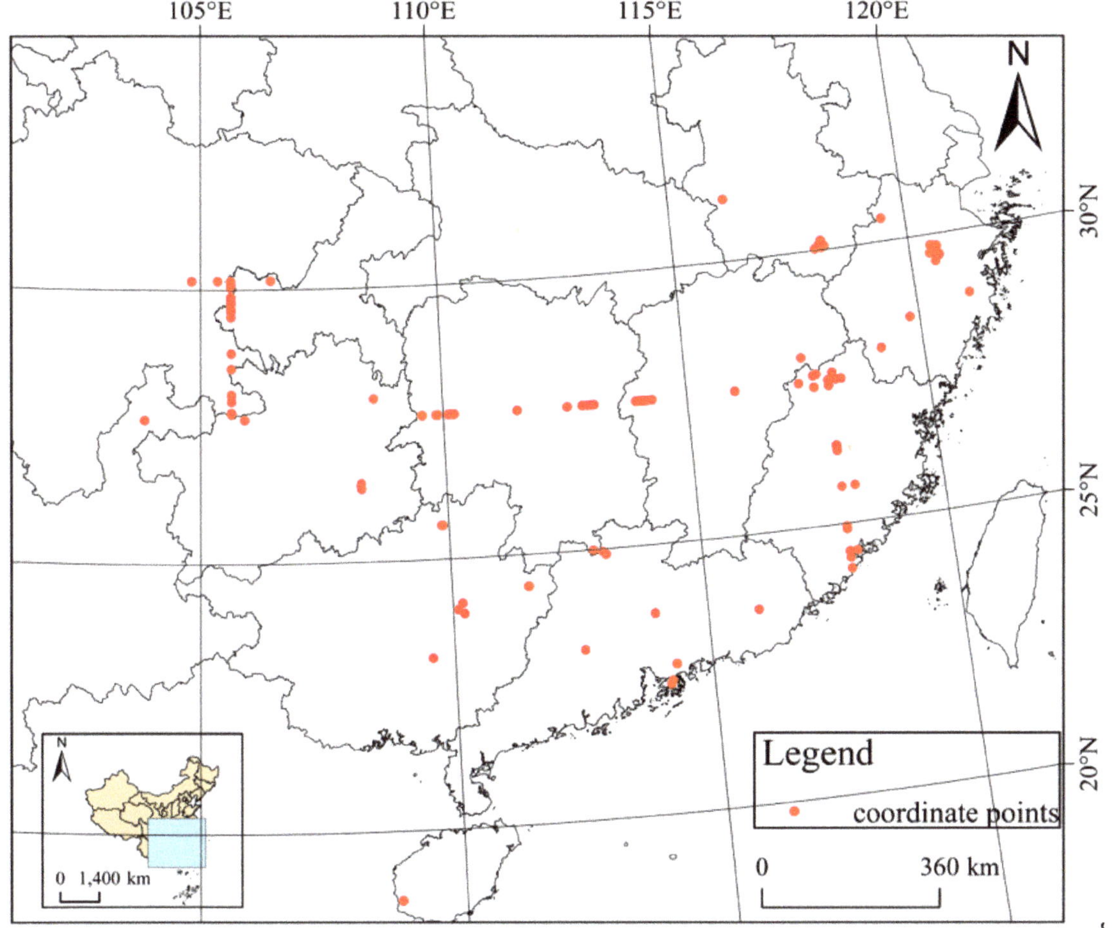

Figure 1. Confirmed coordinate distribution of *Quasipaa spinosa* in China.

In order to minimize the bias fitting of the model, the ENMTools package (Package) was used to perform correlation analysis on 21 environmental factors [33]. When the correlation was greater than |0.8|, the variables with small contribution rates were removed. Finally, 10 significant factors were screened out (Table 1).

Table 1. Various parameters of the main environmental variables of *Q. spinosa*.

Environmental Variables	Description	Unit	PC (%)	PI (%)
ele	Elevation	m	5.5	8.2
bio2	Mean Diurnal Range	°C	3.9	3.2
bio3	Isothermality	-	3.4	20.0
bio4	Temperature Seasonality	-	2.6	40.9
bio5	Max Temperature of Warmest Month	°C	0.6	0
bio6	Min Temperature of Coldest Month	°C	1.9	9.6
bio14	Precipitation of Driest Month	mm	79.3	11.7
bio15	Precipitation Seasonality	-	1.6	2.9
bio18	Precipitation of Warmest Quarter	mm	0.7	1.9
people	Human Foot	-	0.5	1.7

2.2. Species Distribution Modeling, Optimization, and Evaluation

Optimized maxent model was implemented in the R programming environment using the package ENMeval [34]. We set 5 regularization multiplier values (β, 1, 2, 3, 4, and 5) and adopted 6 features (H, L, LQ, LQH, LQHP, and LQHTP), among which L, Q, H, P, and T were linear, quadratic, hinge, product, and threshold, respectively [35,36]. Finally, the best model parameter combination with the lowest Akaike Information Criterion Correction (AICc) value (delta. AICc = 0) was obtained for the establishment of MaxEnt model.

Q. spinosa distribution occurrence data and selected environmental variables were loaded into the MaxEnt model. The data were randomly divided: 75% of the location point data were used as a training model, with the remaining 25% used for validating the MaxEnt model [37,38]. By default, the maximum number of iterations was set as 10,000, the model was repeated for 10 times, and the predicted results were output in "Cloglog" format and "asc" file type [39]. The maximum number of iterations after optimization is 420. The prediction effect was tested by AUC under the receiver operating characteristic curve (ROC) [40]. AUC value between 0.5 and 0.6 was considered to be unqualified, 0.6~0.7 was poor, 0.7~0.8 was average, 0.8~0.9 was good, and 0.9~1.0 was excellent [34]. When the TSS value was <0.4, the predictive power of the model was considered "poor", while 0.4–0.8 was considered "good" and 0.8–1 was considered "excellent" [41].

2.3. Classification of Suitable Living Grade of Q. spinosa

The average value of MaxEnt simulation results in each period was imported into ArcGIS 10.2 software, and the model distribution area was divided into suitability grades and visualized [42]. The habitat suitability index was divided into four levels: high-suitability area, medium-suitability area, low-suitability area, and non-suitable area by the Nature Breaks (Jenks) method. They were represented by different colors: red is high-suitability area, yellow is medium-suitability area, green is low-suitability area, and white is non-suitable area. We used ArcGIS raster tools to count the area of suitable areas.

3. Results

3.1. Model Optimization and Accuracy Evaluation

The MaxEnt model was used to simulate the potential distribution areas in different scenarios currently and in future (2050s and 2070s) according to the AICc (the minimum information criterion) for the screened 130 *Q. spinosa* distribution points and 10 environmental variable layers. The optimal parameter combination of FC = LQHP and RM = 3 is obtained after the optimization of the ENMeval program package. At this time, AICc is the smallest and delta.AICc = 0. The optimized AICc and the difference between the AUC value of the training set and the AUC of the test set are both lower than the default settings (Table 2).

Table 2. Evaluation metrics of MaxEnt model generated by ENMeval.

Type	FC	β	delta. AICc	avg. diff. AUC
Default	LQHPT	1	101.9003	0.1211
Optimized	LQHP	3	0	0.1008

FC: feature combination; RM: regulatory multiplier; AICc: LQPH: linear features (L) + quadratic features (Q) + product features (P) + hinge features (H); LQH: linear features (L) + quadratic features (Q) + hinge features (H); delta. AICc: the minimum information criterion AICc value; avg. diff. AUC: difference between the AUC value.

The MaxEnt simulations for modern and future (2050s and 2070s) time periods with optimal parameters are performed, and the AUC results are shown in Figure 2. The values of the area under the receiver operating characteristic curve were all greater than 0.95 (Table 3). The true skill statistics were all greater than 0.8 (Table 3). In conclusion, the optimized MaxEnt model has high reliability in the prediction results of the potential suitable areas of *Q. spinosa*. The 10 environmental factors used for modeling included 8 climate factors, 1 terrain factor, and 1 anthropogenic factor (Table 1).

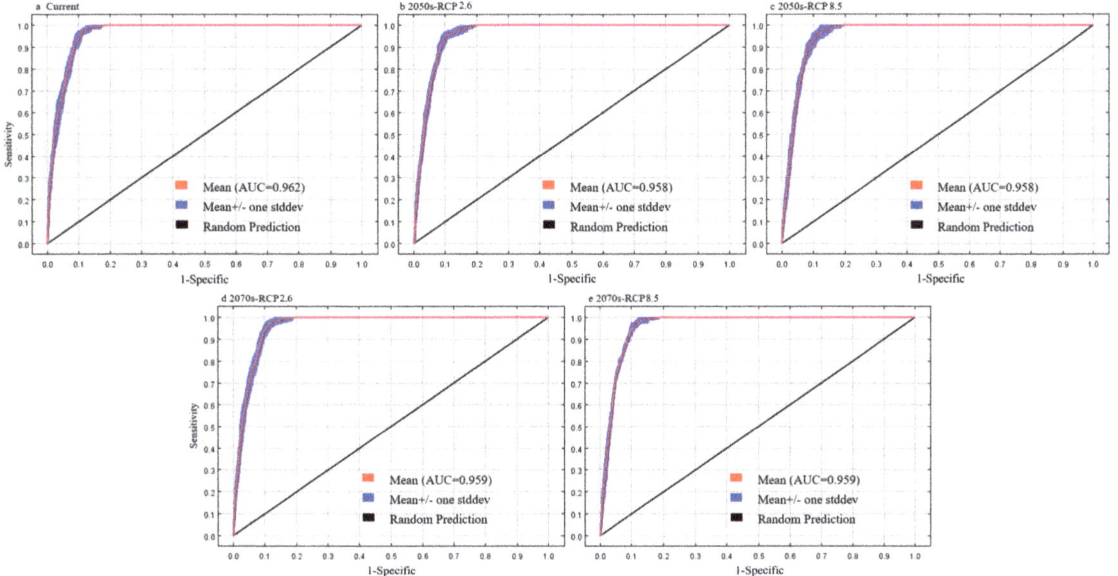

Figure 2. AUC value of *Q. spinosa* predicted by MaxEnt model.

Table 3. Evaluate the AUC and TSS index values of the MaxENT model.

	Current	50s-RCP2.6	50s-RCP8.5	70s-RCP2.6	70s-RCP8.5
AUC	0.962 ± 0.0073	0.958 ± 0.0073	0.958 ± 0.0073	0.959 ± 0.0072	0.959 ± 0.008
TSS	0.817 ± 0.014	0.813 ± 0.009	0.804 ± 0.010	0.822 ± 0.015	0.818 ± 0.015

Note: The values in the table are expressed as "mean ± standard deviation".

3.2. The Importance of Environmental Variables

Percent contribution (PC) and permutation importance (PI) are the main indicators for evaluating the importance of environmental variables. The larger the index value, the higher the importance of environmental variables. The top 5 environmental variables with percent contributions are precipitation of the driest month (bio14, 79.3%), elevation (ele, 5.5%), mean diurnal range (bio2, 3.9%), isothermality (bio3, 3.4%), and temperature seasonality (bio4, 2.6%), accounting for 94.7% in total. The top 5 environmental variables with permutation importance are temperature seasonality (bio4, 40.9%), isothermality (bio3, 20.0%), precipitation of the driest month (bio14, 11.7%), minimum temperature of the coldest month (bio6, 9.6%), and elevation (ele, 8.2%), accounting for 90.4% of the total.

The Jackknife test results show (Figure 3F–H) that test gain, regularized training gain, and AUC are basically the same without variables and with all variables. When only variables are used, the top five factors with the largest test gain, regularized training gain, and AUC are precipitation of the driest month (bio14), mean diurnal range (bio2), minimum temperature of the coldest month (bio6), precipitation of the warmest quarter (bio18), and temperature seasonality (bio4), indicating that these five environmental factors have the greatest impact on the distribution of *Q. spinosa*. In conclusion, the precipitation of the driest month, temperature seasonality, elevation, isothermality, and minimum temperature of the coldest month are the dominant variables affecting the distribution of *Q. spinosa*.

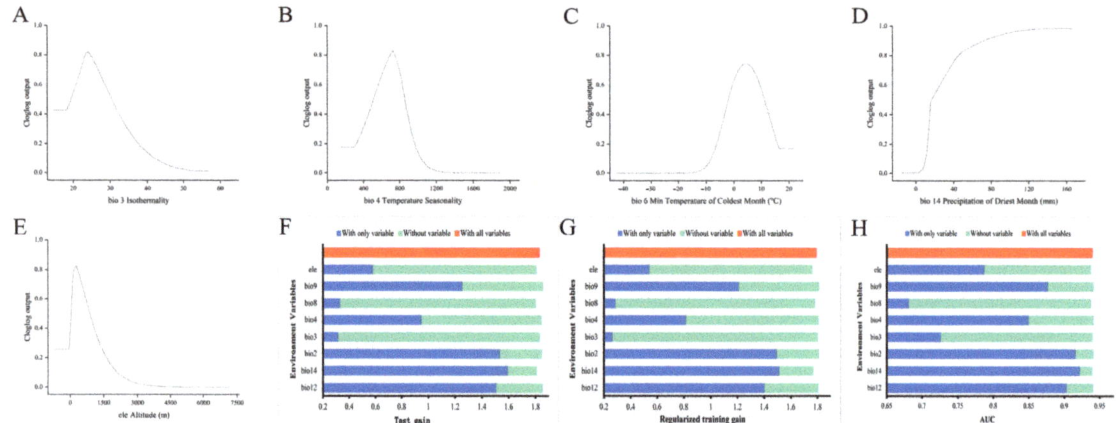

Figure 3. Response curves and Jackknife test of the environment variables. (**A–E**) Respectively represent the response curves of isothermality, temperature seasonality, min temperature of coldest month, precipitation of driest month and Altitude. (**F–H**) The contribution of each environmental factor to each scenario using the Jacknife test in In test gain, regularized training gain and AUC, respectively.

The environmental variables (>0.63) were the most suitable for the survival of *Q. spinosa* when the precipitation of the driest month was 28.4–166.1 mm, the altitude was 79.6–635.8 m, the minimum temperature of the coldest quarter was 0.1–9.0 °C, the temperature seasonality was 584.0–821.3, and the isothermality was 21.2–28.0 (Figure 3A–E).

3.3. Current and Future Potential Suitable Areas and Their Spatiotemporal Changes

The distribution range of the potential suitable habitats of *Q. spinosa* in the future is generally the same, mainly in Hunan, Hubei, Guangxi, Fujian, Zhejiang, Anhui, Jiangxi, Guizhou, Guangdong, Sichuan, Taiwan, Hong Kong, and other provinces (Figure 4a–e). According to the MaxEnt simulation results, the modern potential total suitable area of *Q. spinosa* is 1.65×10^6 km^2, accounting for 17.19% of the total land area in China, including the high-suitability area of 4.9×10^5 km^2, accounting for 5.10%; the medium-suitability area of 6.3×10^5 km^2, accounting for 6.56%; and the low-suitability area of 5.3×10^5 km^2, accounting for 5.52% (Table 4; Figure 4f). In the future 2050s, the proportion of the total suitable area for the frog under high-concentration (RCP8.5) and low-concentration (RCP2.6) emission scenarios will be 17.92% and 18.33%, respectively. In the future 2070s, the proportions of the total suitable area of *Q. spinosa* under the high-concentration (RCP8.5) and low-concentration (RCP2.6) emission scenarios will be 17.29% and 17.71%, respectively, which are higher than the modern 17.19% (Table 4). In the future 2050s and 2070s, the proportion of suitable areas under the high-concentration emission scenario (RCP2.6) is larger than that under the low-concentration emission scenario (RCP 8.5).

Table 4. Area statistics of suitable areas for *Q. spinosa* under different climate scenarios.

Circumstances	Low Suitability		Medium Suitability		High Suitability		All	
	Area ($\times 10^4$ km^2)	Percentage (%)	Area ($\times 10^4$ km^2)	Percentage (%)	Area ($\times 10^4$ km^2)	Percentage/%	Area ($\times 10^4$ km^2)	Percentage (%)
Current	53	5.52	63	6.56	49	5.10	165	17.19
2050s-RCP 2.6	67	6.98	58	6.04	47	4.90	172	17.92
2050s-RCP 8.5	66	6.88	67	6.98	43	4.48	176	18.33
2070s-RCP 2.6	61	6.35	64	6.67	41	4.27	166	17.29
2070s-RCP 8.5	64	6.67	61	6.35	45	4.69	170	17.71

Note: The area percentages are the ratios of the suitable areas of different grades to the national land area (960×10^4 km^2) under different climate scenarios.

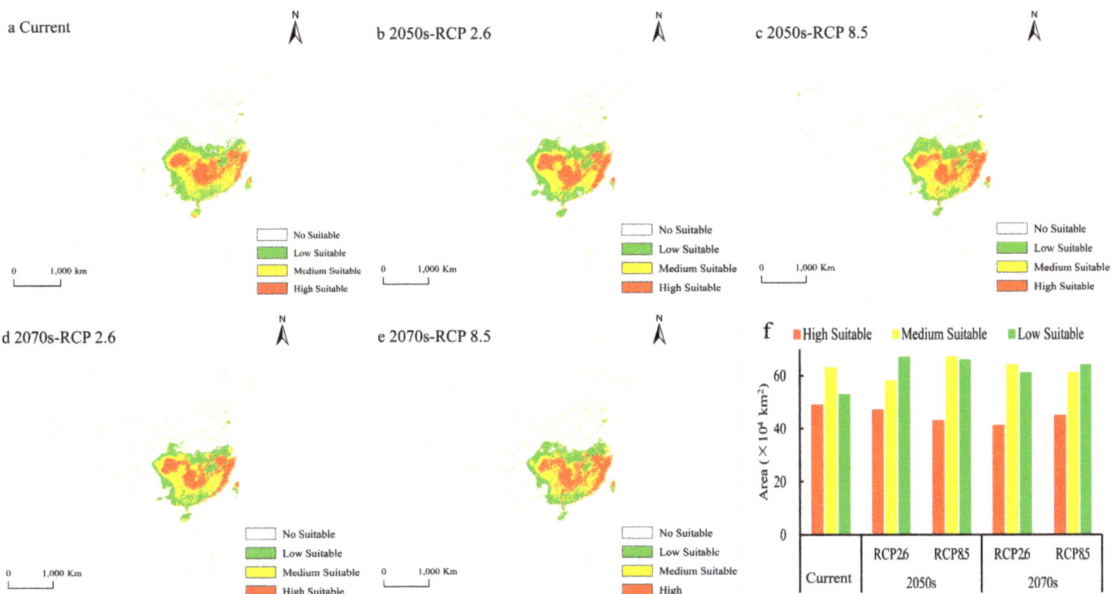

Figure 4. Potential distribution area of *Q. spinosa* under different scenarios. (**a–e**) Predicted distribution map of different suitable areas of *Q. spinosa* at current, 2050s and 2070s. (**f**) Statistical maps of different suitable areas of *Q. spinosa* in China at different periods.

Compared with the current time period, the proportion of high-suitability areas in the 2050s and 2070s was reduced. Under the future low-concentration emission scenario (RCP2.6), the area of the low-suitability area first increased and then decreased, but was higher than that of the present time; the area of the medium-suitability area was the smallest in 2050s and the largest in 2070s; and the high-suitability area showed a decreasing trend. Under the future high-concentration emission scenario (RCP8.5), the areas of medium- and low-suitability areas both increased first and then decreased but were higher than that of the current generation; and the areas of high-suitability areas decreased first and then increased, but were lower than that of the current generation.

Compared with the current potential distribution range of *Q. spinosa*, the newly added potentially suitable areas of 2050s-RCP2.6, 2050s-RCP8.5, 2070s-RCP2.6, and 2070s-RCP8.5 were all larger than the lost areas, increasing by 60,000, 20,000, 130,000, and 80,000 km², respectively, accounting for 3.64%, 1.21%, 7.88%, and 4.85% of the modern area, respectively (Table 5). In the future, the potential suitable area of *Q. spinosa* showed a trend for migration to the northwest and north, and the potential suitable area in the south gradually decreased. In the future 2050s and 2070s, the loss areas are mainly in Yunnan, Guizhou, Guangdong, Taiwan, Jiangsu, and other provinces, and the newly added areas are mainly in Sichuan, Gansu, and Qinling (Figure 5). Under the future low-concentration emission scenario (RCP 2.6), Yunnan is both the main increase area and the main loss area of suitable habitat. Under the high-concentration emission scenario (RCP 8.5), more suitable habitats of *Q. spinosa* were lost in Yunnan, and the loss area increased with time (Figure 5).

Table 5. Spatial variations in suitable habitat for *Q. spinosa* in different periods.

Circumstances	Area ($\times 10^4$ km^2)			Rate of Change (%)		
	Gain	Loss	Change	Gain	Loss	Change
2050s-RCP2.6	14	8	6	8.48	4.85	3.64
2050s-RCP8.5	9	7	2	5.45	4.24	1.21
2070s-RCP2.6	18	5	13	10.91	3.03	7.88
2070s-RCP8.5	14	6	8	8.48	3.64	4.85

Note: The rate of change is the percentage of the area of each period and the area of the contemporary suitable area. The current *Q. spinosa* area of the potential suitable habitat is 165×10^4 km^2.

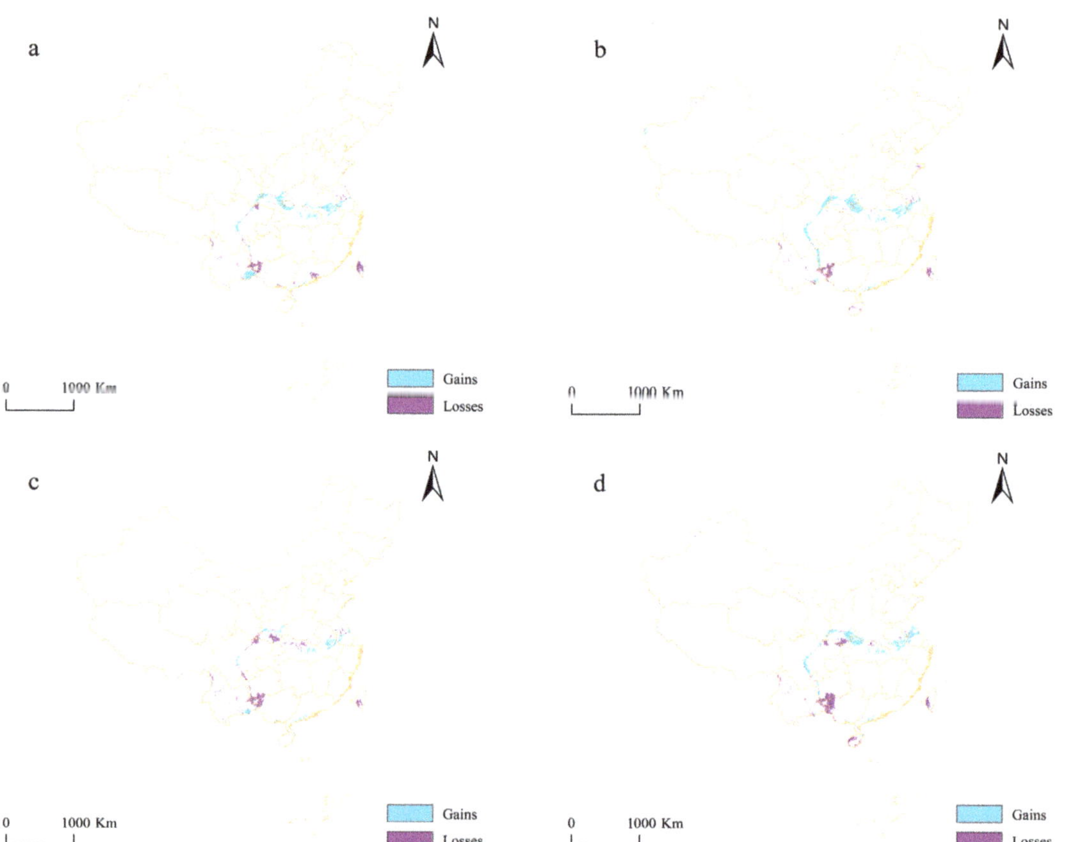

Figure 5. Spatial transformation pattern of *Q. spinosa* suitable area in different periods. (**a**) Greenhouse gas emission concentration is lowest in 2041–2060; (**b**) greenhouse gas emission concentration is highest in 2041–2060; (**c**) greenhouse gas emission concentration is lowest in 2061–2080; (**d**) greenhouse gas emission concentration is highest in 2061–2080. The increase and loss of suitable areas are derived and compared to current area.

Under the low-concentration emission scenario (RCP2.6), the increase in the area of suitable areas increased from 3.64% in the 2050s to 7.88%. Under the high-concentration emission scenario (RCP8.5), the area ratio increases from 1.21% in the 2050s to 4.45% in the 2070s. Under the low-concentration emission scenario in the future 2070s, the area of suitable habitats will increase the most (Table 5).

4. Discussion

4.1. Rationality of Model

Currently, among the reported species distribution models, the MaxEnt model has better stability and higher accuracy, and has less distortion in dealing with group temperature and precipitation factors [43–45]. Amphibians are ectothermic animals, and the external environment, especially temperature, precipitation, and so on, is the main restricting factor for their growth and development [46–48]. In this paper, temperature, altitude, and precipitation related to the characteristics of amphibians are selected as the main climatic factors, and climatic extreme values are selected to make related predictions. This study uses the ENMeval program package to optimize MaxEnt to reduce the degree of overfitting and sampling bias, improving the prediction accuracy [49,50]. The model prediction results shows that the AUC and TSS values of the current and future time periods (2050s and 2070s) are all greater than 0.95 and 0.80, respectively, indicating the high accuracy and distinguishing ability of the prediction results. The results of the optimized MaxEnt model showed that the potential suitable habitats of the vulnerable species *Q. spinosa* were mainly in the provinces south of the Qinling Mountains in China, which was basically consistent with the reported distribution range of *Q. spinosa* [51–54].

4.2. Main Environmental Factors Affecting the Distribution of Q. spinosa

Q. spinosa is a water-dwelling frog [55], which mainly inhabits the areas beside streams in mountainous areas, and temperature and water sources have an important impact on its survival [56]. Altitude is closely related to temperature changes to a certain extent. The results of the Jackknife, AUC, PC, and PI tests indicated that the importance of precipitation in the driest month, temperature seasonality, elevation, isothermality, and the mean temperature of the coldest quarter played a major role in affecting the distribution of *Q. spinosa*. This is similar to *Rana hanluica* [57] and *Buergeria oxycephala* [8], but the precipitation factor has a lower gain on the distribution model of *Buergeria oxycephala*, which may be related to the lower water dependence of its adult frog habitat. The distribution of *Odrrana hainanensis* is less affected by temperature, humidity, and sunshine, which may be related to the fact that its study area is located in the equatorial tropics [2]. While *Q. spinosa* is a cold-water frog with a small suitable temperature range and a distribution area mainly located in the subtropical region, large changes in temperature and precipitation have a great impact on its distribution. The main factors influencing their potential distribution varied among the five species of *Scutiger* [6]. The precipitation of the warmest quarter and temperature seasonality are the main factors affecting the potential distribution of *Rana heaviness* [7]. In summary, different species have different ecological requirements, and the main environmental factors affecting their potential distribution are also different.

This study is basically consistent with the findings of Zou et al. [58]. The results show that the most suitable habitat altitude of *Q. spinosa* is 79.6–635.8 m. However, Liu [55] found that the distribution range of *Q. spinosa* was 150–1000 m above sea level, while Fei et al. [59] found that it was 300–1500 m, and Liang et al. [60] found that it was 700–800 m. The reason may be related to climate change in recent years, sample bias, and different sampling intensities. When the altitude was lower than 79.6 m or higher than 635.8 m, the suitability of *Q. spinosa* decreased dramatically. The lower elevations in the south are mainly plains, where human dwellings congregate, and water is plentiful, but summer temperatures are high. Higher altitudes, however, are not conducive to the survival of *Q. spinosa* due to the low temperatures and low biodiversity combined with the lack of water and food. From the 1950s to 2007, the annual precipitation in the central region of South China decreased year by year, and the temperature showed an upward trend [61]. The changes in temperature and precipitation were related to the altitude [62–64], which led to the change in the habitat altitude of *Q. spinosa*. When the mean diurnal range is 3.5–7.7 °C and the max temperature of the warmest month is 29.5–34.3 °C (Figure 6), it is most suitable for the survival of *Q. spinosa*, which is basically consistent with the reports [16,51]. Southern China is dominated by a subtropical monsoon climate, with rain and heat in the same period. The *Q. spinosa* is optimally distributed in areas with a warmest quarter precipitation of 549.5–850.9 mm, which is basically consistent with the precipitation climate in southern China [61].

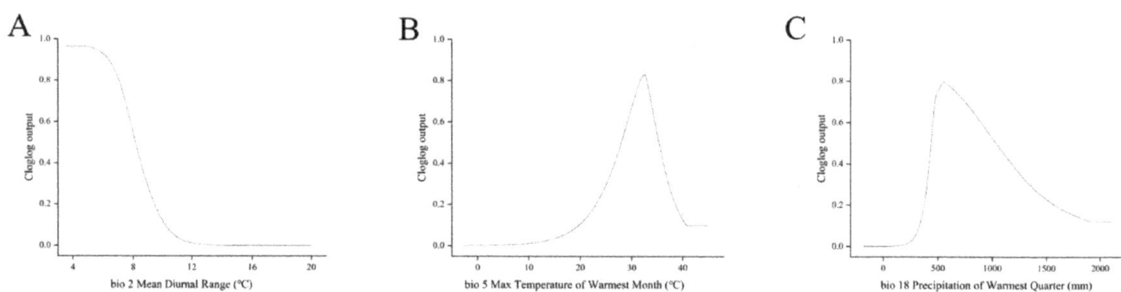

Figure 6. Response curve of the partial environment variables. (**A–C**) Respectively represent the response curves of mean diurnal range, max temperature of warmest month and precipitation of warmest quarter.

In this study, the human footprint refers to the pressure humans put on the environment. This factor has little effect on the distribution of *Q. spinosa*, which may be due to the fact that the human capture factor is not included in this environmental data and the living area of *Q. spinosa* is mainly near mountain streams.

4.3. Changes in Potential Suitable Areas

Compared with the current time period, under the two emission scenarios in the future, the area of low-suitability areas increased more than that of high-suitability areas, and the area of potentially suitable areas showed an overall upward trend. Under the low-concentration emission scenario, the area of suitable habitat for *Q. spinosa* showed an increasing trend, and maintaining the low-concentration emission scenario would expand the suitable habitat distribution range of *Q. spinosa*; however, the high-suitability area gradually decreased. Under the high-concentration emission scenario, the size of the suitable area first increased and then decreased. Although the area of the high-suitability area increased in the 2070s, it was still smaller than the current time period.

The prediction results of the optimization model showed that the distribution area of *Q. spinosa* tended to expand to the north and northwest in the future, which is mainly reflected in the increase in the potentially suitable areas in Qinling, Sichuan, Gansu, and other provinces. During the past century of climate change in China, the eastern region showed a decreasing trend for precipitation as a whole [65], the accumulated temperature in the southern region showed an increasing trend, the high-accumulated-temperature region gradually decreased from the southeast to the northwest, and the annual precipitation in the central region of the south decreased year by year [61]. *Q. spinosa* is a cold-water frog with an optimal survival temperature of 22–26 °C, and 80% of its life cycle is in water. Mountainous areas have good water retention, with the reduction of ice and snow providing water replenishment, which explains the expansion direction of the future suitable area for *Q. spinosa* and the reduction in high-suitability areas.

4.4. Resource Conservation of Q. spinosa

Based on the goal of striving to achieve peak carbon by 2030 and carbon neutrality by 2060 (referred to as the "dual carbon" goal) [66], China's energy structure is continuing to transition to green energy. Although carbon emissions continued to grow for the fourth consecutive year, with an increase of 0.6%, the carbon emission intensity decreased by 1%, and the overall carbon emission per unit of GDP showed a downward trend and is now only 0.7 (2018) [67]. For carbon emissions trading, relevant policies have been issued to standardize the operation of the carbon market [68]. From the perspective of economic and social development status and energy structure, China's carbon neutrality goal still faces many challenges and uncertainties [69]; however, China is committed to improving the ecological environment and promoting harmonious coexistence between man and nature. China's carbon peak by 2030 and carbon neutrality by 2060 can be expected.

Global warming, the increase in harsh climates, and man-altered mountains and rivers have greatly affected the ecological balance of biological habitats. *Q spinosa* feeds on live insects [27,70], and climate change may indirectly affect the survival and distribution of its prey organisms, leading to the frog migrating to more suitable habitats. Comprehensive analysis shows that the changes in environmental climate since modern times have caused some changes in the habitat of *Q. spinosa*, and it tends to be in areas with an altitude of about 600 m, small temperature differences, and excellent water sources.

The following suggestions should be helpful for the conservation of *Q. spinosa* resources. First of all, under the "dual carbon" goal, the state should further strengthen the coupled management of resources and establish and improve a sustainable forest resource management mechanism, which is of great significance to further improve the natural ecosystem. Secondly, the state should put *Q. spinosa* on the list of protected animals in China, prohibiting the sale and capture of wild *Q. spinosa*. Thirdly, the state should further clarify the prohibited types and prohibited areas of pesticides, or limit the amount of use. Finally, we need to take multiple measures to protect the natural population of *Q. spinosa*, such as in situ protection, ex situ protection, proliferation and release, etc. Priority protection can only be achieved when the whole of society works together.

5. Conclusions

In this study, the MaxEnt model for *Q. spinosa* provided satisfactory results. The prediction results can provide an important reference for wild *Q. spinosa* development and protection. Frog diversity conservation and biogeographical research are enriched by our results. *Q. spinosa* is mainly distributed around the streams in the mountains of southern China. Climate is one of the important factors for the distribution of *Q. spinosa*. In the future, maintaining a low-carbon life could alleviate climate deterioration and protect the habitat of spiny chestnut frogs. In addition, improving mountain and wildlife protection strategies is also important.

Author Contributions: J.H. conceived the study. J.X. designed the study. J.H. processed the data, performed the analyses, analyzed the results, and wrote the manuscript. J.X., D.L. and X.L. edited the manuscript. All authors have read and agreed to the published version of the manuscript.

Funding: This research was funded by [National Natural Science Foundation of China] grant/award number [31772832]; [National Key Research and Development Program of China] grant/award number [2019YFD0900603]; [Hunan Provincial Modern Agricultural Research System] grant/award number [2019-105].

Institutional Review Board Statement: Not applicable.

Informed Consent Statement: Not applicable.

Data Availability Statement: All data generated by this study are available from the corresponding author upon reasonable request.

Acknowledgments: Thanks to the reviewers for their constructive and valuable comments and the editors for their assistance in refining this article.

Conflicts of Interest: The authors declare no conflict of interest.

References

1. Soberon, J.; Peterson, A.T. Interpretation of models of fundamental ecological niches and species' distributional areas. *Biodivers. Inform.* **2005**, *2*, 1–10. [CrossRef]
2. Huang, Y.; Lu, J.; Wang, F.; Lin, Y.; Liu, L.; Mi, H.; Fangxu, M.; Fang, J.; Jialing, L. Predicting the potential geographical distribution of Hainan odorous frog (*Odorrana hainanensis*) in Hainan province. *Chin. J. Zool.* **2017**, *1*, 30–41. [CrossRef]

3. Ward, D.F. Modelling the potential geographic distribution of invasive ant species in New Zealand. *Biol. Invasions* **2007**, *9*, 723–735. [CrossRef]

4. Zhu, G.; Bu, W.; Gao, Y.; Liu, G. Potential geographic distribution of brown marmorated stink bug invasion (*Halyomorpha halys*). *PLoS ONE* **2012**, *7*, e31246. [CrossRef] [PubMed]

5. Pearson, R.G.; Raxworthy, C.J.; Nakamura, M.; Townsend Peterson, A.J.J.o.b. ORIGINAL ARTICLE: Predicting species distributions from small numbers of occurrence records: A test case using cryptic geckos in Madagascar. *J. Biogeogr.* **2007**, *34*, 102–117. [CrossRef]

6. Zhao, Z.; Xiao, N.; Liu, G.; Li, J. Prediction of the potential geographical distribution of five specices of *Scutiger* in the south of hrngduan Mountains Biodiversity Conservation Priority Zone. *Acta Ecol. Sin.* **2022**, *42*, 2636–2647.

7. Zhang, K.; Yang, K.; Wo, Y.; Tong, H.; Jin, Y.; Jin, Y. Suitability evaluation of potential geographic distribution for *Rana zhenhaiensis* based on MaxEnt. *Chin. J. Ecol.* **2018**, *37*, 164–170. [CrossRef]

8. Mu, S.; Qi, X.; Xie, L.; Qian, T.; Zhou, Z.; Lu, Y.; Mo, Y.; Li, P. MaxEnt-based prediction on the geographical distribution of Hainan stream treefrog (*Buergeria oxycephala*) in Hainan island. *Chin. J. Wildl.* **2021**, *42*, 809–816. [CrossRef]

9. Zhou, W.W.; Zhang, B.L.; Chen, H.M.; Jin, J.Q.; Yang, J.X.; Wang, Y.Y.; Jiang, K.; Murphy, R.W.; Zhang, Y.P.; Che, J. DNA barcodes and species distribution models evaluate threats of global climate changes to genetic diversity: A case study from *Nanorana parkeri* (Anura: Dicroglossidae). *PLoS ONE* **2014**, *9*, e103899. [CrossRef]

10. Nottingham, S.; Pelletier, T.A. The impact of climate change on western *Plethodon* salamanders' distribution. *Ecol. Evol.* **2021**, *11*, 9370–9384. [CrossRef]

11. Zhao, Z.; Xiao, N.; Shen, M.; Li, J. Comparison between optimized MaxEnt and random forest modeling in predicting potential distribution: A case study with *Quasipaa boulengeri* in China. *Sci. Total Env.* **2022**, *842*, 156867. [CrossRef]

12. Long, J.; Hou, J.; Zhou, W.; Xiang, J.; Pan, W. Analysis of karyotype in *Quasipaa spinosa*. *J. Anhui Agric. Sci.* **2021**, *49*, 95–97+103.

13. Long, J.; Xiang, J.; He, T.; Zhang, N.; Pan, W. Gut microbiota differences during metamorphosis in sick and healthy giant spiny frogs (*Paa spinosa*) tadpoles. *Lett. Appl. Microbiol.* **2020**, *70*, 109–117. [CrossRef] [PubMed]

14. Xiang, J.G.; He, T.Y.; Wang, P.P.; Xie, M.; Xiang, J.; Ni, J.J. Opportunistic pathogens are abundant in the gut of cultured giant spiny frog (*Paa spinosa*). *Aquac. Res.* **2018**, *49*, 2033–2041. [CrossRef]

15. Ye, C. *Rare and Economical Amphibians of China*; Sichuan Science and Technology Press: Sichuan, Chian, 1993.

16. Zhou, W.; Li, J.; Hou, J.; Xiang, J.; Pan, W. Culture technology of *Quasipaa spinosa* (first)—Site selection and design. *Curr. Fish.* **2021**, *46*, 80–81.

17. Liang, Z.; Xu, Q.; Jiang, Y.; Qin, J.; Deng, W. Breeding situation and development strategy of *Rana spinosa* in Yongfu County. *Guangxi J. Anim. Husb. Vet. Med.* **2013**, *29*, 244–246.

18. Chan, H.K.; Shoemaker, K.T.; Karraker, N.E. Demography of *Quasipaa* frogs in China reveals high vulnerability to widespread harvest pressure. *Biol. Conserv.* **2014**, *170*, 3–9. [CrossRef]

19. Gao, S. *Annals of Animal Medicine in China*; Jilin Science and Technology Publishing House: Jilin, China, 1996.

20. Mei, Y.; Zheng, R.; Zheng, S.; Yan, H.; Liu, Z.; Hong, Y. Gonad differentiation and the effects of temperature on sex determination in *Quasipaa spinosa*. *Acta Ecol. Sin.* **2018**, *38*, 4809–4816.

21. Yu, Z.; Ma, B. Status of the *Quasipaa spinosa* industry in Jiangxi. *Jiangxi Fish. Sci. Technol.* **2012**, *130*, 4–8. [CrossRef]

22. Hou, J.; Long, J.; Xiang, J.; Pan, W.; Li, D.; Liu, X. Ontogenetic characteristics of the intestinal microbiota of Quasipaa spinosa revealed by 16S rRNA gene sequencing. *Lett. Appl. Microbiol.* **2022**, *75*, 1182–1192. [CrossRef]

23. Yu, S.S.; Zhao, Z.H.; Gong, X.F.; Fan, X.L.; Lin, Z.H.; Chen, J. Antimicrobial and immunomodulatory activity of beta-defensin from the Chinese spiny frog (*Quasipaa spinosa*). *Dev. Comp. Immunol.* **2022**, *126*, 104264. [CrossRef] [PubMed]

24. Lei, X.P.; Yi, G.; Wang, K.Y.; OuYang, P.; Chen, D.F.; Huang, X.L.; Huang, C.; Lai, W.M.; Zhong, Z.J.; Huo, C.L.; et al. Elizabethkingia miricola infection in Chinese spiny frog (*Quasipaa spinosa*). *Transbound Emerg. Dis.* **2019**, *66*, 1049–1053. [CrossRef] [PubMed]

25. Liu, Z.P.; Gu, W.B.; Wang, S.Y.; Wang, L.Z.; Zhou, Y.L.; Dong, W.R.; Shu, M.A. Functional differences of three CXCL10 homologues in the giant spiny frog Quasipaa spinosa. *Dev. Comp. Immunol.* **2020**, *109*, 103719. [CrossRef] [PubMed]

26. Wang, X.D.; Xie, Y.G.; Hu, W.; Wei, Z.Y.; Wei, X.Y.; Yuan, H.; Yao, H.Y.; Dunxue, C. Transcriptome characterization and SSR discovery in the giant spiny frog Quasipaa spinosa. *Gene* **2022**, *842*, 146793. [CrossRef]

27. Hou, J.; Zhou, W.; Li, J.; Xiang, J.; Pan, W. Culture technology of *Quasipaa spinosa* (V)—Commercial frog farming. *Curr. Fish.* **2021**, *6*, 68.

28. Zhou, W.; Li, J.; Hou, J.; Xiang, J.; Pan, W. Breeding technology of *Quasipaa spinosa* (III)—Tadpole breeding. *Curr. Fish.* **2021**, *46*, 80+82.

29. Li, J.; Zhou, W.; Hou, J.; Xiang, J.; Pan, W. Breeding technology of *Quasipaa spinosa* (IV)—Feeding and management during metamorphosis. *Curr. Fish.* **2021**, *46*, 75–79+81.

30. Global Biodiversity Information Facility (GBIF) Occurrence Download. Available online: https://www.gbif.org/occurrence/download/0173717-210914110416597 (accessed on 6 March 2022).

31. Venter, O.; Sanderson, E.W.; Magrach, A.; Allan, J.R.; Beher, J.; Jones, K.R.; Possingham, H.P.; Laurance, W.F.; Wood, P.; Fekete, B.M.; et al. *Last of the Wild Project, Version 3 (LWP-3): 2009 Human Footprint, 2018 Relisades*; NASA Socioeconomic Data and Applications Center (SEDAC): Palisades, NY, USA, 2018. [CrossRef]

32. Shi, W.; Zhu, E.; Wang, Y.; Ma, F.; He, Q.; Yi, C. Prediction of potentially suitable distribution area of *Propomacrus davidi Deyrolle* in China based on MaxEnt model. *Chin. J. Ecol.* **2021**, *40*, 2836–2944. [CrossRef]

33. Warren, D.L.; Glor, R.E.; Turelli, M. ENMTools: A toolbox for comparative studies of environmental niche models. *Ecography* **2010**, *33*, 607–611. [CrossRef]
34. Phillips, S.J.; Dudik, M. Modeling of species distributions with Maxent: New extensions and a comprehensive evaluation. *Ecography* **2008**, *31*, 161–175. [CrossRef]
35. Carlson, C.J.; Burgio, K.R.; Dougherty, E.R.; Phillips, A.J.; Bueno, V.M.; Clements, C.F.; Castaldo, G.; Dallas, T.A.; Cizauskas, C.A.; Cumming, G.S.; et al. Parasite biodiversity faces extinction and redistribution in a changing climate. *Sci. Adv.* **2017**, *3*, e1602422. [CrossRef]
36. Chen, Q.H.; Yin, Y.J.; Zhao, R.; Yang, Y.; da Silva, J.A.T.; Yu, X.N. Incorporating local adaptation into species distribution modeling of paeonia mairei, an endemic plant to China. *Front. Plant Sci.* **2020**, *10*, 1717. [CrossRef] [PubMed]
37. Rabinowitz, T.; Polsky, A.; Golan, D.; Danilevsky, A.; Shapira, G.; Raff, C.; Basel-Salmon, L.; Matar, R.T.; Shomron, N. Bayesian-based noninvasive prenatal diagnosis of single-gene disorders. *Genome Res.* **2019**, *29*, 428–438. [CrossRef] [PubMed]
38. Yuan, Y.D.; Tang, X.G.; Liu, M.Y.; Liu, X.F.; Tao, J. Species distribution models of the *Spartina alterniflora* loisel in its origin and invasive country reveal an ecological niche shift. *Front. Plant Sci.* **2021**, *12*, 738–769. [CrossRef] [PubMed]
39. Li, X.; Huang, Y.; Ruan, T.; Wei, W. Maxent model-based evaluation og habitat suitability of Chinese red panda in Qionglai mountains. *J. Guizhou Norm. Univ. Nat. Sci.* **2022**, *40*, 34–39. [CrossRef]
40. Tin, A.; Marten, J.; Kuhns, V.L.H.; Li, Y.; Wuttke, M.; Kirsten, H.; Sieber, K.B.; Qiu, C.X.; Gorski, M.; Yu, Z.; et al. Target genes, variants, tissues and transcriptional pathways influencing human serum urate levels. *Nat. Genet.* **2019**, *51*, 1459–1474. [CrossRef]
41. Allouche, O.; Tsoar, A.; Kadmon, R. Assessing the accuracy of species distribution models: Prevalence, kappa and the true skill statistic (TSS). *J. Appl. Ecol.* **2006**, *43*, 1223–1232. [CrossRef]
42. Wang, W.; Li, Z.J.; Zhang, Y.L.; Xu, X.Q. Current Situation, Global Potential Distribution and Evolution of Six Almond Species in China. *Front. Plant Sci.* **2021**, *12*, 619883. [CrossRef]
43. Elith, J.; Graham, C.H.; Anderson, R.P.; Dudík, M.; Ferrier, S.; Guisan, A.; Hijmans, R.J.; Huettmann, F.; Leathwick, J.R.; Lehmann, A.J.E. Novel methods improve prediction of species' distributions from occurrence data. *Ecography* **2006**, *29*, 129–151. [CrossRef]
44. Merow, C.; Smith, M.J.; Silander, J.A. A practical guide to MaxEnt for modeling species' distributions: What it does, and why inputs and settings matter. *Ecography* **2013**, *36*, 1058–1069. [CrossRef]
45. Phillips, S.J.; Anderson, R.P.; Schapire, R.E. Maximum entropy modeling of species geographic distributions. *Ecol. Model.* **2006**, *190*, 231–259. [CrossRef]
46. Bull, C.M.; Burzacott, D. Changes in climate and in the timing of pairing of the Australian lizard, *Tiliqua rugosa*: A 15-year study. *J. Zool.* **2002**, *256*, 383–387. [CrossRef]
47. Lin, W. *Studies on the Physiology and Biochemistry of Juvenile Quasipaa Spinosa and the Growth and Breeding Technology of Year-Round Phenology*; Fuzhou University: Fuzhou, China, 2017.
48. Soares, C.; Brito, J.C. Environmental correlates for species richness among amphibians and reptiles in a climate transition area. *Biodivers. Conserv.* **2007**, *16*, 1087–1102. [CrossRef]
49. Kong, W.; Li, X.; Zou, H. Optimizing MaxEnt model in the prediction of species distribution. *Chin. J. Appl. Ecol.* **2019**, *6*, 2116–2128. [CrossRef]
50. Ye, X.; Zhang, M.; Lai, W.; Yang, M.; Fan, H.; Zhang, G.; Chen, S.; Liu, B. Prediction of potential suitable distribution of *Phoebe bournei* based on MaxEnt optimization model. *Acta Ecol. Sin.* **2021**, *41*, 8135–8144.
51. Chen, Q. *Physiological and Ecological Studies and the Annual Growth and Culture Techniques of Rana spinosa Tadpoles*; Fuzhou University: Fuzhou, China, 2017.
52. Kong, S. *Interspecific Hybridization of Quasipaa spinosa and Hybrid Tadpole Fitness Research*; Zhejiang Normal University: Jinhua, China, 2017.
53. Xie, Y.; Wei, Z.; Wei, X.; Luo, H.; Chen, D. Acute Toxicity of four conventional disinfectants in aquaculture on tadpoles of *Quasipaa spinosa*. *Fish. Sci.* **2021**, 1–10. [CrossRef]
54. Zhang, D. *Molecular Basis of Local Adaptation and Speciation of Quasipaa spinosa*; Zhejiang Normal University: Jinhua, China, 2020. [CrossRef]
55. Liu, C.; Hu, S. *The Tailless Amphibians of China*; Science Press: Beijing, China, 1961.
56. Tao, Z.; Ma, B.; Yu, Z.; Huang, J.; Zhang, H.; Zhang, A.; Li, Z.; Zhu, G. Effects of environmental factors on the growth of *Rana spinosa* tadpole. *Hunan Agric. Sci.* **2015**, *2*, 55–56+59.
57. Xia, X.; Li, Y.; Yang, D.; Pi, Y. Potential geographical distribution of *Rana hanluica* in China under climate change. *Chin. J. Appl. Ecol.* **2021**, *32*, 4307–4314. [CrossRef]
58. Zou, M.; Zhong, Y. Ecological investigation and artificial trial culture observation of *Quasipaa spinosa* in northwest Fujian. *Chin. J. Zool.* **1986**, *3*, 4–8. [CrossRef]
59. Fei, L.; Ye, C.; Jiang, J. *Color Map of Amphibians and Their Distribution in China*; Sichuan Science and Technology Press: Sichuan, Chian, 2012.
60. Liang, R.; Dong, Y. Ecological survey of *Rana spinosa*. *J. Anhui Norm. Univ. Nat. Sci.* **1984**, *1*, 30–38.
61. Zhao, J.; Yang, X.; Liu, Z.; Cheng, D.; Wang, W.; Chen, F. The possible effect of global climate changes on cropping systems boundary in China II. The characteristics of climatic variables and the possible effect on northern limits of cropping systems in south China. *Sci. Agric. Sin.* **2010**, *43*, 1860–1867. [CrossRef]

62. Guo, Y.; Zhao, Y.; Zhou, Y.; Huang, Q.; Yu, Z.; Gu, Z. Diurnal variation of summer precipitation and its relationship with altitude in Tianshan Mountains of Xinjiang. *Arid Land Geogr.* **2022**, *1*, 57–65.

63. Miao, W.; Liu, S.; Zhu, Y.; Duan, S.; Han, F. Spatio-temporal differentiation and altitude dependence of temperature and precipitation in Meili Snow Mountains. *Clim. Chang. Res.* **2022**, *18*, 328–342.

64. Yang, Q.; Ma, Z.; Chen, L. A preliminary study on the relationship between precipitation trend and altitude in China. In Proceedings of the 28th Annual Meeting of the Chinese Meteorological Society—S17 3rd Annual Meeting of Graduate, Xiamen, China, 1 November 2011; p. 138.

65. Zhang, H.; Zhang, Q.; Liu, Q.; Yan, P. Analysis on variation characteristics and differences of the Climate drying degree between South and North of China. *Plateau Meteorol.* **2016**, *35*, 1339–1351. [CrossRef]

66. Zhao, R.; Huang, X.; Yun, W.; Wu, K.; Chen, Y.; Wang, S.; Lu, H.; Fang, K.; Li, Y. Key issues in natural resource management under carbon emission peak and carbon neutrality targets. *J. Nat. Resour.* **2022**, *37*, 1123–1136. [CrossRef]

67. Tian, S. Study on impact of "Operation Green Fence" on China's carbon emission. *Jiangxi Univ. Financ. Econ.* **2021**. [CrossRef]

68. Yin, Y.; Chang, X. Does China's carbon emission trading policy promote regional green total factor productivity? *Financ. Econ.* **2022**, *3*, 60–70. [CrossRef]

69. Normile, D. Can China, the world's biggest coal consumer, become carbon neutral by 2060? *Science* **2020**. [CrossRef]

70. Wang, S. Comparative study on tadpole growth and development of *Quasipaa spinosa* introduced from different places and local species. *Mod. Agric. Sci. Technol.* **2010**, 278–279.

 *biology*

Article

Larval Dispersal Modeling Reveals Low Connectivity among National Marine Protected Areas in the Yellow and East China Seas

Jiaying Lu, Yuanjie Chen, Zihan Wang, Feng Zhao, Yisen Zhong, Cong Zeng and Ling Cao *

School of Oceanography, Shanghai Jiao Tong University, Shanghai 200240, China
* Correspondence: caoling@sjtu.edu.cn

Simple Summary: In this study, a biophysical model was developed to assess the ecological connectivity of national marine protected areas in the Yellow and East China Seas. The results showed that ocean dynamics, pelagic larval duration, and distribution patterns had significant effects on larval dispersal. The existing national marine reserves in the Yellow and East China Seas did not form a well-connected network, and nearly 30% of them were isolated. The only connections were mostly from north to south. Four marine protected areas (all in coastal Zhejiang) emerged as key nodes for ensuring multi-generational connectivity. Due to the selection of study species with weak to strong potential dispersal, the results of this study can be applied to other organisms with similar life history characteristics, and can provide scientific reference for future reserve planning in coastal China.

Abstract: Marine protected areas (MPAs) are vital for protecting biodiversity, maintaining ecosystem integrity, and tackling future climate change. The effectiveness of MPA networks relies on connectivity, yet connectivity assessments are often skipped in the planning process. Here we employed a multi-species biophysical model to examine the connectivity patterns formed among the 21 national MPAs in the Yellow and East China Seas. We simulated the potential larval dispersal of 14 oviparous species of five classes. Larvae of non-migratory species with pelagic larval duration (PLD) were assumed to be passive floating particles with no explicit vertical migration. A total of 217,000 particles were released according to spawning period, living depth, and species distribution, and they were assumed to move with currents during the PLD. Most larvae were dispersed around the MPAs (0–60 m isobaths) and consistent with the currents. Larval export increased with PLD and current velocity, but if PLD was too long, few larvae survived due to high daily mortality during pelagic dispersal. The overall connectivity pattern exhibited a north-to-south dispersal trend corresponding to coastal currents. Our results indicated that the national MPAs in the Yellow and East China Seas did not form a well-connected network and nearly 30% of them were isolated. These MPAs formed three distinct groups, one in the Yellow Sea ecoregion and two in the East China Sea ecoregion. Four MPAs (all in coastal Zhejiang) emerged as key nodes for ensuring multi-generational connectivity. Under the pressure of future climate change, high self-recruitment and low connectivity present significant challenges for building well-connected MPA networks. We suggest adding new protected areas as stepping stones for bioecological corridors. Focused protection of the Yellow Sea ecoregion could have a good effect on the southern part of the population recruitment downstream. Conservation management should be adjusted according to the life cycles and distributions of vulnerable species, as well as seasonal changes in coastal currents. This study provides a scientific basis for improving ecological connectivity and conservation effectiveness of MPAs in the Yellow and East China Seas.

Keywords: marine protected areas; connectivity; larval dispersal; biophysical modeling; network

Citation: Lu, J.; Chen, Y.; Wang, Z.; Zhao, F.; Zhong, Y.; Zeng, C.; Cao, L. Larval Dispersal Modeling Reveals Low Connectivity among National Marine Protected Areas in the Yellow and East China Seas. *Biology* **2023**, *12*, 396. https://doi.org/10.3390/biology12030396

Academic Editor: John R. Turner

Received: 10 December 2022
Revised: 25 February 2023
Accepted: 27 February 2023
Published: 2 March 2023

1. Introduction

Marine ecosystems are the entities on which thousands of marine creatures depend for survival and reproduction. Human activities and climate change have put increasing pressure on marine ecosystems, leading to a decrease in marine biodiversity [1]. Marine protected areas (MPAs) are frequently established and powerful tools for conserving species. However, global marine biodiversity has continued to deteriorate [2], indicating a great need for more efficient MPA management. Building and improving ecological connectivity between MPAs can effectively assist in forming a well-connected MPA network [3]. An expanding body of empirical evidence has demonstrated the potential benefits of incorporating connectivity into conservation management [4,5]. Both the Aichi Biodiversity Target 11 (https://www.cbd.int/aichi-targets/target/11) (accessed on 3 August 2021) and the 2030 action target three of the Convention on Biological Diversity have called for ecological connectivity of MPAs as an essential criterion when evaluating the effectiveness of MPAs and accomplishing biodiversity goals [6].

The overall connectivity of a protected area network is reflected by studying the ecological connectivity of species in the protected areas, which is primarily affected by larval dispersal [7]. Many marine organisms have a pelagic larval phase and a relatively stationary adult phase, while the larval phase can be quite dispersive. Due to the fluidity and continuity of the ocean, protected species will move about and disperse between different habitats. The dispersal of larvae can greatly affect population dynamics [8]. Linkages between local populations are often maintained only through larval exchange between habitat patches. This is a fundamental ecological process that structures marine populations and confers ecosystems with resilience, and is thus important for planning MPAs [9,10]. Pelagic larval duration (PLD), often known as the time larvae spend as plankton and drift with currents, is the period between spawning and the juvenile stage of marine life history. In general, the longer the PLD, the greater the dispersal potential of the species, and the wider the distribution range; therefore, the larger the area to be protected [7].

Larval labeling [11,12], otolith microchemical analysis [13–15], genetic parentage methods [4,16,17], and biophysical modeling [18,19] are popular methods for assessing ecological connectivity in MPAs by studying larval dispersal. In the context of smaller spatiotemporal scales, larval labels, genetic parental analysis, otolith microchemical analysis, and landscape analysis are efficient techniques. However, these methods require intensive sampling and are usually expensive. Biophysical models can avoid such limitations and allow for accurate descriptions of larval dispersal on spatial and temporal scales [7,8,20–23]. At the beginning of MPA design, other methods are usually not applicable due to data and cost constraints. Biophysical models are therefore the preferred method for assessing the ecological connectivity of MPAs, especially at larger temporal and spatial scales [24]. The effects of larval behavior and topography on larval dispersal along the coast as well as population recruitment of the larval stages of species and connectivity between MPAs have been demonstrated by a number of realistic numerical models of coastal oceans [10,23,25,26]. These are ecologically pivotal phases of the dispersal process, but they are all correlated to some degree by the interactions of the characteristics of larvae (e.g., PLD) and the velocity fields of a particular ocean region.

MPAs have been in place in China for nearly six decades. As of 2021, 273 MPAs had been established, although the effectiveness of China's MPAs has been impeded by accumulated issues over the years [27]. Most existing marine reserves were designed and managed according to limited local systems, and as such only few MPAs have involved ecological connectivity in marine conservation [28]. The connection patterns of existing protected areas are still unclear. Although connectivity was not considered in the initial construction of the reserves, it is likely that these MPAs are connected due to the movement of planktonic larvae under the influence of strong coastal currents. Due to the rich biodiversity and marine resources, MPA establishment in the Yellow and East China Seas has made significant progress. The Yellow and East China Seas initially formed a national MPA system with marine biodiversity protection as the core goal [29]. National MPAs have the highest protection level and represent the highest conservation value for biodiversity, but there are few studies on the connectivity assessment among these MPAs.

In this study, we assessed the ecological connectivity among the national MPAs of the Yellow and East China Seas, and examined whether the existing reserves have formed a well-connected network. By simulating potential larval dispersal patterns of species with different life histories and distribution characteristics, we investigated the importance of individual MPAs in strengthening ecological connectivity. This study incorporated multi-species and ecological connectivity into MPA planning, aiming to provide a scientific basis for improving the effectiveness of marine biodiversity conservation.

2. Materials and Methods

2.1. Study Region

Based on the consideration of comparing the connectivity between the Yellow Sea ecoregion and the East China Sea ecoregion, this study primarily focused on the continental shelf waters of the Yellow and East China Seas. The study area ranged from Lianyungang, Jiangsu Province, in the north to Dongshan County, Fujian Province, in the south (Figure 1). The study focused on 21 national MPAs (accounting for 27.3% of total national MPAs in China) and adjacent waters (24° N–36° N), involving fourteen special marine protected areas (SMPAs) and seven marine nature reserves (MNRs). MNR, in which extractive activities are highly restricted, and SMPA, including marine parks, in which multiple resource use is allowed. The ocean currents in the Yellow and East China Seas are from coastal currents (the Yellow Sea Coastal Current, the Yangtze Diluted Water, the Zhejiang–Fujian Coastal Current, and the Taiwan Warm Current) and the Kuroshio Current and its branches.

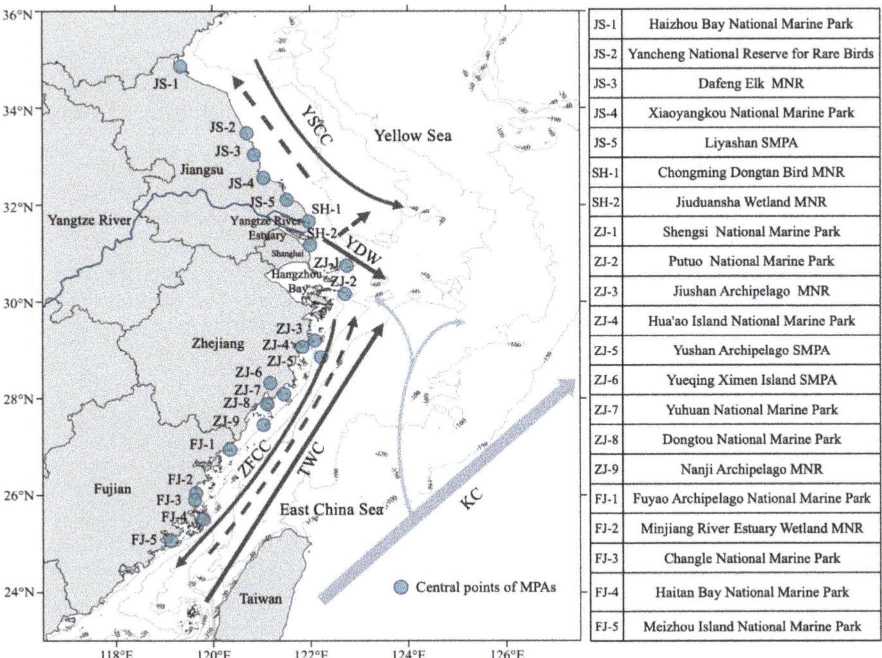

Figure 1. Distribution of 21 national MPAs and major currents in the Yellow and East China Seas. The blue points are the central points of MPAs. Arrowhead lines show the current directions. Dashed arrowhead lines are the possible directions of YSCC, YDW, and ZFCC in summer [30–33]. YSCC: the Yellow Sea Coastal Current; YDW: Yangtze Diluted Water; ZFCC: Zhejiang–Fujian Coastal Current; TWC: Taiwan Warm Current; KC: Kuroshio Current. MNR and SMPA in right table were marine nature reserves and special marine protected areas, respectively.

2.2. Ocean Model Configuration

Larval dispersal between MPAs is strongly influenced by ocean currents and is also a function of species' life history [22]. For this reason, this study developed a biophysical modeling approach based on Regional Ocean Modeling System (ROMS) to quantify potential larval dispersal between MPAs in the Yellow and East China Seas. ROMS is a free-surface, terrain-following, three-dimensional nonlinear baroclinic ocean model widely used for larval dispersal in connectivity modeling [23,34,35]. This model solves the Reynolds-Averaged Navier–Stokes equations based on hydrostatics and Boussinesq approximation in sigma terrain-following coordinates (S-levels) and a curvilinear orthonormal Arakawa C grid over the vertical and horizontal axes, respectively [35]. The modeled area of this study followed Chen (2020) [31] and was located at 117° E–135° E, 24° N–42° N, including the Bohai Sea, the Yellow Sea, the East China Sea, the Korean Strait, part of the Sea of Japan, and the Northwest Pacific Ocean [31]. The model grid mesh spanned 730 × 438 cells in the horizontal direction (Figure S6) and contained 20 vertical layers. Horizontal grid density was enhanced near the coast of mainland China. The horizontal resolution varied from ~1 km near the Yangtze Estuary to ~6 km near the east open boundary.

The bathymetric data came from two sources: (1) digital chart data with high accuracy were used in the Yangtze Estuary, Hangzhou Bay, and Subei Shoal; (2) ETOPO1 data with a resolution of 1/60° were used in other sea areas. Since this study focused on the continental shelf area, the minimum depth of water was taken as 5 m. The maximum depth was 2000 m. The model northern and western boundaries were closed, and the eastern and southern boundaries were open boundaries employing the Chapman scheme [36] and the Flather scheme [37] for the free-surface and two-dimensional velocities, respectively. The atmospheric forcing and boundary forcing data were acquired from the ECMWF (The European Centre for Medium-Range Weather Forecasts) ERA-Interim reanalysis product and the Copernicus global ocean analysis product. The tidal harmonic constants on the open boundary were taken from the Global Tidal Database of Oregon State University (http://volkov.oce.orst.edu/tides/) (accessed on 7 March 2019) comprising 13 constituents (M2, S2, N2, K2, K1, O1, P1, Q1, M4, MS4, MN4, MM, and MF). The time step for the inner mode was 120 s, and the time step for the outer mode was 6 s. The model was continuously integrated from the initial stationary state. The simulation time ranged from 1 January to 31 December 2016, because the particle dispersal dynamic in 2016 was close to the average field in recent eight years (see Supplementary Materials). The verification results showed that the model could simulate the dynamic characteristics of the currents in the Yellow and East China Seas [31].

2.3. Lagrangian Particle Dispersion Model

A Lagrangian particle dispersion model-based ROMS was used to simulate larval dispersal of different species. In the model, particles that moved passively with ocean currents represented the larvae of species. These particles were released at specific locations, times, and water depths according to their life history characteristics, regardless of active swimming and vertical migration. To analyze the ecological connectivity of different species in the Yellow and East China Sea national marine reserves, this study focused on marine organisms distributed in the abovementioned areas with high conservation or fishery value. Representative species for connectivity assessment were selected from (1) the species assessed as critically endangered (CR), endangered (EN), or vulnerable (VU) in the IUCN Red List of Threatened Species, and (2) the main economic species in The Atlas of Main economic species in the East China Sea. A species could be selected if it met the criteria of (1) living in one of the 21 MPAs in the Yellow and East China Seas, (2) belonging to marine groups, (3) having a planktonic larval stage, and (4) having exhaustive life history information.

The biological parameters of the model included life history, distribution information, pelagic larval duration (PLD), living depth, and spawning window [38] that have been demonstrated to play important roles in ecological connectivity. The data were collected from published articles and public databases (Table S1). To make the research more general, study species were selected to represent other organisms with similar life history characteristics (pelagic larval duration, living depth, and spawning season) and covering as diverse a range of each biological parameter as possible. The process was completed on 8 June 2022, resulting in 14 species (Table 1) in five classes (Actinopterygii, Actinozoa, Crustacea, Holothuroidea, Cephalopoda). These species represented a range of dispersal phenotypes (from shorter to longer dispersal distances). The PLD of the 14 species ranged from 7 to 75 days and was set as the floats' release duration in the model. Timing of spawning was defined as the time when spawning events happened during the entire year.

Table 1. The biological parameters of species included in the model.

Scientific Name	Classification	Category	PLD (Days)	Depth (m)	Spawning Window
Nemipterus virgatus	Actinopterygii	Threatened	75	25	April–May
Sepiella maindroni	Cephalopoda	Economic	45	10	April–May
Epinephelus bruneus	Actinopterygii	Threatened	45	20	May–June
Larimichthys crocea	Actinopterygii	Threatened	33	15	April–May
Argyrosomus argentatus	Actinopterygii	Economic	33	50	June–August
Argyrosomus japonicus	Actinopterygii	Threatened	33	100	January–March
Evynnis cardinalis	Actinopterygii	Threatened	30	45	November–January
Apostichopus japonicus	Holothuroidea	Threatened	20	10	May–June
Scomberomorus niphonius	Actinopterygii	Economic	19	20	May–June
Portunus Trituberculatus	Malacostraca	Economic	17	20	April–June
Epinephelus akaara	Actinopterygii	Threatened	15	25	April–June
Anguilla japonica	Actinopterygii	Threatened	10	3	November–December
Penaeus japonicus	Crustacea	Economic	10	20	December–March
Acropora solitaryensis	Anthozoa	Threatened	7	5	April–May

Note: PLD: pelagic larval duration, Depth: experimental living depth, Spawning window: the species' spawning period in the Yellow and East China Seas. Species were classified into endangered and economic species according to the number of distribution sites. The scientific names were displayed in order of PLD length.

All of these species only had one spawning window per year. The particles were released at 12:00 a.m. in the middle day of the spawning window. The particles were released according to the distribution of species [39]. Once a species was likely to occur in any MPA, the central point (with a resolution of 0.0001°) of the MPA was set as the starting point. For one species, 1000 particles were released from each starting point. Passive particles with a constant size were released without vertical migration, ontogenetic changes, or swimming capabilities between the depth of 5 and 100 m in terms of their habitats. The position of particles was output every six hours. Both central points of MPAs and the number of particles released for each species are displayed in Table 2. For example, for *Anguilla japonica* that occurred in 21 MPAs, 21,000 particles were released from MPAs and floated passively for 10 days. Although the influence of mortality on connectivity was widely recognized [7], we had little information about the exact effects (i.e., sea temperature) on larval mortality since the field-based data are limited. The mortality was usually set at a fixed value in the studies of MPA connectivity [23,40]. It is known that the first stages of larval life are characterized by a high daily mortality rate that ranges from 10% to 20% [40,41]. Thus, this study included mortality rates, and 15% of the particles of each species were removed daily and at random to obtain effective connectivity information during model output analysis.

Based on the data obtained from the abovementioned models, species dispersal trajectories were drawn to determine whether the established MPA network could cover the species' corridors. The dispersal fluxes of 14 species to each protected area were calculated to represent the accessibility of species to the protected area (taking the actual boundary of MPAs as the range). The dispersal flux d(i) was defined as the particle abundance of each species at 21 national MPAs. Dispersal fluxes of some species in MPA i were calculated by the fraction of the cumulative quantity of larvae particles entering MPA i, $r(i)$ to the total quantity of particles released from all k starting points.

$$d(i) = r(i) / \sum_x r(i) \tag{1}$$

Table 2. The occurrence of study species and the number of released particles in each MPA.

MPA	Area (km²)	*Acropora solitaryensis*	*Apostichopus japonicus*	*Epinephelus akaara*	*Epinephelus bruneus*	*Nemipterus virgatus*	*Scomberomorus niphonius*	*Argyrosomus japonicus*	*Larimichthys crocea*	*Evynnis cardinalis*	*Portunus Trituberculatus*	*Anguilla japonica*	*Argyrosomus argentatu*	*Sepiella maindroni*	*Penaeus japonicus*
JS-1	514.55								✓	✓	✓	✓	✓	✓	✓
JS-2	2472.6							✓	✓	✓	✓	✓	✓	✓	✓
JS-3	780								✓	✓	✓	✓	✓	✓	✓
JS-4	47.1								✓	✓	✓	✓	✓	✓	✓
JS-5	15.46								✓	✓	✓	✓	✓	✓	✓
SH-1	241.55									✓	✓	✓	✓	✓	✓
SH-2	423.2									✓	✓	✓	✓	✓	✓
ZJ-1	549		✓	✓	✓	✓	✓	✓	✓	✓	✓	✓	✓	✓	✓
ZJ-2	218.4		✓	✓	✓	✓	✓	✓	✓	✓	✓	✓	✓	✓	✓
ZJ-3	484.78					✓	✓	✓	✓	✓	✓	✓	✓	✓	✓
ZJ-4	44.19					✓	✓	✓	✓	✓	✓	✓	✓	✓	✓
ZJ-5	57		✓	✓	✓	✓	✓	✓	✓	✓	✓	✓	✓	✓	✓
ZJ-6	30.8					✓	✓	✓	✓	✓	✓	✓	✓	✓	✓
ZJ-7	306.69			✓	✓	✓	✓	✓	✓	✓	✓	✓	✓	✓	✓
ZJ-8	311.04			✓	✓	✓	✓	✓	✓	✓	✓	✓	✓	✓	✓
ZJ-9	201.06		✓	✓	✓	✓	✓	✓	✓	✓	✓	✓	✓	✓	✓
FJ-1	67.83			✓	✓	✓	✓	✓	✓	✓	✓	✓	✓	✓	✓
FJ-2	22.6			✓	✓	✓	✓	✓	✓	✓	✓	✓	✓	✓	✓
FJ-3	24.44			✓	✓	✓	✓	✓	✓	✓	✓	✓	✓	✓	✓
FJ-4	34.9	✓	✓	✓	✓	✓	✓	✓	✓	✓	✓	✓	✓	✓	✓
FJ-5	69.11	✓		✓	✓	✓	✓	✓	✓	✓	✓	✓	✓	✓	✓
Distribution points		2	5	11	11	14	14	15	19	21	21	21	21	21	21
Number of Released particles		2000	5000	11,000	11,000	14,000	14,000	15,000	19,000	21,000	21,000	21,000	21,000	21,000	21,000

However, higher larval abundance in one national MPA could only indicate that the accessibility of the area for that species was better, and this would not explain population recruitment of species and connection probability $c(i,j)$ between MPAs. Therefore, further analysis of self-recruitment rate selfr(i) and subsidy recruitment rate subr(i,j) of the protected area needed to be included. Here, each protected area was regarded as a node in the network. The connection probability was the proportion of larvae departing from MPA i that arrived in MPA j [42]. The self-recruitment rate was defined as the proportion of larvae that originated from a MPA and reached the same MPA. The subsidy recruitment rate of a single MPA was the proportion of larval replenishment coming from other MPAs summed over all MPAs and species [8]. The connectivity probability revealed the relative contribution of every MPA to surrounding protected areas in terms of recruitment, i.e., sources and sinks of the population recruitment. Protected areas with high source intensity could promote biodiversity conservation of both protected species and economically important species, and thus they were crucial for the entire MPA network [43].

$$\mathrm{selfr}(i) = c(i,i) / \sum_j c(i,j) \tag{2}$$

$$\mathrm{subr}(i,j) = \sum_{j \neq i} c(i,j) / \sum_j c(i,j) \tag{3}$$

Patterns of larval abundance at the end of the pelagic larval duration demonstrated the potential larval dispersal. Since MPAs' size, location, time since establishment, fishing restrictions, and regulation enforcement are important factors affecting larval dispersal [44,45], larval export was likely to vary among MPAs. Because of the lack of data, this study only involved MPA size, and the MPAs in the pattern were drawn with weights in terms of their sizes.

Graph theory was used to quantify each national MPA's ecological connectivity within the network [8,43]. Graph theory is widely used to measure and visualize connectivity patterns between reserves in a lot of related research [7,8,20,21,23]. Nodes and edges of the graph were defined by MPAs and larval trajectories, respectively [46]. Two network node importance metrics were illustrated: (1) Degree centrality of node i was the number of nodes connected to node i. The assumption was that the crucial MPAs were those that had many connections. The more connections an MPA had, the stronger its influence. Since the connection between two protected areas was directional, we used indegree and outdegree of graph theory to quantitatively analyze the role of each MPA in larval recruitment. They referred to the set of MPAs connected immediately upstream (Indegree of nodes) and downstream (Outdegree of nodes). (2) Betweenness centrality of node i was the sum of all the shortest paths connecting nodes x and y across node i, according to Andrello et al. [8] and Treml et al. [43]:

$$bc(i) = \sum_{x \neq y \neq i} \frac{\sigma_{xy}(i)}{\sigma_{xy}} \tag{4}$$

where σ_{xy} is the total number of shortest paths between node x and node y, and $\sigma_{xy}(i)$ is the number of those passing across the node i. When incorporating connectivity into the design of an MPA network, analysis of the effect that every MPA plays in promoting connectivity has been demonstrated to be significant [47]. As an MPA is connected with other MPAs in a multi-step way, each act as an essential node to transfer genes and individuals between MPAs that are not directly connected. The roles of such central MPAs were identified by betweenness centrality [48]. As a result, betweenness centrality demonstrates the significance of each node in acting as a corridor for propagating larvae across the network [46,48] and is related to the landscape ecology concept of "stepping stones" [42]. Degree centrality and betweenness centrality are therefore two complementary node-level metrics that quantify the connectivity of single MPAs in relation to their closest neighbors (degree centrality) or to all nodes within the network (betweenness centrality) [8]. The network connectivity variables were obtained by the equal-proportion addition of threatened species and economically important species. All metrics were calculated by the igraph package in R [49].

3. Results

3.1. Larval Dispersal

The entire flow field simulated by the model (Figure 2) was consistent with the results of previous studies [30,31]. The velocity of most of the study areas was 0.1–0.2 m/s. The dominant current of the Yellow Sea ecoregion (i.e., YSCC) was much weaker than that of the East China Sea ecoregion. Outside the mouth of the Yangtze River, the YDW generally flows to the northeast during the southerly winds in Seasons 2 and 3. In Season 1, the runoff of the Yangtze River decreases greatly, and under the northerly winds the YDW flows southward along the coast. ZFCC was dominant along the coast of Zhejiang and Fujian provinces in the East China Sea ecoregion. Because this flow was located close to the MPAs, it may be the ocean current that has the greatest impact on population connectivity between protected areas in the study region. The upstream area of ZFCC (near the Yangtze River and the Hangzhou Bay) had the highest velocity, especially during Season 4 (October–February) due to the stable and strong northeast winds. The TWC is the current occurring on the east side of the East China Sea coastal current and south of the Yangtze Estuary. This current flowed to the northeast along the coast of Fujian and Zhejiang almost all year round, except in Season 4 when the surface layer was easily affected by the northerly monsoon and the flow direction turned southward.

Most of the particles were scattered within the 40 m isobath and radiated outward from the released points. As a result of a combination of biological characteristics, ocean currents and habitat distribution, the potential dispersal path of different species varied widely (Figures S1–S4). Similar to previous studies, for most species, the longer the planktonic larval stage, the wider the larval dispersal range, and the higher the degree of connectivity between protected areas [21]. However, when the PLD was too long, for example, in *Nemipterus virgatus* with the longest PLD (75 days), due to high daily mortality during the floating period the survival rate of larvae decreased rapidly, and thus the ecological connectivity of this specie was limited in the study area (Figure S1).

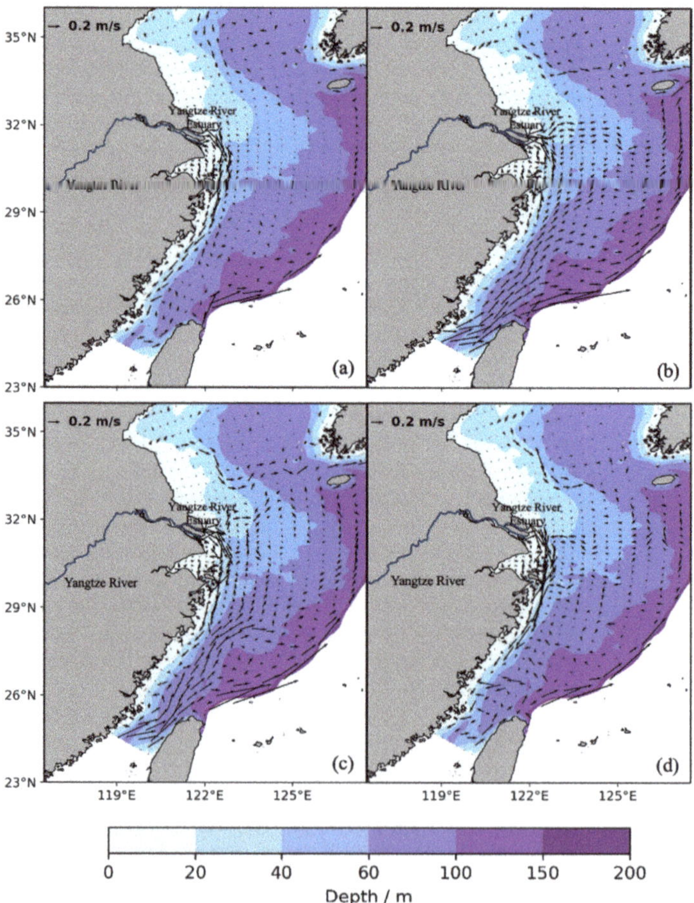

Figure 2. Modeling quarterly mean velocity and direction of ocean currents in Yellow and East China Seas: (**a**) January–March, (**b**) April–June, (**c**) July–September, and (**d**) October–December. The length of the line indicates the average velocity, and the arrows indicate the current direction.

Ocean current directions and the continental shelf appeared to account for the spatial patterns of larval dispersal (Figure 3). Larvae could accumulate even though they floated far from their original location (the Yangtze River estuary), where the continental shelf was large (40–60 m isobath). The directions of ocean currents dominated larval transport. Currents were likely to carry larvae for long distances to areas unsuitable for settlement. Because the velocity of surface water was greater than in deeper layers, species living in shallow water were more likely to reach distant MPAs. Although the larval concentration of different species differed widely in the pelagic part, similar aggregation phenomena existed in the nearshore reserve. The floating trajectory of the particles extended as far as the 60 m isobath sea area, and the diffusion path was highly consistent with the direction and intensity of ocean currents in four seasons.

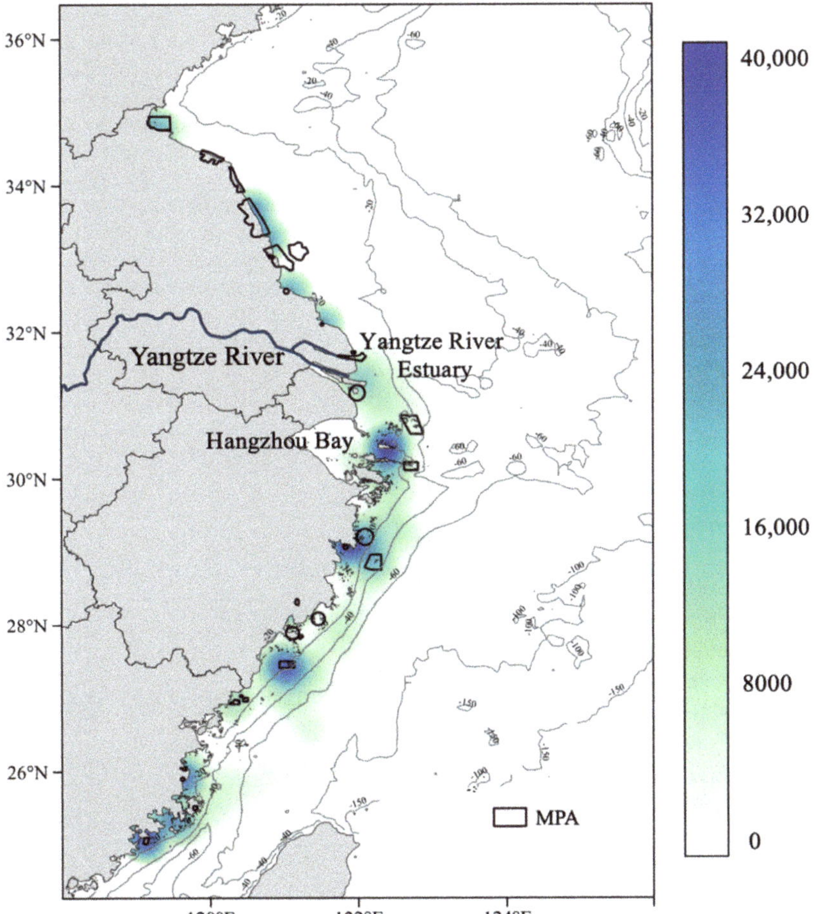

Figure 3. Total larval abundance at the end of pelagic larval duration. The color indicates the density particles. Particles of all 14 species were superimposed and were given the same weight.

Since the national MPAs were usually close to the coast, larval dispersal was primarily affected by coastal currents and the Kuroshio branches. Under the influence of the Yellow Sea Coastal Current and the Taiwan Warm Current as well as the Kuroshio current branches, the particles from the north and south gathered together in the Yangtze River estuary with high abundance. The current velocity near the Yangtze River was relatively high. Because of the Yangtze-diluted water, particles had a clear trajectory from west to east and moved northward from winter to summer. Due to the variation in the direction and velocity of currents in the four seasons, the dispersion patterns of particles released in different spawning seasons showed significant differences. Spawning from October to December may be conducive to larval diffusion and enhanced species communication. Larval export distance increased between Season 1 and Season 2 owing to the strengthening of coastal current velocity (from south to north). Because of the fast flow at the Yangtze River estuary from April to June, even though the PLD of *Anguilla japonica* was short, there was still pronounced particle aggregation near the ocean currents.

The number of distribution points affected larval dispersal in this study. Due to the wide distribution of economically important species, their dispersal fluxes in 21 MPAs were larger than those of endangered species, especially *Penaeus japonicus* and *Anguilla japonica* (Figure 4). Conversely, since *Acropora solitaryensis* only occurred at two MPAs according to the assessment of suitable habitats (Table 2), the distribution characteristic resulted in high concentration in a single area. High larval abundance was found in JS-2, JS-5, ZJ-3, ZJ-9, FJ-3, and FJ-5, indicating that the MPAs mentioned above were focal reserves with higher species diversity. Despite the distribution points, species with short PLD had higher abundance in the MPAs (e.g., *Acropora solitaryensis*, *Epinephelus akaara*, *Penaeus japonicus*, and *Anguilla japonica*). When the PLD was ≥ 45 days (the top three rows in Figure 4), larvae followed the currents for longer periods of time, floating far beyond the MPA boundary limits relative to their size, which reduced the proportions remaining in protected areas. However, the high abundance of particles in a protected area could only indicate that the MPA was highly accessible for certain species, and this could not explain the population recruitment of species and connection among 21 MPAs in the network. Therefore, recruitment rate calculation and directed network analysis were required.

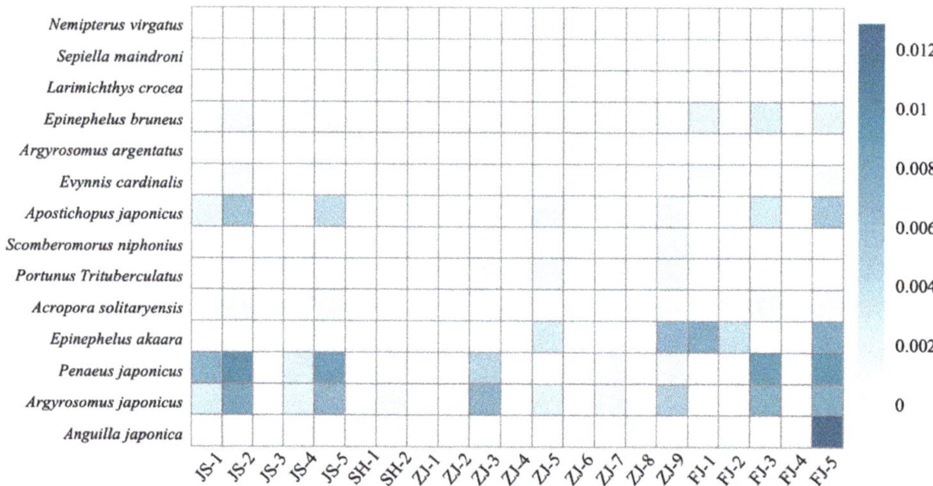

Figure 4. Dispersal fluxes of 14 species in each MPA. The species are arranged from top to bottom according to the length of PLD.

3.2. MPA Ecological Connectivity Analysis

Under comprehensive consideration of different types of representative species in the Yellow and East China Seas, the established MPAs did not form a connected protected area network. The self-recruitment fraction of all protected areas was far higher than the subsidized recruitment fraction and was highly variable among MPAs (Figure S5). Self-recruitment fractions were distributed serially in the range of 0 and 1 (median 0.029, interquartile range 0.0415). Six MPAs relied totally on self-recruitment (JS-1, JS-4, JS-5,

SH-1, SH-2, FJ-5), while seven MPAs had zero self-recruitment (JS-3, ZJ-1, ZJ-2, ZJ-4, ZJ-7, ZJ-8, FJ-4). Total self-recruitment indicated that the majority of larvae tended to stay in the original reserves rather than migrate to adjacent protected areas that showed zero self-recruitment. MPAs ZJ-1, ZJ-2, and ZJ-8 relied only on subsidized recruitment. As a consequence, the ecological connectivity of MPAs in the Yellow and East China Seas was very low. MPA FJ-5 had the highest self-recruitment fraction (0.08). The highest subsidized recruitment fraction was from MPA ZJ-8 to ZJ-9 (0.02). The high dispersal fluxes shown in Figure 4 of MPA JS-5, ZJ-3, FJ-3, and FJ-5 were due to the larvae distributed in these MPAs not leaving the MPAs or returning to the original MPAs, resulting in high rates of self-recruitment.

Consistent with the results in Figure S5, the subsidized recruitment fraction between connected MPAs (Figure 5) was always very low, mostly in the range of 0.0001–0.001 (median 0.00019, interquartile range 0.00077). Among the Yellow Sea ecoregion, JS-3 contributed high larval supplementation to JS-2, where the connection direction was from south to north. No connection was observed between other protected areas in the region. The population replenishment between MPAs located at the coast of Zhejiang in the East China Sea ecoregion was clearly better than those of other provinces. Most larvae were exported from northern protected areas to southern protected areas, and only four connections were from south to north. Among the latter, MPA ZJ-2 to ZJ-6 and ZJ-8 to ZJ-3 and ZJ-9 had the highest recruitment. As to Fujian province in the south of the East China Sea ecoregion, MPA FJ-4 had larvae supplemented to FJ-2 and FJ-3 movement from south to north.

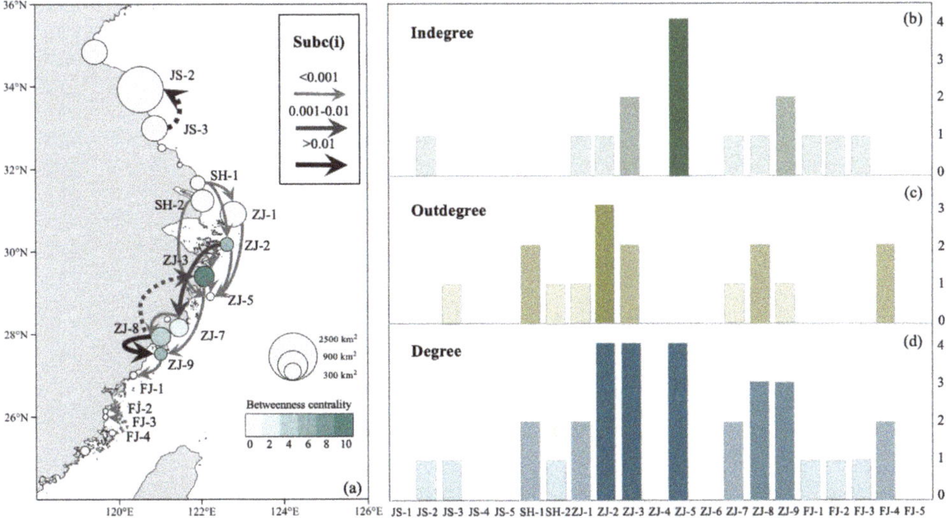

Figure 5. Schematic diagram of connections between MPAs and the network node importance metrics. Circle sizes indicate the relative sizes of protected areas. The directions of the arrows represent the direction of species particle dispersal. Dashed lines indicate from south to north, and solid lines indicate from north to south. Thickness of the lines represents the subsidized recruitment rate between MPAs. The network node importance metrics are shown in (**a**) Betweenness centrality, (**b**) Indegree centrality, (**c**) Outdegree centrality, and (**d**) Degree centrality.

The degree of centrality (indegree and outdegree) and betweenness centrality varied greatly across MPAs (Figure 5), indicating the importance of individual MPAs. The degree pattern clearly showed that MPAs along Zhejiang province had a higher degree (Figure 5d) and betweenness centrality (Figure 5a) among the 21 MPAs, probably due to their central positions where the estuary of the Yangtze River met the coastal currents. MPA ZJ-2,

ZJ-3, and ZJ-5 had the highest degree (which was four), making the most connections to other protected areas. Having the highest outdegree (which was three), MPA ZJ-2 was an important source for the surrounding reserves (Figure 5c). Meanwhile, the degree of MPA ZJ-5 was the same as the indegree (Figure 5d), i.e., its connections were all due to larval supplementation from other protected areas. In particular, due to the frequent southward connections, MPA ZJ-2 and ZJ-3 possessed high betweenness centrality. These MPAs thus were inferred as key nodes to ensure the connectivity between the Yellow Sea and East China Sea ecoregions. The highest betweenness centrality occurred in MPA ZJ-3 (which was ten), showing that it was the most critical MPA in the entire network.

To summarize, MPAs in the Yellow and East China Seas did not form a systematic protection network, as the connections between MPAs were always one-way only. There were three MPA clusters located in Jiangsu, Shanghai, and Zhejiang, and Fujian coastal areas. Jiangsu and Fujian MPA clusters only involved 2–3 MPAs, and all of the connections were from south to north. The ecological connectivity was very low, and MPAs were relatively isolated. The Shanghai and Zhejiang (SZ) MPA cluster covered at least nine MPAs that had larval export or recruitment with other MPAs (Figure 5a). Moreover, the MPAs of this cluster had the highest degree of centrality among all of the studied MPAs, and five MPAs of the cluster were demonstrated to be crucial "stepping stones" (high betweenness centrality), and thus ecological connectivity was highly strengthened by frequent multidirectional connections. In addition, MPA ZJ-9 seemed to be the "hub" of the Fujian MPA cluster and the SZ MPA cluster, as it was the closest geographically to MPA FJ-1.

4. Discussion

MPAs are widely identified as useful tools in biodiversity conservation and sustainable fisheries research, while their effectiveness relies on species larval export and ecological connectivity between MPAs [4,5,26]. Taking representative species with fishery and conservation value as a case study, this paper has revealed that (1) the connectivity between national MPAs in the Yellow and East China Seas was very low, and the system did not form a connected network; (2) larval dispersal and connectivity were primarily affected by coastal currents, PLD, and species occurrence; (3) individual MPAs can be crucial sources or sinks of populations (high recruitment) and can be essential in maintaining the connectivity of entire system (high degree and betweenness centrality).

4.1. Model Strengths and Limitations

Early studies on MPAs employed markers and sampling methods [12,14]. More recently, genetic analysis and microchemical methods have been significant developments and extended to deduce patterns of ecological connectivity and larval dispersal [15,16,26]. These methods incur high costs and involve high-intensity sampling and experiments for analysis, and thus they can barely be used on wider spatial and time scales. In this context, the advantages of biophysical modeling are highlighted due to their low cost, no sampling requirement, and applicability to broad spatial and temporal scales with the support of sufficient biological and hydrodynamic data. This study is the first evaluation of ecological connectivity among several MPAs at the scale of entire seas in China.

Although there is a trend toward increasing the number and coverage of China's MPAs, biodiversity conservation is not working well [2]. One of the most mentioned problems is the decentralization due to the lack of systematic planning and design that demands a connectivity assessment of the whole system and the establishment of an effective MPA network in China [28]. Sustainable population recruitment determined by larval dispersal is more important for conservation diversity than MPA size and shape [20]. The model presented in this paper can be taken as one of the first steps in attempting to achieve this task by biophysical modeling.

Nonetheless, this study has four main limitations. First, although the horizontal grid size of ROMS was narrowed near the coast (the resolution near the Yangtze River estuary is precise to 1 km), larval export and connectivity can be affected by ocean processes at much smaller scales [50]. Models with a higher resolution can better simulate the demographics of larval dispersal [51].

Second, the larval export was predicted under the hypothesis that larvae were passive particles without growth or voluntary movement and behavior, while vertical migration, active movement, and interaction with resident species can change the patterns of larval dispersal [21,22]. The passive floating that only relied on currents may underestimate the probability of recruitment [38]. Future models that fully considered the abovementioned biological factors would be conducive to biological process research [52].

Third, due to the lack of latest data, the life cycle information of studied species was mostly from the literature and research from past decades (Table S1). The biological data from recent years will be beneficial for the prediction results that would be closer to the actual larval export [7]. In addition, this study simulated the larval dispersal over one year. Considering the impact of climate change and other environmental variables on biological behaviors and hydrodynamic modes over decades, it is possible that there may be significant differences in the connectivity patterns from year to year [53].

Finally, species can also spawn and reproduce in unprotected areas or in other MPAs with lower conservation levels. The individuals from these spawning and nursery grounds can also recruit the national MPAs involved in this study and strengthen the larval abundance. From this perspective, the connectivity may be underestimated. Moreover, subsidized recruitment depends on the productivity gap between protected areas and fished areas [23]. Potential larval supply from outside the MPAs is unknown due to the lack of productivity data both within and outside the MPAs. Thus, predicting whether the number of larvae exported is adequate to lead to recruitment supply remains challenging.

In consideration of these restrictions, the present results should be taken with caution. A model with a higher resolution, a better understanding of larval biology, and finer data for productivity, biomass, and intensity of fishery could optimize the analyses and results in this work [51]. This study conducted larval dispersal simulations on species with different dispersal strategies and analyzed the effects of ocean currents, pelagic larval duration, living depth, and spawning month on larval dispersal, taking into account survival rates. Instead of assuming that all species occur in each MPA, simulation for larvae was carried out based on previous suitable habitat studies on related species, with the goal to approximate actual larval dispersal. In spite of all these restrictions, the results remained conservative, as this work considered an optimistic scenario when assessing connectivity. That is, fisheries were effectively confined in all MPAs in the Yellow and East China Seas, and the moderate larval pelagic mortality was taken in the analyses.

4.2. Ecological Connectivity between the MPAs

The main result of this study is that the Yellow and East China Sea MPAs did not form a well-connected network, as there was very low ecological connectivity. Larvae spread to a limited number of adjacent MPAs, and most did not settle in the protected areas. The number of connections generated by 14 species between 21 MPAs was only 33 (degree). In addition, 88% of connections were from northern to southern MPAs, and MPAs in Jiangsu and Fujian only had northerly larval export. Six MPAs were totally isolated, and 70% of the MPAs had connections, while most of these were unidirectional, highlighting the weak connectivity. Moreover, the connectivity in the East China Sea was much better than in the Yellow Sea ecoregion. The notable finding was that Zhejiang MPAs had a relatively tight cluster involving important sinks and sources (indegree and outdegree) compared with the coasts of other provinces. Apart from the small number of connections between MPAs, the connection probabilities were always low. The low connection probability (<0.001) indicated that the population migration caused by larval dispersal was insufficient to influence demography [53] but may be sufficient for genetic differentiation and evolution [8]. In this work, only 30% of the connection probabilities between MPAs in the Yellow and East China Seas were above this value, meaning that these connections can be considered demographically relevant.

Most connections between MPAs were consistent with the direction of major coastal currents in the Yellow and East China Seas. In spite of this, because the distribution data in this study were derived from the results of suitable habitat model simulation, the consequences of larval dispersal and recruitment were supposed to be interpreted as potential settlement availability. Since the three major coastal currents (YSCC, YDW, ZFCC) possibly change their directions during different seasons, the larval export of species with distinct spawning times showed various patterns, especially in winter and summer. The MPAs along the coasts of Zhejiang and Fujian provinces showed frequent bidirectional connections, although the southern direction connections dominated (Figure 5). We attributed this to the fact that most of the studied species spawn from December to June, which are critical period for marine fish breeding in the Yellow and East China Seas. The dominant coastal current, ZFCC, is strong and southward in spring and winter but weak and northward in summer and autumn. During the spawning time, southerly currents are strong. As a result, most larvae flowed from the northern MPAs to the southern MPAs. Less directional connectivity from southern to northern MPAs was observed in Jiangsu and Fujian provinces. It seems the connections were determined by summer coastal currents that were probably due to the greater distribution of species spawning in Season 2 in the MPAs.

MPAs located upstream of the coastal currents (ZJ-3, ZJ-4, ZJ-5) are important larval sources for downstream reserves. Because these currents span long distances, after having received larvae from upstream MPAs, these MPAs could act as sources of larvae for the southern protected areas, ensuring sufficient connectivity in this zone. Larval abundance was generally high in the waters around the Yangtze Estuary that is likely to be an area with rich biodiversity due to the transport of sediments and nutrients by the currents [31]. This result highlights the need for the incorporation of these areas in the current MPA zonation to improve the MPA effectiveness. ZJ-9 could act as a bridge between the Zhejiang and Fujian coasts thanks to relatively high MPA density and the opposite flow of ZFCC in summer and winter, leading to a Zhejiang MPA cluster. Compared with Zhejiang MPAs, the low MPAs density of the other three provinces indicates that connectivity may depend on the number and spatial arrangement of MPAs. In other areas of the globe where MPAs are fewer, connectivity will be likely lower than in the East China Sea.

Populations of almost 50% of the MPAs were only replenished by locally bred larvae, and there was no supply from other MPAs. Previous studies have also found low connectivity and high self-recruitment using biophysical models for other seas [7,38]. Low connectivity and high self-recruitment may lead to detrimental results in the adjustment of nearby populations. A population that depends on self-recruitment can be demographically stable. However, even if new recruits are adequate, the lack of larval supply from other populations can result in inbreeding depression, increasing the risk of population extinction [54]. Moreover, since isolated populations cannot send new adaptive alleles to other populations, weak connectivity may decrease the probability of the population adapting to global climate change [55].

4.3. Future Planning of the MPA Network

China is currently implementing a strategy of "eco-civilization," with MPAs expected to be one approach to achieving sustainable marine ecosystems [56]. More protected areas will be established along coastal China and some isolated nodes are likely to be integrated, which is very promising to form networks of MPAs. MNRs are designed to avoid disturbances by prohibiting human activities strictly, whereas SMPAs allow multiple use of marine resources with limited extractive activities [28]. Ideally, MNRs are preferred to maximize the connectivity, but SMPAs are more operational when considering social and economic development.

Larval dispersal to surrounding areas hinges on MPA areas and spacing between MPAs [57]. Expanding the area of these reserves and setting them closer together would be beneficial in enhancing the ecological connectivity of the southern Yellow and East China Seas. MPA ZJ-2, ZJ-3, and ZJ-5 played crucial roles as sources or sinks of recruits for surrounding MPAs as well as were key gateways in sustaining the connectivity of the entire network (Figure 5). Taking them as the focus of MPA designation can have spillover effects on the surrounding reserves that may effectively strengthen ecological connectivity and improve the effectiveness of marine biodiversity protection in the entire area. In addition, high self-recruitment MPAs should be given priority protection, for example, MPA JS-2, JS-5, FJ-3, and FJ-5. Increasing their protection level with stricter fishing bans can be effective to the conservation of biodiversity. Additional protected areas should be placed in areas identified as being crucial for genetic diversity conservation, paving stepping stones between existing MPAs and enhancing connectivity [25,58].

The need for customized and differentiated conservation strategies for different protected species should be emphasized [59]. The potential ecological corridors of 14 general species types that this study considered are the reference sites from which new protected areas can be established (Figures S1–S4). For species with weak dispersal potential, strong sessile ability, and scattered distributions such as *Acropora solitaryensis* and *Apostichopus japonicus*, relatively small protected areas that cover the habitats they need would be reasonable under the condition of saving conservation costs. Vulnerable species such as corals, which are confined to a single protected area, usually live in the same area for their entire life, and thus enhanced habitat protection and management in the original area can be best. For species with strong dispersal potential and wide distributions such as *Epinephelus akaara*, *Anguilla japonica*, and *Larimichthys crocea*, these circulate in multiple MPAs during the planktonic larval stage, and thus it may be necessary to establish a large network of protected areas to form an ecological corridor.

The population replenishment was generally from the northern protected areas to the southern protected areas. Thus, new protected areas could be built at the ecological corridors along the Yellow Sea ecoregion (Jiangsu coast) first. Focused protection of the northern seas could lead to multi-step connections with protected areas in the south, promoting population connectivity and gene exchange between the north and the south. Moreover, the currents in the Yellow and East China Seas have strong seasonal differences, carrying larvae in different directions. Therefore, more protected areas should be established along the gateways according to the seasonal variation of connectivity and management plans need to be adjusted in different seasons to balance development and conservation. It has been reported that the Yellow and East China Seas have experienced warming at twice the global average since 2011 [60]. Local populations in the northern parts of the Yellow and East China Seas have possibly not adapted to the warmer water. The lack of population recruitment from south, warm-adapted populations can weaken the species sustainability in the north of Yellow and East China Seas. In climate change scenarios, genetic thermal stress resistance is likely to be the basis for species to adapt to ocean warming. These populations with higher resistance should be established in areas that can support more vulnerable communities [55]. Therefore, it is essential to research the effects of ocean warming and variation in current velocity and direction on larval floating and survival in future studies [61].

Increasing numbers of studies have begun to use multiple methods [62,63]. This paper revealed weak connections between the MPAs in the Yellow and East China Seas using oceanographic dynamics and biological information, while approaches that capitalize on the complementary strengths of genetic parentage datasets and biophysical models can produce accurate larval dispersal patterns at regional scales [64]. Connectivity among MPAs in the Yellow and East China Seas has hardly been the object of genetics studies. Thus, it is necessary that future research identifies the hereditary stability of species protected in MPAs to explain the geographic distribution of adaptive genetic variation.

5. Conclusions

We used a biophysical model to simulate larval dispersal in national reserves in the Yellow and East China Seas to assess connectivity between MPAs. Evaluating the larval recruitment of study region and larval supply outside MPAs is challenging due to the lack of information of larval biology, and the productivity and intensity of fishing in MPAs. Our study revealed a high self-recruitment rate and weak connectivity of the present MPAs system in the Yellow and East China Seas, demonstrating potentially deleterious consequences for conservation effectiveness and population persistence. The connectivity patterns are likely to be the results of ocean currents, species occurrence, and life cycle. The future MPA design needs to establish a functioning network with better connectivity within and between MPAs with the consideration of essential factors above, global climate change as well as economic and social costs. Further studies are needed to combine different methods to explore larval connectivity changes and their effects on the population recruitment within and outside MPAs.

Supplementary Materials: The following supporting information can be downloaded at: https://www.mdpi.com/article/10.3390/biology12030396/s1, Table S1: Source of biological parameters for study species; Figure S1: Potential larval export of species with PLD $\geq$ 45: Nemipterus virgatus, Sepiella maindroni, Epinephelus bruneu; Figure S2: Potential larval export of species with PLD (30~44): Larimichthys crocea, Argyrosomus argentatus, Argyrosomus japonicus, Evynnis cardinalis; Figure S3: Potential larval export of species with PLD (16~29): Apostichopus japonicus, Scomberomorus niphonius, Portunus Trituberculatus; Figure S4: Potential larval export of species with short PLD ($\leq$15): Epinephelus akaara, Anguilla japonica, Penaeus japonicus, Acropora solitaryensis; Figure S5: Self-recruitment fraction and subsidy recruitment fraction of MPAs; Figure S6: The computational grid of the ROMS model and the domain of configuration area; Particle stability analysis for recent eight years. Figure S7: The circumcircle areas of the triangles formed by particles for recent eight years (2014–2021). References [7,8,21,23,34,65–75] are cited in the supplementary materials.

Author Contributions: J.L.: investigation, conceptualization, methodology, data analysis and visualization, writing—original draft, writing—review and editing. Y.C.: methodology, data verification. Z.W. and F.Z.: methodology, data analysis. Y.Z.: methodology, data analysis. C.Z.: investigation, conceptualization, methodology, data verification and analysis, writing—review and editing. L.C.: resources, funding acquisition, project administration, supervision, writing—review and editing. All authors have read and agreed to the published version of the manuscript.

Funding: We would like to thank the Ministry of Science and Technology of China (Grant No. 2022YFC3102404), National Natural Science Foundation of China (Grant No. 42142018, 42206082), Shanghai Pilot Program for Basic Research—Shanghai Jiao Tong University (Grant No. 21TQ1400220), the Oceanic Interdisciplinary Program of Shanghai Jiao Tong University (Grant No. SL2021PT101), the Key Laboratory of Marine Ecological Monitoring and Restoration Technologies (Grant No. MEMRT202112), Blue Planet Found (OPF and WWF) (Grant No. PORO001426) and Shanghai Frontiers Science Center of Polar Science (SCOPS) for financial support. Any opinions, findings, and conclusions or recommendations expressed in this material are those of the authors and do not necessarily reflect the views of the funders.

Institutional Review Board Statement: Not applicable.

Informed Consent Statement: Not applicable.

Data Availability Statement: The data and biophysical model configuration are available from the corresponding author upon reasonable request (caoling@sjtu.edu.cn).

Acknowledgments: We gratefully thank Chuning Wang and Shuangzhao Li for help during configuration of the ocean model.

Conflicts of Interest: We declare that we have no known competing financial interests or personal relationships that could have appeared to influence the work reported in this paper.

References

1. Hughes, T.; Baird, A.; Bellwood, D.; Card, M.; Folke, C.; Grosberg, R.; Hoegh-Guldberg, O.; Jackson, C.; Kleypas, J.; Lough, J. Climate change, human impact, and the resilience of coral reefs. *Aust. Inst. Mar. Sci.* **2003**, *301*, 929–933. [CrossRef] [PubMed]
2. Lin, J.; Liu, X. Evaluation of the Management Effectiveness of Marine Protected Areas: A Brief Review of Recent Research Progress. *J. Oceanol. Limnol.* **2019**, *26*, 286–294.
3. Woodley, S.; Bertzky, B.; Crawhall, N.; Dudley, N.; Sandwith, T. Meeting Aichi Target 11: What does success look like for Protected Area systems? *Parks* **2012**, *18*, 23–36.
4. Harrison, H.B.; Williamson, D.H.; Evans, R.D.; Almany, G.R.; Thorrold, S.R.; Russ, G.R.; Feldheim, K.A.; van Herwerden, L.; Planes, S.; Srinivasan, M.; et al. Larval Export from Marine Reserves and the Recruitment Benefit for Fish and Fisheries. *Curr. Biol.* **2012**, *22*, 1023–1028. [CrossRef] [PubMed]
5. Olds, A.; Connolly, R.; Pitt, K.; Maxwell, P. Habitat connectivity improves reserve performance. *Conserv. Lett.* **2012**, *5*, 56–63. [CrossRef]
6. Kalinina, M.U.N. Body Releases Draft Plan to Put Biodiversity on Path to Recovery by 2050. Available online: https://www.pewtrusts.org/en/research-and-analysis/articles/2021/07/12/un-body-releases-draft-plan-to-put-biodiversity-on-path-to-recovery-by-2050 (accessed on 18 June 2022).
7. Assis, J.; Fragkopoulou, E.; Serrão, E.A.; Horta e Costa, B.; Gandra, M.; Abecasis, D. Weak biodiversity connectivity in the European network of no-take marine protected areas. *Sci. Total Environ.* **2021**, *773*, 145664. [CrossRef]
8. Andrello, M.; Mouillot, D.; Beuvier, J.; Albouy, C.; Thuiller, W.; Manel, S. Low Connectivity between Mediterranean Marine Protected Areas: A Biophysical Modeling Approach for the Dusky Grouper Epinephelus marginatus. *PLoS ONE* **2013**, *8*, e68564. [CrossRef]
9. Cowen, R.; Lwiza, K.; Sponaugle, S.; Olson, D. Connectivity of Marine Populations: Open or Closed? *Science* **2000**, *287*, 857–859. [CrossRef]
10. Watson, J.; Hays, C.; Raimondi, P.; Mitarai, S.; Dong, C.; McWilliams, J.; Blanchette, C.; Caselle, J.; Siegel, D. Currents connecting communities: Nearshore community similarity and ocean circulation. *Ecology* **2011**, *92*, 1193–1200. [CrossRef]
11. Ponchon, A.; Aulert, C.; Le Guillou, G.; Gallien, F.; Peron, C.; Gremillet, D. Spatial overlaps of foraging and resting areas of black-legged kittiwakes breeding in the English Channel with existing marine protected areas. *Mar. Biol.* **2017**, *164*, 119. [CrossRef]
12. Thorrold, S.R.; Jones, G.P.; Hellberg, M.E.; Burton, R.S.; Swearer, S.E.; Neigel, J.E.; Morgan, S.G.; Warner, R.R. Quantifying larval retention and connectivity in marine populations with artificial and natural markers. *Bull. Mar. Sci.* **2002**, *70*, 291–308.
13. Gillanders, B.M. Using elemental chemistry of fish otoliths to determine connectivity between estuarine and coastal habitats. *Estuar. Coast. Shelf Sci.* **2005**, *64*, 47–57. [CrossRef]
14. Campana, S.E. Chemistry and composition of fish otoliths: Pathways, mechanisms and applications. *Mar. Ecol. Prog. Ser.* **1999**, *188*, 263–297. [CrossRef]
15. Standish, J.D.; Sheehy, M.; Warner, R.R. Use of otolith natal elemental signatures as natural tags to evaluate connectivity among open-coast fish populations. *Mar. Ecol. Prog. Ser.* **2008**, *356*, 259–268. [CrossRef]
16. Li, Y.; Han, Z.; Song, N.; Gao, T.-X. New evidence to genetic analysis of small yellow croaker (*Larimichthys polyactis*) with continuous distribution in China. *Biochem. Syst. Ecol.* **2013**, *50*, 331–338. [CrossRef]
17. Planes, S.; Jones, G.P.; Thorrold, S.R. Larval dispersal connects fish populations in a network of marine protected areas. *Proc. Natl. Acad. Sci. USA* **2009**, *106*, 5693–5697. [CrossRef]
18. Andrello, M.; Mouillot, D.; Somot, S.; Thuiller, W.; Manel, S. Additive effects of climate change on connectivity between marine protected areas and larval supply to fished areas. *Divers. Distrib.* **2015**, *21*, 139–150. [CrossRef]
19. Xing, Q.; Yu, H.; Yu, H.; Sun, P.; Liu, Y.; Ye, Z.; Li, J.; Tian, Y. A comprehensive model-based index for identification of larval retention areas: A case study for Japanese anchovy Engraulis japonicus in the Yellow Sea. *Ecol. Indic.* **2020**, *116*, 106479. [CrossRef]
20. Roberts, K.; Cook, C.; Beher, J.; Treml, E. Assessing the current state of ecological connectivity in a large marine protected area system. *Conserv. Biol.* **2020**, *35*, 699–710. [CrossRef]
21. Liu, G.; Bracco, A.; Quattrini, A.; Herrera, S. Kilometer-Scale Larval Dispersal Processes Predict Metapopulation Connectivity Pathways for Paramuricea biscaya in the Northern Gulf of Mexico. *Front. Mar. Sci.* **2021**, *8*, 790927. [CrossRef]
22. Ospina-Alvarez, A.; Weidberg, N.; Aiken, C.M.; Navarrete, S.A. Larval transport in the upwelling ecosystem of central Chile: The effects of vertical migration, developmental time and coastal topography on recruitment. *Prog. Oceanogr.* **2018**, *168*, 82–99. [CrossRef]
23. Lopera, L.; Cardona, Y.; Zapata-Ramírez, P.A. Circulation in the Seaflower Reserve and Its Potential Impact on Biological Connectivity. *Front. Mar. Sci.* **2020**, *7*, 385. [CrossRef]
24. Buonaccorsi, V.P.; Kimbrell, C.A.; Lynn, E.A.; Vetter, R.D. Limited realized dispersal and introgressive hybridization influence genetic structure and conservation strategies for brown rockfish, Sebastes auriculatus. *Conserv. Genet.* **2005**, *6*, 697–713. [CrossRef]
25. Crochelet, E.; Roberts, J.; Lagabrielle, E.; Obura, D.; Petit, M.; Chabanet, P. A model-based assessment of reef larvae dispersal in the Western Indian Ocean reveals regional connectivity patterns—Potential implications for conservation policies. *Reg. Stud. Mar. Sci.* **2016**, *7*, 159–167. [CrossRef]
26. Di Franco, A.; Gillanders, B.; Benedetto, G.; Pennetta, A.; De Leo, G.; Guidetti, P. Dispersal Patterns of Coastal Fish: Implications for Designing Networks of Marine Protected Areas. *PLoS ONE* **2012**, *7*, e31681. [CrossRef] [PubMed]

27. Bohorquez, J.J.; Xue, G.; Frankstone, T.; Grima, M.M.; Kleinhaus, K.; Zhao, Y.; Pikitch, E.K. China's little-known efforts to protect its marine ecosystems safeguard some habitats but omit others. *Sci. Adv.* **2021**, *7*, eabj1569. [CrossRef]
28. Zeng, X.; Chen, M.; Zeng, C.; Cheng, S.; Wang, Z.; Liu, S.; Zou, C.; Ye, S.; Zhu, Z.; Cao, L. Assessing the management effectiveness of China's marine protected areas: Challenges and recommendations. *Ocean Coast. Manag.* **2022**, *224*, 106172. [CrossRef]
29. Deng, B.; Ji, H.; He, Y.; Ji, X. The Development of the National Marine Reserve Area in East China Sea District. *Ocean Dev. Manag.* **2017**, *34*, 64–67.
30. Chen, J.; Ma, J.; Xu, K.; Liu, Y.; Cao, W.; Wei, T.; Zhao, B.; Chen, Z. Provenance discrimination of the clay sediment in the western Taiwan Strait and its implication for coastal current variability during the late-Holocene. *Holocene* **2016**, *27*, 110–121. [CrossRef]
31. Chen, Y.; Cheng, P. Numerical modelling study of tidal energy and dissipation in the East China Seas. *J. Xiamen Univ.* **2020**, *59*, 61–70+151–152.
32. Jiang, X. The Numerical Simulation of Circulations, Mesoscale Eddies and Submesoscale Processes in the East China Sea. Ph.D. Thesis, Nanjing University of Information Science & Technology, Nanjing, China, 2022.
33. Zhang, K.; Li, A.; Zhang, J.; Lu, J.; Wang, H. Seasonal variations in the surficial sediment grain size in the East China Sea continental shelf and their implications for terrigenous sediment transport. *J. Oceanogr.* **2020**, *76*, 1–14. [CrossRef]
34. Rassweiler, A.; Ojea, E.; Costello, C. Strategically designed marine reserve networks are robust to climate change driven shifts in population connectivity. *Environ. Res. Lett.* **2020**, *15*, 034030. [CrossRef]
35. Shchepetkin, A.F.; McWilliams, J.C. The regional oceanic modeling system (ROMS): A split-explicit, free-surface, topography-following-coordinate oceanic model. *Ocean Model.* **2005**, *9*, 347–404. [CrossRef]
36. Chapman, D.C. Numerical Treatment of Cross-Shelf Open Boundaries in a Barotropic Coastal Ocean Model. *J. Phys. Oceanogr.* **1985**, *15*, 1060–1075. [CrossRef]
37. Carter, G.S.; Merrifield, M.A. Open boundary conditions for regional tidal simulations. *Ocean Model.* **2007**, *18*, 194–209. [CrossRef]
38. Treml, E.A.; Roberts, J.J.; Chao, Y.; Halpin, P.N.; Possingham, H.P.; Riginos, C. Reproductive Output and Duration of the Pelagic Larval Stage Determine Seascape-Wide Connectivity of Marine Populations. *Integr. Comp. Biol.* **2012**, *52*, 525–537. [CrossRef]
39. Wang, Z.; Zeng, C.; Jiang, Z.; Cao, L. Conservation gap analysis of threatened fish in the East China Sea and adjacent sea areas. *J. Trop. Oceanogr.* **2023**, *46*, 66–82.
40. Schill, S.; Raber, G.; Roberts, J.; Treml, E.; Brenner, J.; Halpin, P. No Reef Is an Island: Integrating Coral Reef Connectivity Data into the Design of Regional-Scale Marine Protected Area Networks. *PLoS ONE* **2015**, *10*, e0144199. [CrossRef]
41. Becker, B.J.; Levin, L.A.; Fodrie, F.J.; McMillan, P.A. Complex larval connectivity patterns among marine invertebrate populations. *Proc. Natl. Acad. Sci. USA* **2007**, *104*, 3267–3272. [CrossRef]
42. Kininmonth, S.J.; De'ath, G.; Possingham, H.P. Graph theoretic topology of the Great but small Barrier Reef world. *Theor. Ecol.* **2010**, *3*, 75–88. [CrossRef]
43. Treml, E.; Halpin, P.; Urban, D.; Pratson, L. Modeling population connectivity by ocean currents, a graph-theoretic approach for marine conservation. *Landsc. Ecol.* **2008**, *23*, 19–36. [CrossRef]
44. Guidetti, P.; Milazzo, M.; Bussotti, S.; Molinari, A.; Murenu, M.; Pais, A.; Spanò, N.; Balzano, R.; Agardy, T.; Boero, F.; et al. Italian marine reserve effectiveness: Does enforcement matter? *Biol. Conserv.* **2008**, *141*, 699–709. [CrossRef]
45. Sala, E.; Ballesteros, E.; Dendrinos, P.; Di Franco, A.; Ferretti, F.; Foley, D.; Fraschetti, S.; Friedlander, A.; Garrabou, J.; Güçlüsoy, H.; et al. The structure of Mediterranean rocky reef ecosystems across environmental and human gradients, and conservation implications. *PLoS ONE* **2012**, *7*, e32742. [CrossRef] [PubMed]
46. Rozenfeld, A.F.; Arnaud-Haond, S.; Hernández-García, E.; Eguíluz, V.M.; Serrão, E.A.; Duarte, C.M. Network analysis identifies weak and strong links in a metapopulation system. *Proc. Natl. Acad. Sci. USA* **2008**, *105*, 18824–18829. [CrossRef]
47. Balbar, A.C.; Metaxas, A. The current application of ecological connectivity in the design of marine protected areas. *Glob. Ecol. Conserv.* **2019**, *17*, e00569. [CrossRef]
48. Bode, M.; Bode, L.; Armsworth, P. Larval dispersal reveals regional sources and sinks in the Great Barrier Reef. *Mar. Ecol. Prog. Ser.* **2006**, *308*, 17–25. [CrossRef]
49. Csardi, G.; Nepusz, T. The Igraph Software Package for Complex Network Research. *InterJ. Complex Syst.* **2005**, *1695*, 1–9.
50. Werner, F.; Cowen, R. Coupled Biological and Physical Models: Present Capabilities and Necessary Developments for Future Studies of Population Connectivity. *Oceanography* **2007**, *20*, 54–69. [CrossRef]
51. Pairaud, I.; Gatti, J.; Bensoussan, N.; Verney, R.; Garreau, P. Hydrology and circulation in a coastal area off Marseille: Validation of a nested 3D model with observations. *J. Mar. Syst.* **2011**, *88*, 20–33. [CrossRef]
52. Cowen, R.; Srinivasan, A. Scaling of Connectivity in Marine Populations. *Science* **2006**, *311*, 522–527. [CrossRef]
53. Treml, E.; Halpin, P. Marine population connectivity identifies ecological neighbors for conservation planning in the Coral Triangle. *Conserv. Lett.* **2012**, *5*, 441–449. [CrossRef]
54. Allendorf, F.; Luikart, G. *Conservation and Genetics of Populations*; Blackwell Publishing: Hoboken, NJ, USA, 2006.
55. Mumby, P.; Edwards, I.; Eakin, C.M.; Skirving, W.; Edwards, H.; Enríquez, S.; Iglesias-Prieto, R.; Cherubin, L.; Stevens, J. Reserve design for uncertain responses of coral reefs to climate change. *Ecol. Lett.* **2011**, *14*, 132–140. [CrossRef]
56. Zeng, X.; Liang, J.; Zeng, J. Evaluating the effectiveness of three national marine protected areas (MPAs) in the Yangtze River delta. *Front. Mar. Sci.* **2022**, *9*, 1393. [CrossRef]

57. López-Sanz, A.; Stelzenmüller, V.; Maynou, F.; Sabatés, A. The influence of environmental characteristics on fish larvae spatial patterns related to a marine protected area: The Medes islands (NW Mediterranean). *Estuar. Coast. Shelf Sci.* **2011**, *92*, 521–533. [CrossRef]
58. Zeng, C.; Rowden, A.; Clark, M.; Gardner, J. Population genetic structure and connectivity of deep-sea stony corals (*O. scleractinia*) in the New Zealand region: Implications for the conservation and management of Vulnerable Marine Ecosystem. *Evol. Appl.* **2017**, *10*, 1040–1054. [CrossRef] [PubMed]
59. Li, Y.; Chen, X.; Chen, C.; Ge, J.; Ji, R.; Tian, R.; Xue, P.; Xu, L. Dispersal and survival of chub mackerel (*Scomber japonicus*) larvae in the East China Sea. *Ecol. Model.* **2014**, *283*, 70–84. [CrossRef]
60. Tang, Y.; Huangfu, J.; Huang, R.; Chen, W. Surface warming reacceleration in offshore China and its interdecadal effects on the East Asia–Pacific climate. *Sci. Rep.* **2020**, *10*, 14811. [CrossRef]
61. Lacroix, G.; Barbut, L.; Volckaert, F. Complex effect of projected sea temperature and wind change on flatfish dispersal. *Glob. Chang. Biol.* **2017**, *24*, 85–100. [CrossRef] [PubMed]
62. Legrand, T.; Di Franco, A.; Ser-Giacomi, E.; Caló, A.; Rossi, V. A multidisciplinary analytical framework to delineate spawning areas and quantify larval dispersal in coastal fish. *Mar. Environ. Res.* **2019**, *151*, 104761. [CrossRef]
63. Gaines, S.; White, C.; Carr, M.; Palumbi, S. Designing Marine Reserve Networks for Both Conservation and Fisheries Management. *Proc. Natl. Acad. Sci. USA* **2010**, *107*, 18286–18293. [CrossRef]
64. Bode, M.; Leis, J.; Mason, L.; Williamson, D.; Harrison, H.; Choukroun, S.; Jones, G. Successful validation of a larval dispersal model using genetic parentage data. *PLoS Biol.* **2019**, *17*, e3000380. [CrossRef]
65. Wu, C.; Dong, Z.; Chi, C.; Ding, F. Reproductive and spawning habits of *Sepiella maindroni* off Zhejiang, China. *Oceanol. Limnol. Sin.* **2010**, *41*, 39–46.
66. Wang, B.; Sun, P.; Zhang, Z.; Ma, J. Biological characteristics of *Epinephelus brunneus* and preliminary experiment of indoor culture. *Fish. Mod.* **2006**, *1*, 28–29.
67. Zhuang, P. *Yangtze Estuary Fishes*; Shanghai Science and Technology Press: Shanghai, China, 2006; Volume 10, pp. 338–340.
68. Wang, B.; Zhang, X.; Qu, X.; Ruan, S.; Qu, Y. Biological characteristics and technologies of seed production of Japanese croaker (*N. japanica*). *Prog. Fish. Sci.* **2002**, *4*, 13–19.
69. Cai, Y.; Chen, Z.; Xu, S.; Zhang, K. Tempo-spatial distribution of Evynnis cardinalis in Beibu Gulf. *South China Fish. Sci.* **2017**, *13*, 1–10.
70. Tang, Q. *Regional Oceanography of China Seas-Fisheries Oceanography*; China Ocean Press: Beijing, China, 2012; Volume 6, pp. 166–169.
71. Oh, C.-W. Population Biology of the Swimming Crab Portunus Trituberculatus (Miers, 1876) (Decapoda, Brachyura) on the Western Coast of Korea, Yellow Sea. *Crustaceana* **2011**, *84*, 1251–1267.
72. Ukawa, M.; Higuchi, M.; Mito, S. Spawning habits and early life history of a serranid fish, Epinephelus akaara (TEMMINCK et SCHLEGEL). *Jpn. J. Ichthyol.* **1966**, *13*, 156–161.
73. Zou, Y. *Special Aquaculture*; China Agriculture Press: Beijing, China, 2002; Volume 7, pp. 30–31.
74. Cai, X.; Lin, Q.; Wan, W. On the Development of the Prawn *Penaeus penicillatus* in Comparison with Both *P. japonicus* and *P. orientalis*. *J. Xiamen Univ. (Nat. Sci.)* **1981**, *2*, 243–252.
75. Zhang, Y.; Huang, H.; Huang, J.; Yuan, T. Experimental cultivation of coral larvae in Xisha Islands. *Ocean. Dev. Manag.* **2013**, *30*, 78–82.

Article

Complete Mitochondrial Genomes of Four *Pelodiscus sinensis* Strains and Comparison with Other Trionychidae Species

Jing Chen, Jinbiao Jiao, Xuemei Yuan, Xiaohong Huang, Lei Huang, Lingyun Lin, Wenlin Yin, Jiayun Yao and Haiqi Zhang *

Agriculture Ministry Key Laboratory of Healthy Freshwater Aquaculture, Key Laboratory of Fish Health and Nutrition of Zhejiang Province, Key Laboratory of Fishery Environment and Aquatic Product Quality and Safety of Huzhou City, Zhejiang Institute of Freshwater Fisheries, Huzhou 313001, China
* Correspondence: zmk407@126.com

Simple Summary: The Chinese soft-shelled turtle (*Pelodiscus sinensis*) is an economically important aquatic reptile species with rich nutrition and medicinal value. However, the wild resources of *P. sinensis* have been depleting due to natural and artificial factors in recent decades. Herein, we report the complete mitochondrial genomes of four *P. sinensis* strains and analyzed the nucleotide composition and variable site for the four mitogenomes. Using Ka/Ks and sliding window analyses, we explored the genetic diversity and selection pressures of different mitochondrial genes in the four *P. sinensis* strains. Through comparative analysis, the present study described the structural variation of 22 tRNAs, replication origin region of the L-strand, and control region, indicating the genetic variations among the four *P. sinensis* strains. Furthermore, the evolutionary relationship of *P. sinensis* strains and other Trionychidae species was determined by phylogenetic analysis. Taken together, our findings provide genetic information and an essential basis for understanding the genetic variations and evolutionary relationship of *P. sinensis* strains, which will play an important role in bioprospecting and conservation of *P. sinensis*.

Citation: Chen, J.; Jiao, J.; Yuan, X.; Huang, X.; Huang, L.; Lin, L.; Yin, W.; Yao, J.; Zhang, H. Complete Mitochondrial Genomes of Four *Pelodiscus sinensis* Strains and Comparison with Other Trionychidae Species. *Biology* **2023**, *12*, 406. https://doi.org/10.3390/biology12030406

Academic Editor: Mingyou Li

Received: 17 January 2023
Revised: 28 February 2023
Accepted: 1 March 2023
Published: 3 March 2023

Abstract: The Chinese soft-shelled turtle (*Pelodiscus sinensis*) is an important aquaculture reptile with rich nutritional and medicinal values. In recent decades, the wild resources of *P. sinensis* have been depleting due to natural and artificial factors. Herein, we report the complete mitochondrial genome of four *P. sinensis* strains, including the Japanese (RB) strain, Qingxi Huabie (HB) strain, Jiangxi (JB) strain, and Qingxi Wubie (WB) strain. The nucleotide composition within the complete mitogenomes was biased towards A + T with a variable frequency ranging from 59.28% (*cox3*) to 70.31% (*atp8*). The mitogenomes of all four strains contained 13 protein-coding genes (PCGs), 22 tRNAs, 2 rRNAs, 1 control region, and a replication origin region of the L-strand replication (OL), which was consistent with most vertebrates. Additionally, the *atp8*, *nad4l*, *nad6*, and *nad3* genes possessed high genetic variation and can be used as potential markers for the identification of these *P. sinensis* strains. Additionally, all PCGs genes were evolving primarily under purifying selection. Through comparative analysis, it was revealed that most of the tRNAs were structurally different in the TψC stem, DHU stem, and acceptor stem. The length of the tandem repeats in the control region was variable in the four *P. sinensis* strains, ranging from 2 bp to 50 bp. Phylogenetic analysis indicated that all *P. sinensis* strains clustered into one branch and were closely related to other Trionychinae species. Overall, this study provides mitochondrial genome information for different *P. sinensis* strains to support further species identification and germplasm resource conservation.

Keywords: *Pelodiscus sinensis*; mitochondrial genome; structural variation; control region; phylogenetic relationship

1. Introduction

The Chinese soft-shelled turtle (*Pelodiscus sinensis*) belongs to the genus *Pelodiscus* of the family Trionychidae, and it is recognized as an important economic species with nutritional value and medicinal value [1–3]. *P. sinensis* is widely distributed in East Asia, including China, Korea, Japan, and Vietnam [4,5]. According to the geographical distribution, *P. sinensis* can be divided into different strains, including the Japanese strain, Yellow River strain, Huaihe River strain, Poyang Lake strain, Taihu Lake strain, Southwest strain, Taiwan strain, etc. [6]. In recent years, two new strains of *P. sinensis* (Qingxi Huabie and Wubie strains) have been identified, which originated from Taihu Lake basin and are unique local strains in Zhejiang Province [7]. Different *P. sinensis* strains have their own traits, such as fast growth rate for the Japanese strain and light-yellow body color for the Yellow River strain [6]. With the continuous improvement in life quality, the consumption demand for *P. sinensis* has also been increasing. The state of world fisheries and aquaculture in 2022 reported that the annual output of *P. sinensis* was 334.3 thousand tons in 2020, accounting for 31.5% of aquatic animals (except finfish, crustaceans, and mollusks) [8]. In the past decades, the scale of artificial breeding of *P. sinensis* has been expanding, and many breeding individuals in non-original wild habitats have escaped to natural waters, causing serious damage to wild germplasm resources [9]. Meanwhile, due to over-fishing and the destruction of the natural environment, the wild *P. sinensis* population has decreased sharply and almost dried up [10,11]. The *P. sinensis* has been listed in the United Nations Red Book on Endangered Species (http://www.iucnredlist.org/details/39620/0) (accessed on 20 October 2022). Until now, the classification of different *P. sinensis* strains has mainly depended on morphological traits, and only a few studies have involved molecular systematics, which plays a crucial role in germplasm resource conservation of *P. sinensis*.

To date, many methods have been developed for species classification and phylogenetic research based on DNA sequencing [12–14]. Among them, mitochondrial DNA (mtDNA) is widely used for its unique advantages; it is rarely affected by gene recombination, it strictly has maternal inheritance in vertebrates, and it has a high mutation rate [15–17]. The rapid evolution of mtDNA implies that many sequence variations can be identified among closely related species, which is an effective tool for species identification [18]. For most vertebrates, the mitochondrial genome contains thirteen protein-coding genes (seven subunits of complex I, one subunit of complex III, three subunits of complex IV, and two subunits of complex V); two rRNAs (12S rRNA and 16S rRNA); one control region (D-loop); and twenty-two tRNAs [19,20]. At present, many studies have reported the phylogenetic relationship and species identification of Trionychidae species by mtDNA analysis. For instance, Zhang et al. reported that *Trionyx sinensis* had a close relationship with *P. axenaria* as determined by the mitochondrial *cytb* gene [21]. Liang et al. amplified the mitochondrial *cox1* gene and found that the Huaihe River strain of *P. sinensis* was more closely related to the Yellow River strain than the Japanese strain [22]. Using the mitochondrial D-loop, Li et al. demonstrated that the Tai Lake population and Hongze Lake population of *P. sinensis* had a close genetic relationship [23]. Chen et al. used the mitochondrial 12S rRNA to identify *P. axenaria* as a new species [24]. Zhang et al. sequenced and analyzed the partial sequences of *nad4*, *cox1*, *nad5*, and *nad6* genes and further confirmed the identification of *P. sinensis* strains [6]. However, most studies only involved some gene fragments of the mitochondrial genome. Increasing evidence has revealed that the complete mitochondrial genome could provide accurate insight into genetic differentiation and species identification [25,26].

In the present study, we amplified and analyzed the complete mitochondrial genome of four *P. sinensis* strains, including the Japanese strain, Jiangxi strain, Qingxi Huabie strain, and Qingxi Wubie strain. To investigate the mitochondrial genome features and genetic variations of the four *P. sinensis* strains, the present study was designed as follows: (1) compare and analyze the length, nucleotide composition, and variable sites of the four mitogenomes; (2) explore the genetic diversity and selection pressures of different mitochondrial genes in the four *P. sinensis* strains; (3) illustrate the structural variation of 22 tRNAs, the origin of L-strand replication, and the control region among the four *P. sinensis* strains; (4) determine the phylogenetic relationship of *P. sinensis* strains and other Trionychidae species.

These results offer abundant genetic information for the four *P. sinensis* strains and will be conducive to the conservation and utilization of germplasm resources of *P. sinensis*.

2. Materials and Methods

2.1. Sample Information

In this study, healthy *P. sinensis* samples with about 150 g weight, including the Japanese (RB) strain, Jiangxi (JB) strain, Qingxi Wubie (WB) strain, and Qingxi Huabie (HB) strain, were collected from a breeding base of Zhejiang province. The liver tissue of these strains was collected and stored at −80 °C. The study was approved by the Institutional Animal Care and Use Committee (IACUC) of the Zhejiang Institute of Freshwater Fisheries.

2.2. DNA Extraction, Sequencing, and Assembly

Genomic DNA was extracted using a genomic DNA extraction kit (Tiangen, Beijing, China). The quantity and quality of DNA was detected by NanoDrop 2000 (Thermo Fisher Scientific, Waltham, MA, USA) and agarose gel electrophoresis. According to the published mitochondrial genome sequence of *P. sinensis* in GenBank (https://www.ncbi.nlm.nih.gov/nuccore/AY687385.1) (accessed on 12 June 2022), the primers used to amplify the complete mitochondrial genome of these strains were designed by Primer premier 5.0 software (Table 1) [27]. The amplification reactions were performed in a total volume of 50 μL, including 19 μL of ddH$_2$O, 25 μL of 2 × Taq Master Mix (Takara, Shiga, Japan), 2 μL of each primer (10 μM/L), and 2 μL of genomic DNA (120 ng/μL). The PCR cycle was an initial denaturation at 95 °C for 3 min; 35 cycles of 95 °C for 30 s, 58 °C for 30 s, and 72 °C for 1 min/Kb (1 min per Kb amplification length); and a final extension at 72 °C for 5 min. PCR amplification products were detected by agarose gel electrophoresis. After purification, the PCR product was sequenced by Sanger sequencing in Tsingke Biological Technology (Beijing, China). The regions containing prominent repeats were amplified into different fragments and then sequenced using the primer walking method. Cluster X 2.0 software [28] and BioEdit 7.0 software [29] were used to align and correct the obtained sequences, and the complete mitochondrial genome sequence was obtained by SeqMan software (DNAStar Inc., Madison, WI, USA) [30].

Table 1. Primers used for amplification of mitochondrial genome.

Forward	Primer Sequence (5′ to 3′)	Reverse	Primer Sequence (5′ to 3′)	Product Length
Mt-1F	AGTGAAAATGCCCTAAAAGTCACATC	Mt-1R	ATACTTATTGTTGCTAGGGGCTATGT	2000 bp
Mt-2F	AATAACAGATGGGGTAAGTCGTAACA	Mt-2R	GTGAAGAAGGCTACAGCAATTAAGAT	2000 bp
Mt-3F	GAGTTCAGACCGGAGC AATCCA	Mt-3R	CAGTTCCTGCGCCTGTTTCAAT	3500 bp
Mt-4F	CTACATGGTTTGATAAGAAGGGGAGT	Mt-4R	ATTGGTGATATTGCGTCTTGAAATCC	2000 bp
Mt-5F	CACTACACCAAACCTGAACCAAAGTA	Mt-5R	GATTGTGAATGGTGCTTCGTAGTATTC	2500 bp
Mt-6F	CACACAACTATCAATGAACATAGCAC	Mt-6R	TGAACTGAAATTGAATGATTGGAAGT	1500 bp
Mt-7F	AGTCTATGGCTCCACATTCTTCGT	Mt-7R	TAGGTTCCAGCATTTAGTCGTTCT	1500 bp
Mt-8F	AGAACCCCTATCACGAAAACGAAC	Mt-8R	GCTATTTTTACGGCGGTTTTTGGT	1500 bp
Mt-9F	AATCTCCTTATAAACCGAGAAGGT	Mt-9R	AGATTTAGTTCGTGGTTTGGCT	1500 bp
Mt-10F	ATCATTGCAGGACTACTAATCTCATCA	Mt-10R	ATTTCATCAGATGGAGATGTTAGATGGA	2000 bp
Mt-11F	GTCAACGCCACAGAATAAGC	Mt-11R	ATTCCGGTTTTGGGGATCGG	1000 bp
Mt-12F	GCCCTATCACCCAAACACTATTCT	Mt-12R	CAGTTTCATTGAGTTGGCAGACAT	1500 bp
Mt-13F	AACCCTTGTTAGTAAGATAC	Mt-13R	CGTTGTTATTGTTGCTTTGG	1500 bp
Mt-14F	TCCATTGACAGTTGGCGTAC	Mt-14R	CTATAACTAAGTCAAGCTTATGC	1500 bp
Mt-15F	ACCAATCTCAAACATAATTG	Mt-15R	GAGATTTACCAACCCTGAATG	2200 bp
Mt-16F	CAGAGCCAGGTAATCAATGC	Mt-16R	CAACTATACCTGCTCAGGCAC	2700 bp
Mt-17F	TCTGAGAAGCATTCTCATCA	Mt-17R	GTAAACTAATAGTTTCGATG	1500 bp
Mt-18F	GAACCACAACCTCTTGGTGC	Mt-18R	GGTAAGAAGGAGTATGGTGATTG	1800 bp
Mt-19F	ACCAATCTCAAACATAATCG	Mt-19R	CTGGCACGAGATTTACCAAC	1500 bp

2.3. Mitogenome Annotation and Sequence Analysis

The sequence annotation was conducted in the ARWEN and MITOS (http://mitos.bioinf.uni-leipzig.de) (accessed on 15 September 2022) online servers [31,32]. The initiation and termination codons of protein-coding genes (PCGs) were identified using other reference sequences of Trionychidae species. The online program OGDraw v1.2 with default parameters was used to map the circular map of the completed mitochondrial genome [33]. Subsequently, the base composition and codon usage of the mitochondrial genome of four *P. sinensis* strains were analyzed by MEGA X software [34], and the AT skew = $(A - T)/(A + T)$ and GC skew = $(G - C)/(G + C)$ were calculated by the method previously reported [35]. Relative synonymous codon usage (RSCU) is an important indicator to judge codon usage preference. The RSCU value represents the ratio of the usage bias of a codon to its expected usage bias in the synonymous codon family (all codons for a particular amino acid are used equally). Codons with an RSCU value > 1.0 have positive codon usage bias, while codons with an RSCU value < 1.0 have negative codon usage bias [36]. The software MEGA X was used to calculate RSCU values. Furthermore, multiple sequence alignment for the mitochondrial genomes of four *P. sinensis* strains was performed using MEGA X software. Then, the PCGs, rRNAs, tRNAs, and control region (D-loop) of four *P. sinensis* strains were analyzed by DnaSPv6.0 software for gene traits and variation sites [37]. The software DnaSPv6.0 was used to calculate the synonymous substitutions per synonymous sites (Ks) and non-synonymous substitutions per non-synonymous sites (Ka). The Ka/Ks ratio was used to assess the selection pressure, Ka/Ks > 1 indicated a positive selection, Ka/Ks = 1 indicated a neutral selection, and Ka/Ks < 1 indicated a purifying (stabilizing) selection [38]. The genetic distance was analyzed by MEGA X software. In addition, the nucleotide diversity (Pi) of 13 PCGs and 2 rRNAs were analyzed by sliding window analysis (500 bp windows every 10 bp) using DnaSPv6.0 software.

2.4. Structural Analyses of Mitogenome and Prediction of Repeat Element

To determine the unique base compositions in the control regions (CRs) of four *P. sinensis* strains, tandem repeats were predicted by the online Tandem Repeats Finder web tool (https://tandem.bu.edu/trf/trf.html) (accessed on 17 October 2022) [39]. For the four *P. sinensis* strains, the stem-loop structures of the origin of L-strand replication were analyzed by the online Mfold web server (http://www.unafold.org/) (accessed on 17 October 2022) [40]. The 22 tRNAs of mitogenomes of the four *P. sinensis* strains were verified in MITOS online server (http://mitos.bioinf.uni-leipzig.de) (accessed on 18 October 2022). Then, the online tRNAscan SE Search Server 2.0 (http://lowelab.ucsc.edu/tRNAscan-SE/) (accessed on 19 October 2022) and RNAstructure software were used to predict the variation in tRNA secondary structure among the mitogenomes of four *P. sinensis* strains [41,42], and their mutation sites were analyzed. The base composition of all components (DHU arm, acceptor stem, TψC arm, an anti-codon arm) were manually checked to distinguish the mutation sites.

2.5. Construction of Phylogenetic Tree

Phylogenetic analysis was performed on the dataset of 13 PCGs, 22 tRNAs, and 2 rRNAs from 19 Trionychidae mitogenomes published in GenBank and four *P. sinensis* mitogenomes sequenced in this study. Sequence alignment of 13 PCGs was performed using MAFFT with default settings in PhyloSuite v1.2.3 [43,44], and the aligned sequences were checked using MEGA X. The best-fit partitioning scheme and corresponding nucleotide substitution models for concatenated nucleotides were selected using PartitionFinder v2.1.1 according to the Bayesian information criterion (BIC) (Table S1) [45]. Phylogenetic trees of concatenated nucleotide sequences were constructed with the best-fit partitioning schemes and nucleotide substitution models using Bayesian inference (BI) and maximum-likelihood (ML) methods in MrBayes v3.2.1 and IQ-tree v2.0.4, respectively [46,47]. Using the best-fit model, the ML analysis was run for each partition with 2000 ultrafast bootstrap (UFB) replicates and performed until a correlation coefficient of at least 0.99 was reached [48,49]. In addition, the BI analysis was run independently using four Markov Chain Monte Carlo (MCMC) chains (three heated chains and one cold chain) starting with a random tree; each chain was run for 2×10^7 generations and sampled every 1000 generations. Convergence of data runs was estimated by the average standard deviation of split frequencies (ASDSF) < 0.01. The phylogenetic trees were visualized in FigTree v1.4.4 [50].

3. Results and Discussion

3.1. Analysis of Mitogenome Features

In this study, we amplified the complete mitochondrial genomes of four *P. sinensis* strains. The result showed that the four mitogenomes were circular molecules with lengths of 17,219 bp, 17,116 bp, 17,235 bp, and 17,182 bp in WB, HB, RB, and JB strains, respectively. These mitogenomes were deposited in GenBank under the accession number OQ236104 for the RB strain, OQ236105 for the HB strain, OQ236106 for the JB strain, and OQ236107 for the WB strain. The mitogenomes of all four strains consisted of 13 protein-coding genes (PCGs), 22 tRNAs, 2 rRNAs, 1 control region, and a replication origin region of the light chain (OL). The arrangement and orientation of these genes were similar to most vertebrates [51–53]. Previous studies had described the length and composition of mitogenomes for other *P. sinensis* strains; for instance, the inked turtle strain of *P. sinensis* was 17,145 bp in length [9], and the Korean soft-shelled turtle was 17,042 bp in length [54]. The mitogenomes of these *P. sinensis* strains had slight differences in length and possessed the same gene arrangement and orientation. However, it was reported that the mitogenome of Mediterranean tortoises contained 23 tRNAs and 2 control regions, and the mitochondrial genes of most Testudoformes species presented a similar arrangement with vertebrates, except *Platysternon megacephalum* and *Malacochersus tornieri* [55,56]. In the mitogenomes of the four *P. sinensis* strains, only the *nad6* gene and eight tRNAs (tRNA-Glu, tRNA-Pro, tRNA-Gln, tRNA-Ala, tRNA-Asn, tRNA-Cys, tRNA-Tyr, and tRNA-Ser) were encoded on the light chain (L-chain), and all other genes were encoded on the heavy chain (H-chain) (Figure 1). Additionally, it was found that there was a slight difference in the length of PCGs, tRNAs, and rRNAs among the four *P. sinensis* strains. The majority of PCGs were initiated by an ATG start codon, with the exception of *cox1*, which was initiated by GTG. Five stop codons (TAG, AGA, TAA, AGG, and T) were found in the mitochondrial genome of four *P. sinensis* strains (Table 2). Among them, the termination codon was TAG for *nad1* and *nad2*; the termination codon was TAA for *cox2, atp8, atp6, nad4l, nad5,* and *cytb*; the termination codon was a single base (T) for *cox3* and *nad4*; and the termination codon was AGA and AGG for *cox1* and *nad6*, respectively. A previous study showed that in the mitogenomes of many animals, AGA and AGG are not used as codons of arginine but are instead used as termination codons [57]. Additionally, for the *nad3* gene, the termination codon TAG was found in the HB and WB strains, and the termination codon T was observed in the JB and RB strains.

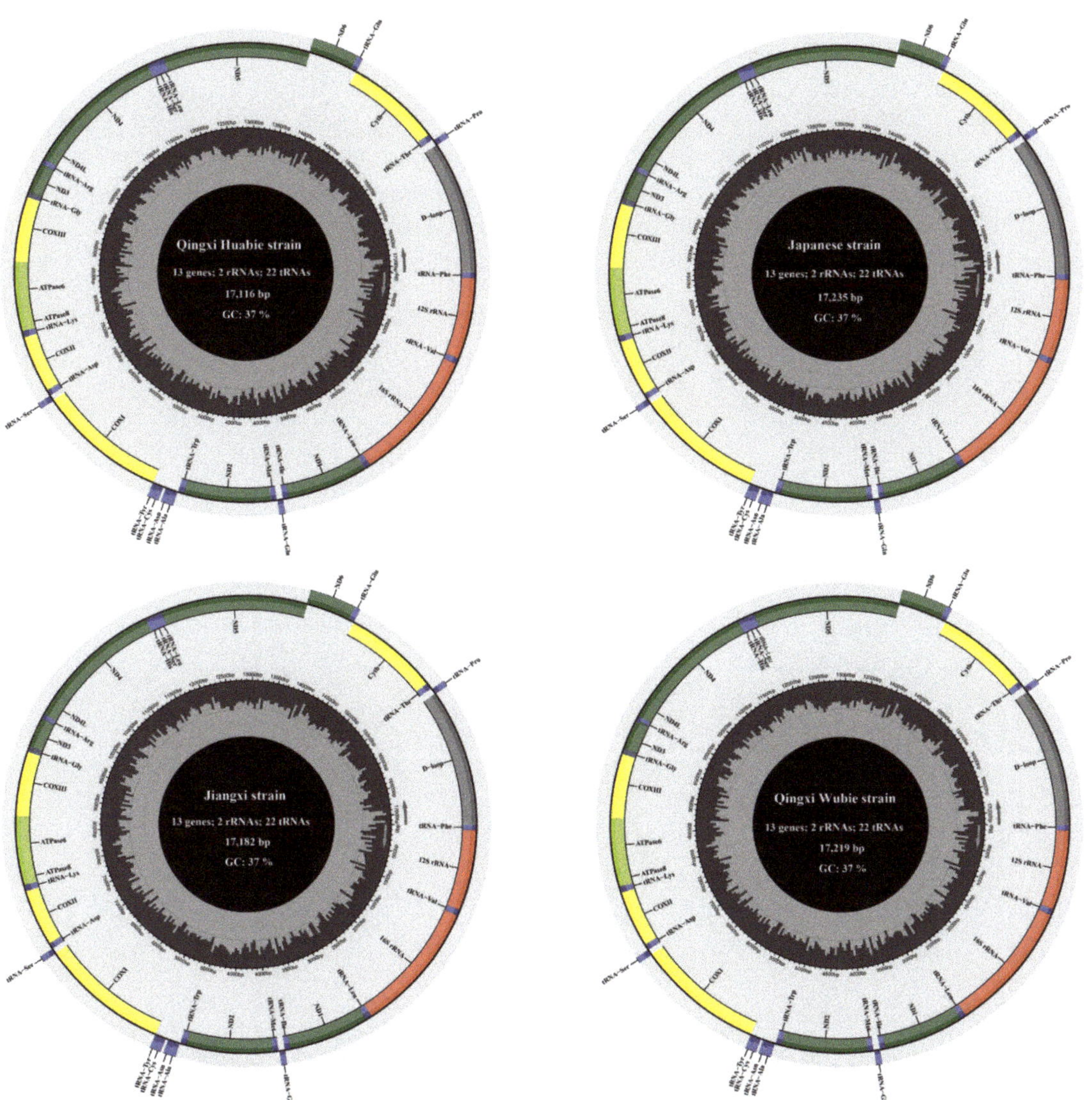

Figure 1. Mitochondrial genome maps of four *P. sinensis* strains. The outer circle represents the light chain (L-chain), and the inner circle represents the heavy chain (H-chain). The dark green box represents mitochondrial complex I (NADH dehydrogenase); the yellow box represents complex IV (cytochrome c oxidase); the light green circle represents ATP synthase; the blue circle represents transfer RNA; the red circle represents ribosomal RNA; and the grey circle represents the control region (D-loop).

Table 2. Gene length, start codon, and stop codon usage of mitochondrial genomes of four *P. sinensis* strains.

Mitochondrial Elements	Qingxi Huabie (HB) Strain				Jiangxi (JB) Strain				Japanese (RB) Strain				Qingxi Wubie (WB) Strain			
	Length (bp)	Intergenic Nucleotides (bp) *	Start Codon	Stop Codon	Length (bp)	Intergenic Nucleotides (bp) *	Start Codon	Stop Codon	Length (bp)	Intergenic Nucleotides (bp) *	Start Codon	Stop Codon	Length (bp)	Intergenic Nucleotides (bp) *	Start Codon	Stop Codon
tRNA^{Phe}	69	0			70	0			70	0			69	0		
12SrRNA	979	0			979	0			980	0			979	0		
tRNA^{Val}	70	0			70	0			70	0			70	0		
16SrRNA	1602	0			1604	0			1605	0			1603	0		
tRNA^{Leu}	77	0			77	0			77	0			77	0		
NAD1	971	0	ATG	TAG	971	0	ATG	TAG	971	0	ATG	TAG	971	0	ATG	TAG
tRNA^{Ile}	70	−1			70	−1			70	−1			70	−1		
tRNA^{Gln}	71	9			71	9			71	9			71	9		
tRNA^{Met}	69	0			69	0			69	0			69	0		
NAD2	1039	0	ATG	TAG	1039	0	ATG	TAG	1039	0	ATG	TAG	1039	0	ATG	TAG
tRNA^{Trp}	73	11			73	11			73	11			73	11		
tRNA^{Ala}	69	1			69	1			69	1			69	1		
tRNA^{Asn}	74	−1			74	−1			74	−1			74	−1		
OL	34	−2			34	−2			34	−2			34	−2		
tRNA^{Cys}	65	0			65	0			65	0			65	0		
tRNA^{Tyr}	66	1			66	1			66	1			66	1		
COX1	1545	−5	GTG	AGA	1545	−5	GTG	AGA	1545	−5	GTG	AGA	1545	−5	GTG	AGA
tRNA^{Ser}	71	1			71	1			71	1			71	1		
tRNA^{Asp}	69	0			69	0			69	0			69	0		
COX2	687	1	ATG	TAA	687	1	ATG	TAA	687	1	ATG	TAA	687	1	ATG	TAA
tRNA^{Lys}	73	1			73	1			73	1			73	1		
ATP8	165	−10	ATG	TAA	165	−10	ATG	TAA	165	−10	ATG	TAA	165	−10	ATG	TAA
ATP6	683	0	ATG	TAA	683	0	ATG	TAA	683	0	ATG	TAA	683	0	ATG	TAA
COX3	784	0	ATG	T	784	0	ATG	T	784	0	ATG	T	784	0	ATG	T
tRNA^{Gly}	70	0			70	0			70	0			70	0		
NAD3	352	−2	ATG	TAG	350	0	ATG	T	350	0	ATG	T	350	0	ATG	TAG
tRNA^{Arg}	70	0			70	0			71	0			70	0		
NAD4L	297	−7	ATG	TAA	297	−7	ATG	TAA	297	−7	ATG	TAA	297	−7	ATG	TAA
NAD4	1381	0	ATG	T	1381	0	ATG	T	1381	0	ATG	T	1381	0	ATG	T
tRNA^{His}	70	0			70	0			70	0			70	0		
tRNA^{Ser}	62	−1			62	−1			62	−1			62	−1		
tRNA^{Leu}	74	0			74	0			74	0			74	0		
NAD5	1779	−5	ATG	TAA	1779	−5	ATG	TAA	1779	−5	ATG	TAA	1779	−5	ATG	TAA
NAD6	525	0	ATG	AGG	525	0	ATG	AGG	525	0	ATG	AGG	525	0	ATG	AGG
tRNA^{Glu}	68	3			68	3			68	3			68	3		
Cytb	1140	3	ATG	TAA	1140	3	ATG	TAA	1140	3	ATG	TAA	1140	3	ATG	TAA
tRNA^{Thr}	74	14			74	14			74	13			74	14		
tRNA^{Pro}	71	0			71	0			71	0			71	0		
D-loop	1597				1660				1711				1699			

* Negative intergenic nucleotides indicate overlapping nucleotides between adjacent elements.

3.2. Nucleotide Composition and Variation Detection

To investigate the nucleotide composition of mitogenomes, we calculated the parameters A + T content, AT-skew, and GC-skew for the four *P. sinensis* strains. The result showed that the average A + T composition was 63.74%, 63.11%, 61.40%, and 64.32% in PCGs, tRNAs, rRNAs, and D-loop, respectively (Table 3). The nucleotide composition within the complete mitogenomes was biased towards A + T with a variable frequency ranging from 59.28% (*cox3*) to 70.31% (*atp8*), which was similar to other Trionychidae species [53]. Previous studies have reported that the mitogenome of the Korean soft-shelled turtle contained 62.6% A + T content and had high nucleotide similarity with the Chinese soft-shelled turtle, and the base composition of the inked turtle strain of *P. sinensis* was 35.5% A, 27.3% T, 11.8% G, and 25.4% C, with an A + T content of 62.8% [9,54]. It was demonstrated that the mitogenomes of these *P. sinensis* strains had a similar A + T content. The average AT- and GC-skew of the complete mitogenome was 0.13 and −0.37, respectively. The AT-skew of all genes varied from −0.02 (*cox1*) to 0.57 (*nad6*), and the GC-skew ranged from −0.68 (*atp8*) to −0.16 (rRNAs) (Table 3). The AT- and GC-skew in most genes indicated that more adenine (A)s/cytosine (C)s than thymine (T)s/guanine (G)s existed in the complete mitogenomes, suggesting no obvious difference among the four *P. sinensis* strains (Figure 2).

Table 3. Base composition statistics of complete mitochondrial genome and each element.

Mitochondrial Elements	Average Base Composition (%)						Average AT-Skew	Average GC-Skew
	T (U)	C	A	G	A + T	G + C		
tRNAs	28.21	21.67	34.90	15.22	63.11	36.89	0.11	−0.17
rRNAs	22.16	22.45	39.24	16.16	61.40	38.61	0.28	−0.16
D-loop	31.78	25.70	32.54	9.99	64.32	35.69	0.01	−0.44
ATP6	30.34	24.71	36.37	8.59	66.71	33.30	0.09	−0.48
ATP8	29.55	25.00	40.76	4.70	70.31	29.7	0.16	−0.68
COX1	30.91	23.53	29.92	15.65	60.83	39.18	−0.02	−0.20
COX2	27.62	23.94	36.64	11.79	64.26	35.73	0.14	−0.34
COX3	27.01	26.05	32.27	14.67	59.28	40.72	0.09	−0.28
Cytb	27.43	28.57	33.22	10.77	60.65	39.34	0.10	−0.45
NAD1	31.48	25.62	31.97	10.93	63.45	36.55	0.01	−0.40
NAD2	25.07	26.51	40.71	7.71	65.78	34.22	0.24	−0.55
NAD3	31.79	25.71	33.21	9.29	65.00	35.00	0.02	−0.47
NAD4	27.75	26.90	36.44	8.91	64.19	35.81	0.14	−0.50
NAD4L	30.89	26.43	34.01	8.67	64.90	35.10	0.05	−0.51
NAD5	26.25	29.06	35.26	9.43	61.51	38.49	0.15	−0.51
NAD6	13.29	31.28	48.50	6.95	61.79	38.23	0.57	−0.64
PCGs	27.64	26.41	36.10	9.85	63.74	36.26	0.13	−0.46
Complete genome	27.23	25.45	35.56	11.75	62.79	37.20	0.13	−0.37

Additionally, the 13 PCGs of the HB strain (11,352 bp length), JB strain (11,350 bp length), RB strain (11,350 bp length), and WB strain (11,352 bp length) were used to estimate the RSCU values. The result showed that these encoded protein sequences contained 21 amino acids; a total of 64 codons were used in the four strains (Figure 3). It was demonstrated that a total of 27 RSCU values computed in this study exceeded 1, indicating that these codons had a high-frequency usage, and most of preference codons ended in purine (A/U) due to the abundance of A/T in the mitogenomes of the four *P. sinensis* strains. Furthermore, we detected and compared the variable sites, singleton variable sites, and parsimony informative sites among the four *P. sinensis* strains. The result showed that a total of 483 variable sites were identified, accounting for 2.83% of the total sites, including 436 singleton variable sites and 47 parsimony informative sites. Moreover, there were 25 variable sites in 22 tRNAs, 113 variable sites in the D-loop, and 291 variable sites in 13 PCGs (Table 4). A previous study reported that the control region is a non-coding sequence in mtDNA, it possesses relatively low selection pressure during evolution, and it has greater polymorphism than other mitochondrial elements [58]. Among all mitochondrial elements, the D-loop, *nad3*, *nad4l*, *nad6*, and *atp8* contained a higher percentage of variable sites than other elements. Therefore, it is inferred that these highly variable elements will play an important role in the classification and identification of the four *P. sinensis* strains.

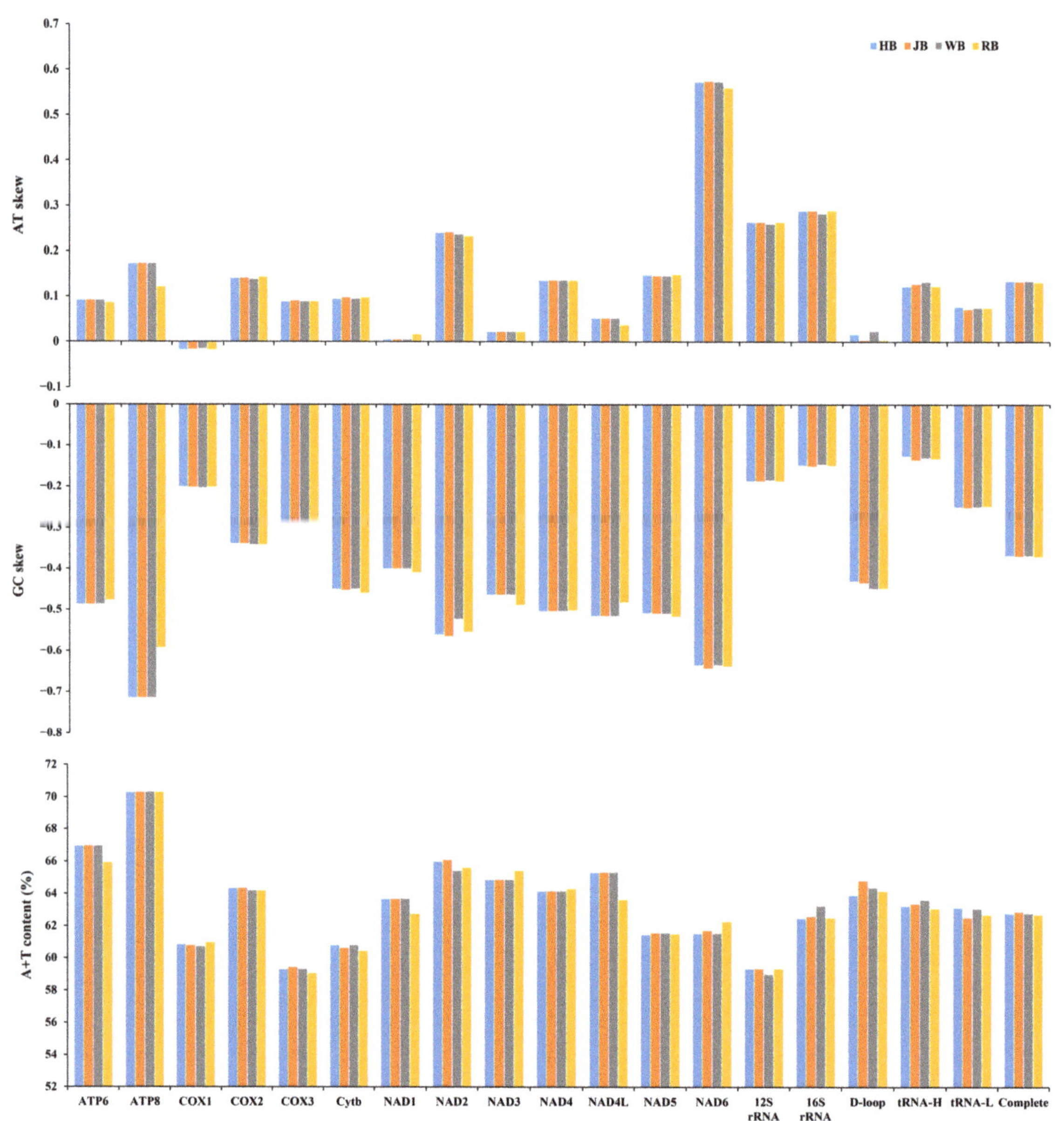

Figure 2. Nucleotide compositions of mitochondrial genome elements of four *P. sinensis* strains.

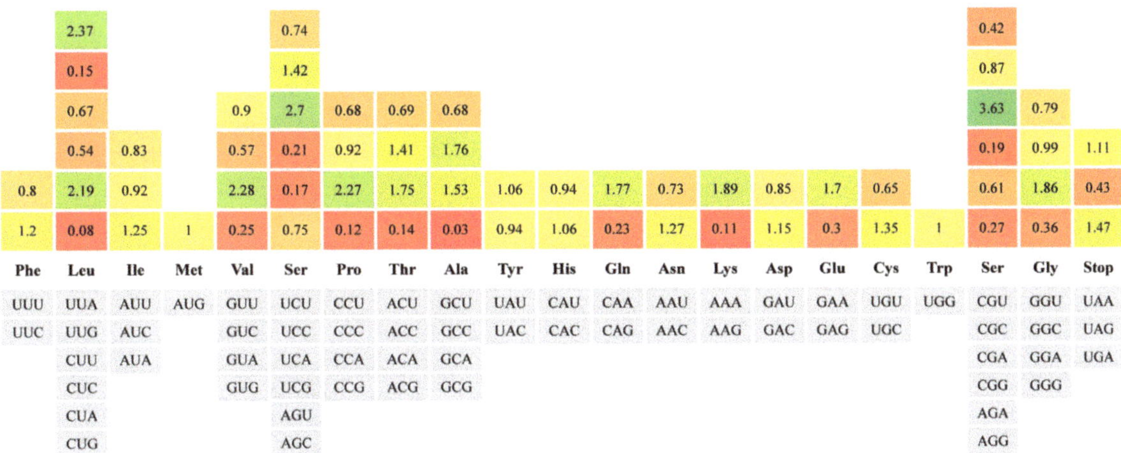

Figure 3. The RSCU values of all protein-coding genes for four *P. sinensis* strains. The red means a low RSCU value, the green indicates a high RSCU value, and the yellow indicates a middle RSCU value; the darker color indicates the largest RSCU value.

Table 4. Comparison of variation loci in mitochondrial genomes of four *P. sinensis* strains.

Mitochondrial Elements	Total Number of Sites	Invariable Sites	Variable Sites	Singleton Variable Sites	Parsimony Informative Sites	Percentage of Variable Sites
Complete genome	17,064	16,581	483	436	47	2.83%
tRNAs	1541	1516	25	22	3	1.62%
rRNAs	2580	2536	44	44	0	1.71%
D-loop	1546	1433	113	74	39	7.31%
ATP6	684	668	16	16	0	2.34%
ATP8	165	159	6	6	0	3.64%
COX1	1545	1512	33	32	1	2.14%
COX2	687	675	12	12	0	1.75%
COX3	784	766	18	17	1	2.30%
Cytb	1140	1113	27	25	2	2.37%
NAD1	972	947	25	25	0	2.57%
NAD2	1041	1013	28	26	2	2.69%
NAD3	350	339	11	11	0	3.14%
NAD4	1381	1340	41	41	0	2.97%
NAD4L	297	286	11	11	0	3.70%
NAD5	1779	1734	45	45	0	2.53%
NAD6	525	507	18	18	0	3.43%

3.3. Nucleotide Diversity and Selection Pressures

The nucleotide diversity of two rRNAs and 13 PCGs was explored using the sliding window analysis. The result showed that the nucleotide diversity levels of these genes were different. The *nad4l* gene (Pi = 0.019) had the highest nucleotide diversity compared with other PCGs genes, followed by *atp8* (Pi = 0.018), *nad6* (Pi = 0.017), and *nad3* (Pi = 0.016). Moreover, *cox2* (Pi = 0.009), *cox1* (Pi = 0.011), *cox3* (Pi = 0.012), and *cytb* (Pi = 0.012) were the most conserved genes across the PCGs and exhibited low nucleotide diversity levels (Figure 4). Compared with many other PCGs, 12S rRNA (Pi = 0.006) and 16S rRNA (Pi = 0.010) had low-level nucleotide diversity with reduced variability. To investigate the evolutionary selection constraints of the four *P. sinensis* strains, we performed Ka/Ks analysis for the 13 PCGs of mitochondrial genomes. The result showed that the Ka/Ks ratios for all PCGs were less than 1, indicating that these genes were evolving primarily under purifying selection. Among them, the lowest Ka/Ks value (0.000) for the *cox2* gene indicates the strongest purifying selection (neutral selection), whereas the highest Ka/Ks value (0.854) for the *nad6* gene showed a highly relaxed purifying selection (Figure 5A). Therefore, the Ka/Ks analysis indicated that the evolution of the four *P. sinensis* strains' mitogenomes has been dominated by purifying selection. Similarly, Kundu et al. reported that most of the PCGs in the mitogenomes of 13 Trionychidae species showed Ka/Ks values of <1, indicating a strong purifying selection among these Trionychidae species [53]. It has been reported that advantageous alleles can be retained by either positive selection or balancing selection, while deleterious alleles are removed through purifying selection [59]. Therefore, it is inferred that purifying selection has played an important role in the elimination of deleterious mutations in the four *P. sinensis* strains during evolution. These findings provide new insights for understanding the natural selection that influences the evolution of Trionychidae species. Previous studies have revealed that purifying selection reduces genetic diversity, and the population differentiation value generated by the gene locus with strong purifying selection is lower than that of the gene locus with a fast evolution rate [60,61]. We speculate that with the continuous progress of technology, excellent varieties of *P. sinensis* have been widely promoted in recent years, and there has been a serious trend of variety simplification, which has resulted in the reduction of genetic diversity.

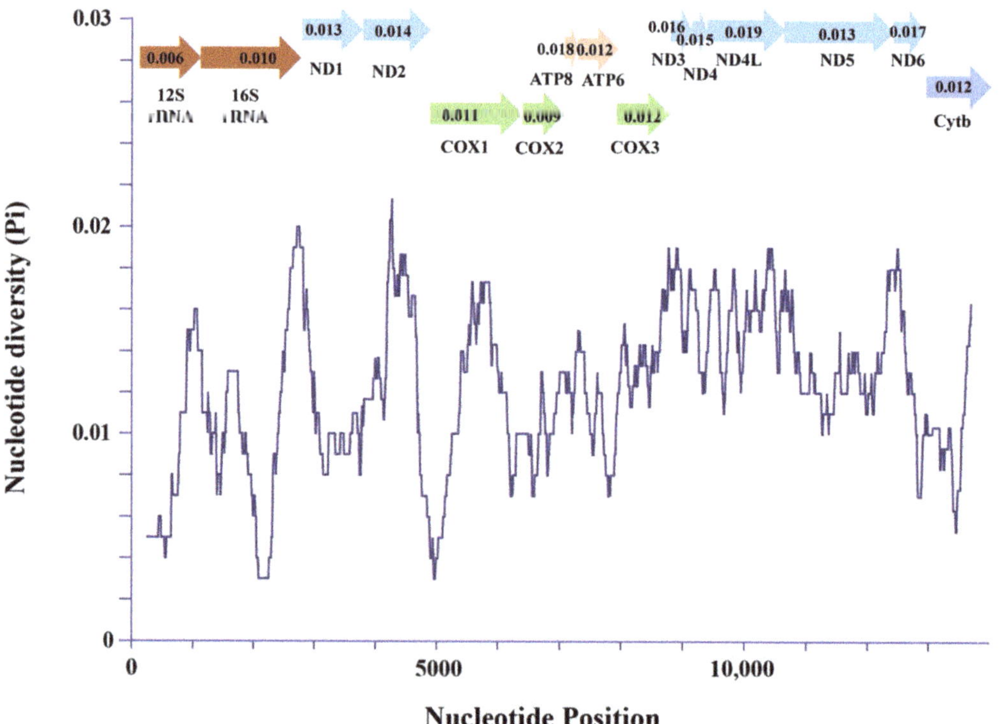

Figure 4. Sliding window analysis of 2 rRNA genes and 13 PCGs among the four *P. sinensis* strains.

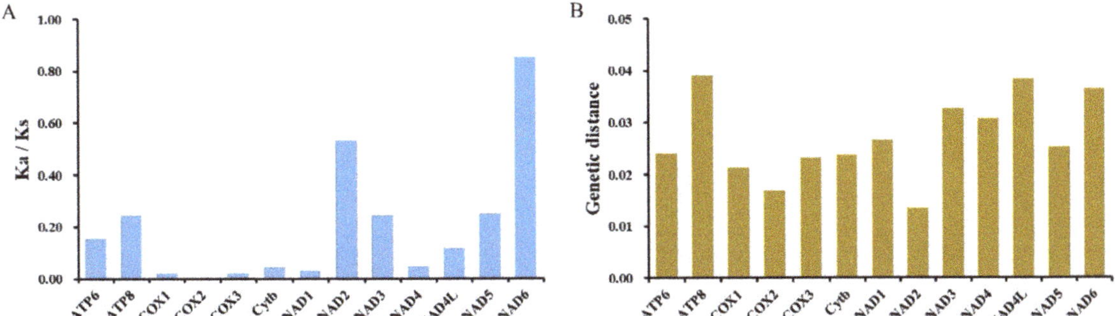

Figure 5. Ka/Ks (**A**) and genetic distance (**B**) analyses of 13 PCGs among four *P. sinensis* strains. Genetic distance indicates the overall mean distances of 13 PCGs among the four strains using the Kimura 2-parameter model of MEGA X software.

Furthermore, we investigated the genetic variation of 13 PCGs for four *P. sinensis* strains using genetic distance analysis. The result showed that *nad2* (average 0.014) and *cox2* (average 0.017) had low genetic distance, while *atp8* (average 0.039) possessed the largest genetic distance value, followed by *nad4l* (average 0.038), *nad6* (average 0.037), and *nad3* (average 0.033) (Figure 5B), which were the most variable genes. Taken together, the above findings reveal that the *nad3*, *nad4l*, *nad6*, and *atp8* genes possess high genetic variation and can be used as potential markers for the identification of these *P. sinensis* strains.

3.4. Comparative Analysis of tRNA Secondary Structure

Extensive comparison of tRNAs is crucial for understanding the structural and functional features of the mitogenomes. Herein, we compared and analyzed the tRNA secondary structure of four *P. sinensis* strains. The result revealed that most of the tRNAs were folded into classic clover-leaf secondary structures; only the DHU stem of tRNA-Ser was missing. Interestingly, similar characteristics were also found in many other Trionychidae species [53]. Through comparative analysis, the result showed that the most variable base pairing was observed in tRNA-Phe, tRNA-Gly, tRNA-Arg, and tRNA-Pro, while unchanged base pairing was detected in tRNA-Val, tRNA-Met, tRNA-Trp, tRNA-His, tRNA-Thr, tRNA-Ser, and tRNA-Asn, which presented a similar secondary structure in the four strains (Figure 6). Additionally, it was found that most tRNAs were structurally different in the TψC stem, DHU stem, and acceptor stem. Variation in the TψC stem occurred in tRNA-Phe, tRNA-Ile, tRNA-Gly, tRNA-Pro, tRNA-Cys, and tRNA-Gln. Variation in the TψC stem and the DHU stem coexisted in tRNA-Phe and tRNA-Pro. Variation in the DHU stem was observed in tRNA-Phe, tRNA-Leu, tRNA-Asp, tRNA-Arg, tRNA-Leu, and tRNA-Pro. Furthermore, structural variation in the acceptor stem was found in tRNA-Phe, tRNA-Lys, tRNA-Arg, tRNA-Pro, tRNA-Glu, and tRNA-Tyr. Similarly, a previous study compared the structural variation of tRNAs among 13 Trionychidae species and found that most tRNAs were different in the structure of the stem and loop [53]. Moreover, it was demonstrated that the variations of 22 tRNAs were mainly base substitutions among the four *P. sinensis* strains, and a few variations involved the change in base number.

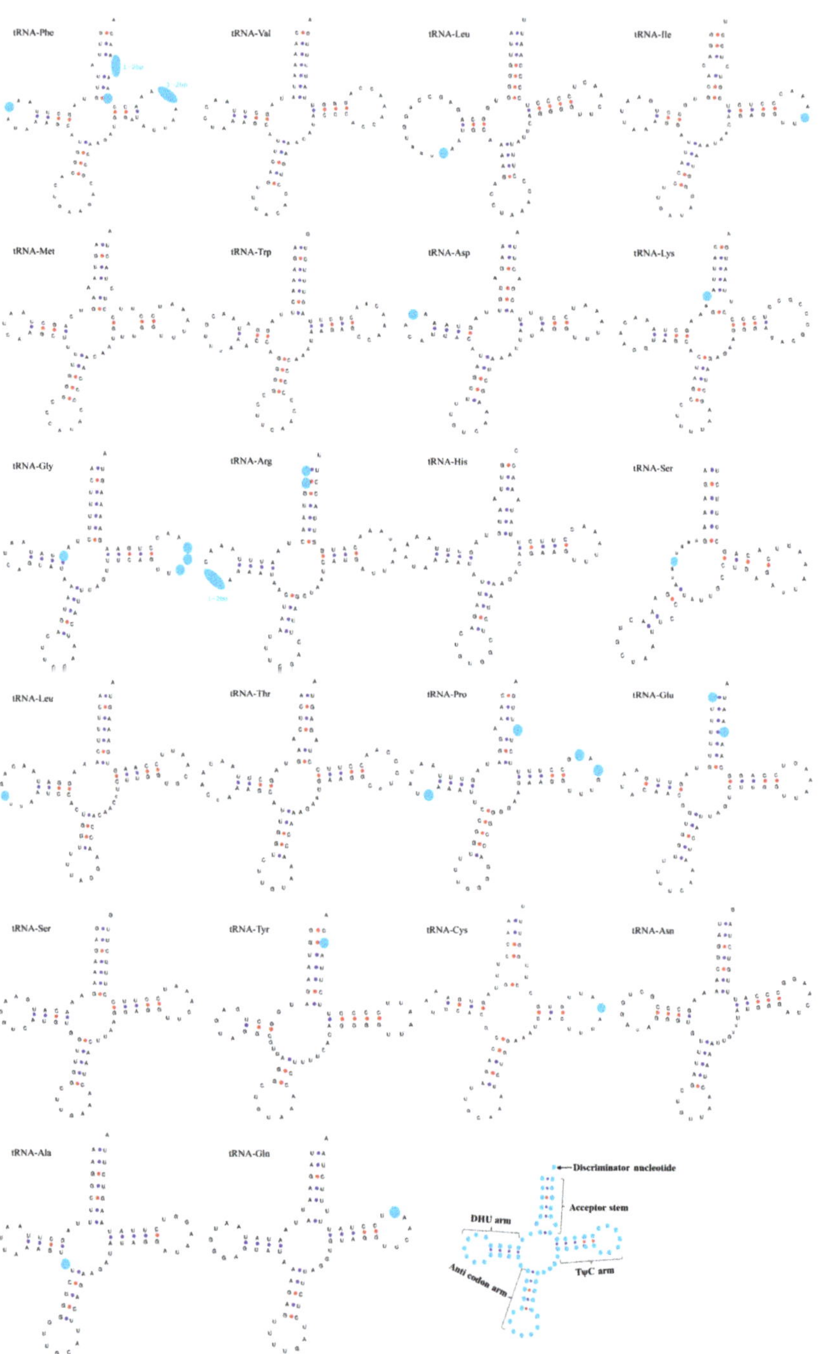

Figure 6. **Secondary structures of 22 transfer RNAs (tRNAs) indicating the structural variation among the four *P. sinensis* strains.** The light blue dots indicate variation sites in the tRNA secondary structure.

3.5. Comparative Analysis of Control Region

In this study, we found the origin of L-strand replication in the mitochondrial genome of four *P. sinensis* strains and detected the conserved sequences of 5′-AAAAT and AACCA-3′ in all four strains. Moreover, a stable stem-loop structure was observed in HB, JB, and WB strains with a length of 11 bp. However, due to the substitution of an adenosine with a thymidine in the RB strain, the stem-loop structure of the RB strain was only 9 bp. A previous study reported the stem-loop structure of the origin of L-strand replication in Cryptodira species but not in *Pelomedusa subrufa* of the suborder Pleurodira [62]. We speculate that this characteristic can be employed to distinguish Cryptodira and Pleurodira species, but further evidence and large-scale investigations are required for the Testudoformes species.

The control region of the mitochondrial genome contains tandem repeats [63]. Herein, we observed that the length of the tandem repeats in the control region was variable in the four *P. sinensis* strains and ranged from 2 bp to 50 bp. It was demonstrated that all four *P. sinensis* strains contained four variable number tandem repeats (VNTRs); the first tandem repeat (50 bp) was found in the four strains. Moreover, the second tandem repeat of HB, JB, RB, and WB strains was $(ACACAT)_{52}$, $(CATACA)_{48}$, $(CACATG)_{47}$, and $(CACATA)_{57}$, respectively. Furthermore, the four strains exhibited different AT contents and CR length, and HB, RB, and WB strains had variable third or fourth tandem repeats. For the HB strain, the length of the CR was 1596 bp, the A + T content was 63.87%, and the third tandem repeat was $(TA)_{37}/(ATATATATATC)_6/(24bp)_3$. The CR length of the RB strain was 1710 bp, the A + T content was 63.71%, and the third tandem repeat was $(TA)_{34}/(TATATATCATA)_6$. The CR length of the WB strain was 1698 bp, the A + T content was 64.45%, and the third and fourth tandem repeats were $(TA)_{40}/(TATATATATCA)_8$ and $(AT)_{21}/(TATCATATA)_6$, respectively (Figure 7). Unlike other strains, the CR of the JB strain contained four single tandem repeats with 1659 bp length and 64.58% A + T content. The CR length of vertebrates varies greatly, and it was suggested that the length difference of the mitochondrial genome may be mainly attributed to the change in CR length [64]. Moreover, the third and fourth tandem repeats of the JB strain were $(TATATATCATA)_6$ and $(TATCATATA)_7$, respectively. Our findings demonstrated that the frequency of tandem repeats was higher at the 3′ end of the CR for all four *P. sinensis* strains, and a single short tandem repeat (TA) was found in HB, RB, and WB strains. It has been reported that the CR of the mitochondrial genome is the most variable fragment as a result of the lack of coding restrictions in most species [65].

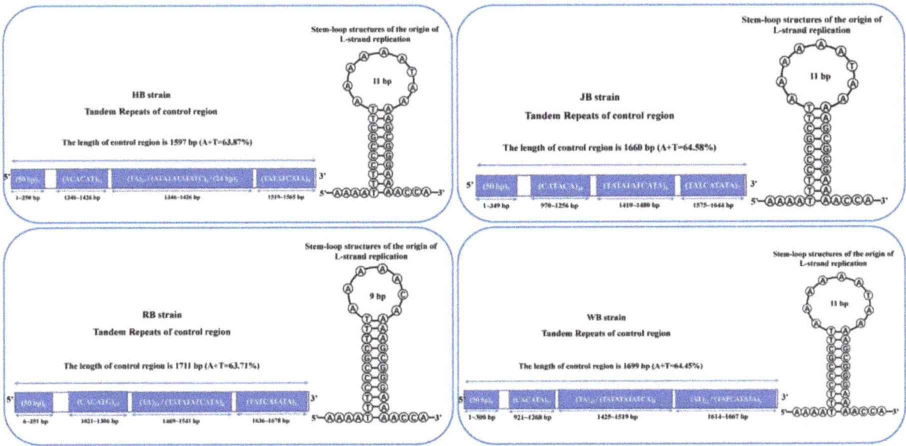

Figure 7. Comparison of length, nucleotide composition in control regions (CRs), and stem-loop structures of origin of L-strand replication for four *P. sinensis* strains.

3.6. Phylogenetic Analyses

To investigate the phylogenetic relationships of *P. sinensis* strains and other Trionychidae species, we conducted a phylogenetic analysis for Trionychidae species, including the subfamily Trionychinae (*P. sinensis, Palea steindachneri, Apalone ferox, A. spinifera, Rafetus swinhoei, Trionyx triunguis, Pelochelys cantorii, Chitra indica, Dogania subplana,* and *Nilssonia nigricans*) and the subfamily Cyclanorbinae (*Lissemys punctata* and *L. scutata*). We performed the phylogenetic analysis with concatenated nucleotides (13 PCGs, 22 tRNAs, and 2 rRNAs) of 23 mitogenomes by BI and ML methods, and the phylogenetic trees constructed by these two methods had almost identical topologies. For the four strains sequenced in this study, we found that the HB and WB strains had the closest genetic relationship, with the JB strain being more closely related to them and the RB strain being farther related to other three strains (Figure 8). The HB and WB strains both originate from the Taihu Lake basin, and it is inferred that they may possess small genetic differences. Overall, the *P. sinensis* strains divided into two main branches: one containing the HB strain, WB strain, JB strain, inked turtle strain (MG431983), and *P. sinensis* from Anhui Province (AY687385) and the other containing the RB strain, the Korean soft-shelled turtle (AY962573), and *P. sinensis* from Shanghai (NC068236). It was revealed that the inked turtle strain and WB strain had a close genetic affinity with the HB strain, while the RB strain, the Korean soft-shelled turtle (AY962573), and *P. sinensis* from Shanghai (NC068236) showed a distant relationship with other *P. sinensis* strains. Using a partial 12S rRNA sequence, Xu et al. reported that the WB strain and Anhui *P. sinensis* strains clustered into one branch and then clustered with the Korean soft-shelled turtle [66]. Therefore, our findings further clarified the phylogenetic relationship of different *P. sinensis* strains and provided molecular evidence for the germplasm resource conservation of *P. sinensis*.

Additionally, the phylogenetic trees showed that all Trionychinae species were clustered together and were retrieved as a sister clade of Cyclanorbinae species, which supported the previous phylogeny of Trionychidae species [67,68]. Moreover, we observed that the species of the same genus gathered into one branch. For instance, all *Apalone* species clustered into one branch, showing a close phylogenetic relationship with *R. swinhoei*. In addition, *D. subplana* and other Trionychinae species had a distant phylogenetic relationship. However, Kundu et al. reported that *T. triunguis, P. cantorii,* and *C. indica* were clustered into one branch and had a distant relationship with other Trionychinae species [53]. Consequently, more sequencing data from different taxonomic ranks of Trionychidae are essential to better understand the phylogenetic and evolutionary relationships among Trionychidae species.

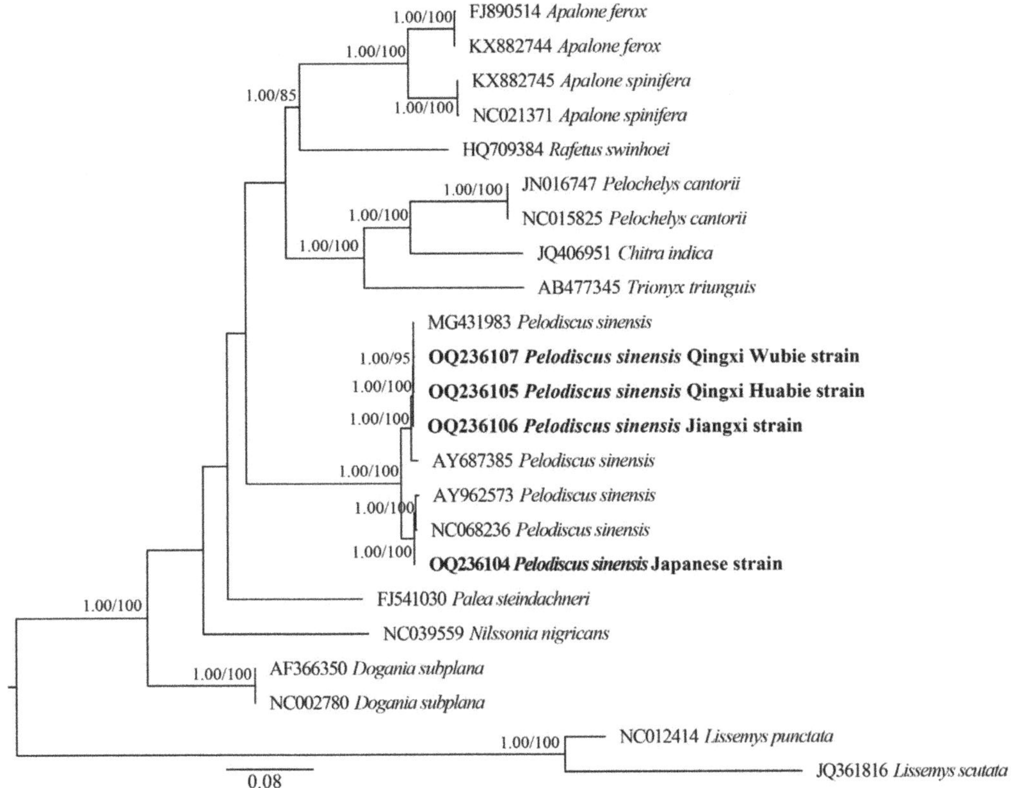

Figure 8. Molecular phylogenetic analysis of Trionychidae species using BI and ML methods based on concatenated mtDNA datasets. The phylogenetic tree nodes were considered well supported when the Bayesian posterior probability (BPP) of the node was ≥0.95 and ultrafast bootstrap was ≥80%. Numbers beside the nodes represent BPP (left) and bootstrap values (right). The bold fonts indicate the four *P. sinensis* strains sequenced in this study.

4. Conclusions

The present study described the complete mitochondrial genomes of four *P. sinensis* strains, including RB, HB, JB, and WB strains. Through comparative analysis, the following conclusions can be drawn: (1) the mitochondrial genomes of the four *P. sinensis* strains consisted of 13 PCGs, 22 tRNAs, 2 rRNAs, 1 control region, and a replication origin region of L-strand replication (OL); (2) most preference codons ended in purine (A/U) due to the abundance of A/T, and the evolution of the four *P. sinensis* strains' mitogenomes has been dominated by purifying selection; (3) the *atp8*, *nad4l*, *nad6*, and *nad3* genes can be used as the potential markers for the identification of these *P. sinensis* strains; (4) most of the tRNAs were folded into classic clover-leaf secondary structures (except for tRNA-Ser), and the structural variation mainly involved the TψC stem, DHU stem, and acceptor stem; (5) the stem-loop structure of the OL in the RB strain was different from the other three strains, and the length of the tandem repeats in the control region was variable in the four *P. sinensis* strains and ranged from 2 bp to 50 bp; and (6) the present study further confirmed phylogenetic relationships of *P. sinensis* strains and other Trionychidae species. Therefore, our findings provide insights for the classification and evolutionary research of different *P. sinensis* strains and offer valuable genetic information for the germplasm resource conservation of Trionychidae species.

Supplementary Materials: The following supporting information can be downloaded at: https://www.mdpi.com/article/10.3390/biology12030406/s1, Table S1: Best partitioning scheme and model selected by PartitionFinder for phylogenetic analyses.

Author Contributions: Conceptualization, H.Z.; Formal analysis, X.Y., X.H., L.H. and L.L.; Funding acquisition, H.Z.; Investigation, J.C., J.J., X.Y., L.L., W.Y. and J.Y.; Methodology, J.C. and J.J.; Software, L.H.; Writing—original draft, J.C.; Writing—review and editing, J.Y. and H.Z. All authors have read and agreed to the published version of the manuscript.

Funding: This work was supported by the Key Scientific and Technological Grant of Zhejiang for Breeding New Agricultural Varieties (No: 2021C02069-8), Exploratory Disruptive Innovation Project of Zhejiang Institute of Freshwater Fisheries (No: 2022TSX01 and 2021TSX02), and "San Nong Liu Fang" Science and Technology collaboration Project of Zhejiang Province (No: 2021SNLF026).

Institutional Review Board Statement: The animal experiments in this study were approved by the Institutional Animal Care and Use Committee (IACUC) of Zhejiang Institute of Freshwater Fisheries.

Informed Consent Statement: Not applicable.

Data Availability Statement: The sequencing data of mitochondrial genomes involved in this study were deposited in GenBank under the accession numbers: OQ236104, OQ236105, OQ236106, and OQ236107.

Conflicts of Interest: The authors declare that they have no competing interests.

References

1. Wang, J.; Qi, Z.; Yang, Z. Evaluation of the protein requirement of juvenile Chinese soft-shelled turtle (*Pelodiscus sinensis*, Wiegmann) fed with practical diets. *Aquaculture* **2014**, *433*, 252–255. [CrossRef]

2. Wu, D.; Huang, L.; Chen, J.; Zhang, Y.; Wang, J.; He, J. Gut microbiota of homologous Chinese soft-shell turtles (*Pelodiscus sinensis*) in different habitats. *BMC Microbiol.* **2021**, *21*, 142. [CrossRef] [PubMed]

3. Liang, Q.; Li, W.; Guo, N.; Tong, C.; Zhou, Y.; Fang, W.; Li, X. Identification and functional analysis of interleukin-1β in the Chinese Soft-Shelled Turtle *Pelodiscus sinensis* . *Genes* **2016**, *7*, 18. [CrossRef] [PubMed]

4. Feng, R.; Zhang, Z.; Guan, Y. Physiological and transcriptional analysis of Chinese soft-shelled turtle (*Pelodiscus sinensis*) in response to acute nitrite stress. *Aquat. Toxicol.* **2021**, *237*, 105899. [CrossRef]

5. Fritz, U.; Gong, S.; Auer, M.; Kuchling, G.; Schneeweiß, N.; Hundsdörfer, A.K. The world's economically most important chelonians represent a diverse species complex (*Testudines: Trionychidae: Pelodiscus*). *Org. Divers. Evol.* **2010**, *10*, 227–242. [CrossRef]

6. Zhang, C.; Xu, X.J.; Zhang, H.Q.; Mu, C.K.; He, Z.Y.; Wang, C.L. PCR-RFLP identification of four Chinese soft-shelled turtle *Pelodiscus sinensis* strains using mitochondrial genes. *Mitochondr. DNA* **2014**, *26*, 538–543. [CrossRef]

7. Li, Y.L.; Zhang, H.Q.; Lv, S.J.; Lin, F.; Liu, L.; Yuan, X.M.; Su, S.Q. Isolation and identification of pathogen causing "head-shaking syndrome" of *Pelodiscus sinensis* nigrum and drug susceptibility analysis. *Ocean. Limn. Sinica* **2020**, *51*, 405–414. (In Chinese)

8. FAO. *Fisheries and Aquaculture*; FAO Yearbook Fishery and Aquaculture Statistics 2022; Food and Agriculture Organization of the United Nations: Rome, Italy, 2022.

9. Zhang, J.; Zhou, Q.; Yang, X.; Yu, P.; Zhou, W.; Gui, Y.; Ouyang, X.; Wan, Q. Characterization of the complete mitochondrial genome and phylogenetic analysis of *Pelodiscus sinensis*, a mutant Chinese soft-shell turtle. *Conserv. Genet. Resour.* **2019**, *11*, 279–282. [CrossRef]

10. Zeng, D.; Li, X.; Wang, X.Q.; Xiong, G. Development of SNP markers associated with growth-related genes of *Pelodiscus sinensis* . *Conserv. Genet. Resour.* **2020**, *12*, 87–92. [CrossRef]

11. Zhang, J.; Wang, F.; Jiang, Y.L.; Hou, G.J.; Cheng, Y.S.; Chen, H.L.; Li, X. Modern greenhouse culture of juvenile soft-shelled turtle, *Pelodiscus sinensis* . *Aquacult. Int.* **2017**, *25*, 1607–1624. [CrossRef]

12. Dong, C.; Jia, Y.; Han, M.; Chen, W.; Mou, D.; Feng, C.; Jia, J.; Liu, X. Phylogenetic analysis of eight species of Anomopoda based on transcriptomic and mitochondrial DNA sequences. *Gene* **2021**, *787*, 145639. [CrossRef] [PubMed]

13. Paul, B.; Raj, K.K.; Murali, T.S.; Satyamoorthy, K. Species-specific genomic sequences for classification of bacteria. *Comput. Biol. Med.* **2020**, *123*, 103874. [CrossRef] [PubMed]

14. Severn-Ellis, A.A.; Scheben, A.; Neik, T.X.; Saad, N.S.M.; Pradhan, A.; Batley, J. Genotyping for Species Identification and Diversity Assessment Using Double-Digest Restriction Site-Associated DNA Sequencing (ddRAD-Seq). *Methods Mol. Biol.* **2020**, *2107*, 159–187. [PubMed]

15. Ling, F.; Yoshida, M. Rolling-Circle Replication in Mitochondrial DNA Inheritance: Scientific Evidence and Significance from Yeast to Human Cells. *Genes* **2020**, *11*, 514. [CrossRef] [PubMed]

16. Jia, M.; Li, Q.; Zhang, T.; Dong, B.; Liang, X.; Fu, S.; Yu, J. Genetic Diversity Analysis of the Chinese Daur Ethnic Group in Heilongjiang Province by Complete Mitochondrial Genome Sequencing. *Front. Genet.* **2022**, *13*, 919063. [CrossRef]

17. Wang, H.; Chen, M.; Chen, C.; Fang, Y.; Cui, W.; Lei, F.; Zhu, B. Genetic Background of Kirgiz Ethnic Group from Northwest China Revealed by Mitochondrial DNA Control Region Sequences on Massively Parallel Sequencing. *Front. Genet.* **2022**, *13*, 729514. [CrossRef]
18. Changbunjong, T.; Bhusri, B.; Sedwisai, P.; Weluwanarak, T.; Nitiyamatawat, E.; Chareonviriyaphap, T.; Ruangsittichai, J. Species identification of horse flies (*Diptera*: *Tabanidae*) in Thailand using DNA barcoding. *Vet. Parasitol.* **2018**, *259*, 35–43. [CrossRef]
19. Yang, X.; Wen, H.; Luo, T.; Zhou, J. Complete mitochondrial genome of *Triplophysa nasobarbatula* . *Mitochondrial DNA B Resour.* **2020**, *5*, 3771–3772. [CrossRef]
20. Zhou, S.B.; Zhang, Z.B.; Zhang, Z.H.; Liu, X.Y.; Guan, P.; Qu, B. The complete mitochondrial genome sequence of *Sinomicrurus peinani* (*Serpentes*: *Elapidae*). *Mitochondrial DNA B Resour.* **2022**, *7*, 964–966. [CrossRef]
21. Zhang, H.; Chen, Z.; Li, G.; Tang, Y.; Wen, Y.; Li, X.; Wang, M.; Liu, J.; Peng, L.; Xiao, Y.; et al. Cloning and genetic diversity analysis of mitochondrial cytochrome b in Hanshou *Trionyx sinensis* . *Acta Laser Biol. Sinica* **2018**, *27*, 359–366. (In Chinese)
22. Liang, H.; Cao, L.; Luo, X.; Zhu, C.; Cui, F.; Zou, G. Genetic diversity of three *Pelodiscus sinensis* strains based on COI gene sequence. *Genom. Appl. Biol.* **2021**, *40*, 2908–2915. (In Chinese)
23. Li, L.; Tan, S.; Wang, B.; Xu, J.; Han, X. Genetic Diversity of Three different populations of soft-shelled turtle *Trionyx sinensis* using mitochondrial D-loop gene. *Chin. J. Fish.* **2020**, *33*, 7–11. (In Chinese)
24. Chen, H.G.; Liu, W.B.; Zhang, X.J. Comparative analysis of mitochondrial DNA 12S rRNA region between *Pelodiscus sinensis* and *Pelodiscus axenaria* and their molecular marker for identification. *Chin. J. Fish.* **2005**, *29*, 318–322. (In Chinese)
25. Cao, J.; Guo, X.; Guo, C.; Wang, X.; Wang, Y.; Yan, F. Complete mitochondrial genome of *Malenka flexura* (*Plecoptera*: *Nemouridae*) and phylogenetic analysis. *Genes* **2022**, *13*, 911. [CrossRef] [PubMed]
26. Li, H.; Yu, R.; Ma, P.; Li, C. Complete mitochondrial genome of *Cultellus attenuatus* and its phylogenetic implications. *Mol. Biol. Rep.* **2022**, *49*, 8163–8168. [CrossRef]
27. Singh, V.K.; Mangalam, A.K.; Dwivedi, S.; Naik, S. Primer premier: Program for design of degenerate primers from a protein sequence. *Biotechniques* **1998**, *24*, 318–319. [CrossRef]
28. Larkin, M.A.; Blackshields, G.; Brown, N.P.; Chenna, R.; McGettigan, P.A.; McWilliam, H.; Valentin, F.; Wallace, I.M.; Wilm, A.; Lopez, R.; et al. Clustal W and Clustal X version 2.0. *Bioinformatics* **2007**, *23*, 2947–2948. [CrossRef]
29. Hall, T.A. BioEdit: A User-Friendly Biological Sequence Alignment Editor and Analysis Program for Windows 95/98/NT. *Nucleic Acids Sym. Ser.* **1999**, *41*, 95–98.
30. Burland, T.G. DNASTAR's Lasergene sequence analysis software. *Methods Mol. Biol.* **2000**, *132*, 71–91.
31. Laslett, D.; Canbäck, B. ARWEN: A program to detect tRNA genes in metazoan mitochondrial nucleotide sequences. *Bioinformatics* **2008**, *24*, 172–175. [CrossRef]
32. Bernt, M.; Donath, A.; Jühling, F.; Externbrink, F.; Florentz, C.; Fritzsch, G.; Pütz, J.; Middendorf, M.; Stadler, P.F. MITOS: Improved de novo metazoan mitochondrial genome annotation. *Mol. Phylogenet. Evol.* **2013**, *69*, 313–319. [CrossRef] [PubMed]
33. Lohse, M.; Drechsel, O.; Bock, R. OrganellarGenomeDRAW (OGDRAW): A tool for the easy generation of high-quality custom graphical maps of plastid and mitochondrial genomes. *Curr. Genet.* **2007**, *52*, 267–274. [CrossRef] [PubMed]
34. Kumar, S.; Stecher, G.; Li, M.; Knyaz, C.; Tamura, K. MEGA X: Molecular Evolutionary Genetics Analysis across Computing Platforms. *Mol. Biol. Evol.* **2018**, *35*, 1547–1549. [CrossRef] [PubMed]
35. Perna, N.T.; Kocher, T.D. Patterns of nucleotide composition at fourfold degenerate sites of animal mitochondrial genomes. *J. Mol. Evol.* **1995**, *41*, 353–358. [CrossRef]
36. Sharp, P.M.; Li, W.H. Codon usage in regulatory genes in Escherichia coli does not reflect selection for 'rare' codons. *Nucleic Acids Res.* **1986**, *14*, 7737–7749. [CrossRef]
37. Rozas, J.; Ferrer-Mata, A.; Sánchez-DelBarrio, J.C.; Guirao-Rico, S.; Librado, P.; Ramos-Onsins, S.E.; Sánchez-Gracia, A. DnaSP 6: DNA Sequence Polymorphism Analysis of Large Data Sets. *Mol. Biol. Evol.* **2017**, *34*, 3299–3302. [CrossRef]
38. Zheng, H.; Zhong, Z.; Shi, M.; Zhang, L.; Lin, L.; Hong, Y.; Fang, T.; Zhu, Y.; Guo, J.; Zhang, L.; et al. Comparative genomic analysis revealed rapid differentiation in the pathogenicity-related gene repertoires between *Pyricularia oryzae* and *Pyricularia penniseti* isolated from a Pennisetum grass. *BMC Genom.* **2018**, *19*, 927. [CrossRef]
39. Benson, G. Tandem repeats finder: A program to analyze DNA sequences. *Nucleic Acids Res.* **1999**, *27*, 573–580. [CrossRef]
40. Zuker, M. Mfold web server for nucleic acid folding and hybridization prediction. *Nucleic Acids Res.* **2003**, *31*, 3406–3415. [CrossRef]
41. Lowe, T.M.; Chan, P.P. tRNAscan-SE On-line: Integrating search and context for analysis of transfer RNA genes. *Nucleic Acids Res.* **2016**, *44*, W54–W57. [CrossRef]
42. Reuter, J.S.; Mathews, D.H. RNAstructure: Software for RNA secondary structure prediction and analysis. *BMC Bioinform.* **2010**, *11*, 129. [CrossRef] [PubMed]
43. Katoh, K.; Standley, D.M. MAFFT multiple sequence alignment software version 7: Improvements in performance and usability. *Mol. Biol. Evol.* **2013**, *30*, 772–780. [CrossRef] [PubMed]
44. Zhang, D.; Gao, F.; Jakovlić, I.; Zou, H.; Zhang, J.; Li, W.X.; Wang, G.T. PhyloSuite: An integrated and scalable desktop platform for streamlined molecular sequence data management and evolutionary phylogenetics studies. *Mol. Ecol. Resour.* **2020**, *20*, 348–355. [CrossRef] [PubMed]
45. Lanfear, R.; Frandsen, P.B.; Wright, A.M.; Senfeld, T.; Calcott, B. PartitionFinder 2: New Methods for Selecting Partitioned Models of Evolution for Molecular and Morphological Phylogenetic Analyses. *Mol. Biol. Evol.* **2017**, *34*, 772–773. [CrossRef] [PubMed]

46. Ronquist, F.; Teslenko, M.; Van der Mark, P.; Ayres, D.L.; Darling, A.; Höhna, S.; Larget, B.; Liu, L.; Suchard, M.A.; Huelsenbeck, J.P. MrBayes 3.2: Efficient Bayesian phylogenetic inference and model choice across a large model space. *Syst. Biol.* **2012**, *61*, 539–542. [CrossRef] [PubMed]

47. Nguyen, L.T.; Schmidt, H.A.; Von Haeseler, A.; Minh, B.Q. IQ-TREE: A fast and effective stochastic algorithm for estimating maximum-likelihood phylogenies. *Mol. Biol. Evol.* **2015**, *32*, 268–274. [CrossRef]

48. Minh, B.Q.; Nguyen, M.A.; Von Haeseler, A. Ultrafast approximation for phylogenetic bootstrap. *Mol. Biol. Evol.* **2013**, *30*, 1188–1195. [CrossRef]

49. Hoang, D.T.; Chernomor, O.; Von Haeseler, A.; Minh, B.Q.; Vinh, L.S. UFBoot2: Improving the Ultrafast Bootstrap Approximation. *Mol. Biol. Evol.* **2018**, *35*, 518–522. [CrossRef]

50. Rambaut, A.; Drummond, A.J.; Xie, D.; Baele, G.; Suchard, M.A. Posterior Summarization in Bayesian Phylogenetics Using Tracer 1.7. *Syst. Biol.* **2018**, *67*, 901–904. [CrossRef]

51. Zhong, L.; Wang, M.; Li, D.; Tang, S.; Zhang, T.; Bian, W.; Chen, X. Complete mitochondrial genome of freshwater goby *Rhinogobius cliffordpopei* (Perciformes, Gobiidae): Genome characterization and phylogenetic analysis. *Genes Genom.* **2018**, *40*, 1137–1148. [CrossRef]

52. Liu, Y.; Wu, P.D.; Zhang, D.Z.; Zhang, H.B.; Tang, B.P.; Liu, Q.N.; Dai, L.S. Mitochondrial genome of the yellow catfish *Pelteobagrus fulvidraco* and insights into Bagridae phylogenetics. *Genomics* **2019**, *111*, 1258–1265. [CrossRef] [PubMed]

53. Kundu, S.; Kumar, V.; Tyagi, K.; Chakraborty, R.; Singha, D.; Rahaman, I.; Pakrashi, A.; Chandra, K. Complete mitochondrial genome of Black Soft-shell Turtle (*Nilssonia nigricans*) and comparative analysis with other Trionychidae. *Sci. Rep.* **2018**, *8*, 17378. [CrossRef] [PubMed]

54. Jungt, S.O.; Lee, Y.M.; Kartavtsev, Y.; Park, I.S.; Kim, D.S.; Lee, J.S. The complete mitochondrial genome of the Korean soft-shelled turtle *Pelodiscus sinensis* (Testudines, Trionychidae). *DNA Seq.* **2006**, *17*, 471–483. [CrossRef] [PubMed]

55. Parham, J.F.; Macey, J.R.; Papenfuss, T.J.; Feldman, C.R.; Türkozan, O.; Polymeni, R.; Boore, J. The phylogeny of Mediterranean tortoises and their close relatives based on complete mitochondrial genome sequences from museum specimens. *Mol. Phylogenet. Evol.* **2006**, *38*, 50–64. [CrossRef]

56. Peng, Q.L.; Pu, Y.G.; Wang, Z.F.; Nie, L.W. Complete Mitochondrial Genome Sequence Analysis of Chinese Softshell Turtle (*Pelodiscus sinensis*). *Chin. J. Biochem. Mol. Biol.* **2005**, *21*, 591–596. (In Chinese)

57. Osawa, S.; Ohama, T.; Jukes, T.H.; Watanabe, K. Evolution of the mitochondrial genetic code. I. Origin of AGR serine and stop codons in metazoan mitochondria. *J. Mol. Evol.* **1989**, *29*, 202–207. [CrossRef]

58. Turanov, S.V.; Lee, Y.H.; Kartavtsev, Y.P. Structure, evolution and phylogenetic informativeness of eelpouts (*Cottoidei: Zoarcales*) mitochondrial control region sequences. *Mitochondrial DNA A DNA Mapp. Seq. Anal.* **2019**, *30*, 264–272. [CrossRef]

59. Gupta, M.K.; Vadde, R. Genetic Basis of Adaptation and Maladaptation via Balancing Selection. *Zoology* **2019**, *136*, 125693. [CrossRef]

60. Maruki, T.; Kumar, S.; Kim, Y. Purifying selection modulates the estimates of population differentiation and confounds genome-wide comparisons across single-nucleotide polymorphisms. *Mol. Biol. Evol.* **2012**, *29*, 3617–3623. [CrossRef]

61. Cvijović, I.; Good, B.H.; Desai, M.M. The Effect of Strong Purifying Selection on Genetic Diversity. *Genetics* **2018**, *209*, 1235–1278. [CrossRef]

62. Zardoya, R.; Meyer, A. Complete mitochondrial genome suggests diapsid affinities of turtles. *Proc. Natl. Acad. Sci. USA* **1998**, *95*, 14226–14231. [CrossRef]

63. Chen, J.Y.; Chang, Y.W.; Zheng, S.Z.; Lu, M.X.; Du, Y.Z. Comparative analysis of the *Liriomyza chinensis* mitochondrial genome with other Agromyzids reveals conserved genome features. *Sci. Rep.* **2018**, *8*, 8850. [CrossRef] [PubMed]

64. Huang, Z.H.; Tu, F.Y. Characterization and evolution of the mitochondrial DNA control region in Ranidae and their phylogenetic relationship. *Genet. Mol. Res.* **2016**, *15*, 1–9. [CrossRef] [PubMed]

65. Crochet, P.A.; Desmarais, E. Slow rate of evolution in the mitochondrial control region of gulls (*Aves: Laridae*). *Mol. Biol. Evol.* **2000**, *17*, 1797–1806. [CrossRef]

66. Xu, X.J.; Zhang, H.Q.; He, Z.Y. Sequence Composition of mitochondrial 12S rRNA genes between two varieties of *Pelodiscus sinensis* . *J. Econ. Anim.* **2012**, *16*, 163–167. (In Chinese)

67. Wang, L.; Zhou, X.; Nie, L.; Xia, X.; Liu, L.; Jiang, Y.; Huang, Z.; Jing, W. The complete mitochondrial genome sequences of *Chelodina rugosa* and *Chelus fimbriata* (Pleurodira: Chelidae): Implications of a common absence of initiation sites (O(L)) in pleurodiran turtles. *Mol. Biol. Rep.* **2012**, *39*, 2097–2107. [CrossRef] [PubMed]

68. Escalona, T.; Weadick, C.J.; Antunes, A. Adaptive patterns of mitogenome evolution are associated with the loss of shell scutes in Turtles. *Mol. Biol. Evol.* **2017**, *34*, 2522–2536. [CrossRef]

 *biology*

Article

Comparative Mitogenome Analyses Uncover Mitogenome Features and Phylogenetic Implications of the Parrotfishes (Perciformes: Scaridae)

Jiaxin Gao [1,2,3,4,5], Chunhou Li [1,2,3,4], Dan Yu [5], Teng Wang [1,2,3,4,*], Lin Lin [1,2,3,4], Yayuan Xiao [1,2,3,4], Peng Wu [1,2,3,4] and Yong Liu [1,2,3,4,*]

1 Key Laboratory of South China Sea Fishery Resources Exploitation and Utilization, Ministry of Agriculture and Rural Affairs, South China Sea Fisheries Research Institute, Chinese Academy of Fishery Sciences, Guangzhou 510300, China
2 Scientific Observation and Research Station of Xisha Island Reef Fishery Ecosystem of Hainan Province, Key Laboratory of Efficient Utilization and Processing of Marine Fishery Resources of Hainan Province, Sanya Tropical Fisheries Research Institute, Sanya 572018, China
3 Guangdong Provincial Key Laboratory of Fishery Ecology Environment, Guangzhou 510300, China
4 Observation and Research Station of Pearl River Estuary Ecosystem, Guangzhou 510300, China
5 Key Laboratory of Aquatic Biodiversity and Conservation, Institute of Hydrobiology, Chinese Academy of Sciences, Wuhan 430072, China
* Correspondence: wangteng@scsfri.ac.cn (T.W.); liuyong@scsfri.ac.cn (Y.L.)

Simple Summary: Parrotfishes are among the most colorful and diverse inhabitants of the coral reefs and sea grass beds and are ecologically important in these habitats. Here, we presented the complete mitogenome sequences from twelve parrotfish species and conducted comparative analysis of mitogenome features among the seven published species for the first time. The comparative analysis revealed both the conserved and unique characteristics of parrotfish mitogenomes. The mitogenome structure, organization, gene overlaps, putative secondary structures of transfer RNAs, and codon usage were relatively conserved among all the analyzed species. However, the base composition and the intergenic spacers varied largely among species. All of the protein-coding genes were under purifying selection. Phylogenetic analysis revealed that the parrotfishes could be divided into two clades with distinct ecological adaptations. Early divergence of these two clades was probably related to the expansion of sea grass habitat, and later diversifications were likely associated with the geomorphology alternation since the closing of the Tethys Ocean. This work offered fundamental materials for further studies on the evolution and conservation of parrotfishes.

Abstract: In order to investigate the molecular evolution of mitogenomes among the family Scaridae, the complete mitogenome sequences of twelve parrotfish species were determined and compared with those of seven other parrotfish species. The comparative analysis revealed that the general features and organization of the mitogenome were similar among the 19 parrotfish species. The base composition was similar among the parrotfishes, with the exception of the genus *Calotomus*, which exhibited an unusual negative AT skew in the whole mitogenome. The PCGs showed similar codon usage, and all of them underwent a strong purifying selection. The gene rearrangement typical of the parrotfishes was detected, with the *tRNA^{Met}* inserted between the *tRNA^{Ile}* and *tRNA^{Gln}*, and the *tRNA^{Gln}* was followed by a putative *tRNA^{Met}* pseudogene. The parrotfish mitogenomes displayed conserved gene overlaps and secondary structure in most tRNA genes, while the non-coding intergenic spacers varied among species. Phylogenetic analysis based on the thirteen PCGs and two rRNAs strongly supported the hypothesis that the parrotfishes could be subdivided into two clades with distinct ecological adaptations. The early divergence of the sea grass and coral reef clades occurred in the late Oligocene, probably related to the expansion of sea grass habitat. Later diversification within the coral reef clade could be dated back to the Miocene, likely associated with the geomorphology alternation since the closing of the Tethys Ocean. This work provided fundamental molecular data that will be useful for species identification, conservation, and further studies on the evolution of parrotfishes.

Citation: Gao, J.; Li, C.; Yu, D.; Wang, T.; Lin, L.; Xiao, Y.; Wu, P.; Liu, Y. Comparative Mitogenome Analyses Uncover Mitogenome Features and Phylogenetic Implications of the Parrotfishes (Perciformes: Scaridae). *Biology* **2023**, *12*, 410. https://doi.org/10.3390/biology12030410

Academic Editor: M. Gonzalo Claros

Received: 31 January 2023
Revised: 28 February 2023
Accepted: 2 March 2023
Published: 7 March 2023

Keywords: parrotfish; mitogenome; gene rearrangement; phylogeny; divergence time

1. Introduction

The mitochondrial genome (mitogenome) of a vertebrate is a small (16–17 kb), compact, and circular double-stranded molecule, typically encoding 13 protein-coding genes (PCGs), 22 transfer RNA genes (tRNAs), two ribosomal RNA genes (rRNAs), and two non-coding regions (the origin of L-strand replication, O_L, and control region, CR) [1]. The mitochondrial DNA sequences have been extensively employed in a variety of study areas, from phylogeography, which elucidates the spatial arrangement of genetic variation among populations or closely related species [2–4], to phylogenetic studies, which decipher the evolutionary relationships across a wide range of taxa at higher taxonomic levels [5–7]. Compared to single or a few mitochondrial gene-based markers, the complete mitogenome sequences generally provide much finer phylogenetic resolution [8]. Moreover, genome-level characteristics, including nucleotide composition, genome structural arrangement, overlap, and non-coding intergenic spacers between genes, vary largely among different species and might possess evolutionary significance [9–11].

Parrotfishes (Scaridae) are among the most colorful and diverse inhabitants of coral reefs and sea grass beds [12]. Currently, a total of 100 species belonging to 10 genera are recognized, with *Scarus* being the most specious genus (52 species) [13]. These fish are mainly herbivorous, foraging mostly by excavating or scraping surfaces of rocks and carbonate substrate that are encrusted with algae, bacterial mats, and detritus [14]. As such, it is widely recognized that parrotfishes play an important role in marine bioerosion [15,16] and serve as determinants of benthic community structure [17]. For example, parrotfish can exert a top-down control on algal communities to provide more space and resources for coals and promote the attachment and recruitment of coral larvae [18–20]. Therefore, it can help to mitigate the competition between coral reefs and macroalgae and increase the resilience of coral reef ecosystems subjected to anthropogenic or natural disturbances [21,22]. In addition, the excavating and scraping species can break the reef framework into sand-sized sediments and facilitate the cycling of calcium carbonate on reefs, which are also dispensable agents in reef erosion and sediment production and transport [23,24].

Deciphering mitogenome structures and sequences can provide insights into evolutionary processes and contribute to species delimitation and conservation efforts [25,26]. Despite the fact that parrotfishes play an irreplaceable role in coral reef and sea grass bed habitats due to their unique behavioral and ecological characteristics, only a few studies have addressed their mitogenome characteristics [27–29], and comparative analysis is scarce. Although a handful of works have tried to elaborate on the phylogenetic relationships among the parrotfishes [30–32], none of them have addressed this question from a mitogenomic perspective. The deficiency of mitogenome data and comparative works hindered us from understanding the evolution of the parrotfish and establishing proper management and conservation decisions. In the present study, we reported twelve parrotfish mitogenomes for the first time and conducted comparative analysis with the published sequences from other seven species to elaborate the detailed features of the parrotfish mitogenomes. Additionally, we also investigated the phylogenetic relationships among these parrotfishes and estimated divergence times using mitogenome data. We hope that our newly generated data and results will provide some insights into the evolution of the parrotfishes as well as contributions towards the identification and conservation of these fishes.

2. Materials and Methods

2.1. Sampling, DNA Extraction, PCR Amplification, and Sequencing

In the present study, we de novo sequenced twelve parrotfish species (with one specimen each): Calotomus carolinus, Cetoscarus bicolor, Hipposcarus longiceps, Scarus globiceps, Scarus chameleon, Scarus rivulatus, Scarus dimidiatus, Scarus oviceps, Scarus frenatus, Scarus niger, Scarus prasiognathos, and Scarus quoyi. The specimens of parrotfish were obtained from the Xisha Islands (15°46′~17°08′ N, 111°11′~112°54′ E), China, and deposited in the South China Sea Fisheries Research Institute, Chinese Academy of Fishery Sciences. Thirteen published mitogenome sequences from seven parrotfish species (Bolbometopon muricatum, KY235362/NC033901; Calotomus japonicus, AP017568/NC035427; Chlorurus sordidus, AP006567; Scarus forsteni, FJ619271/NC011928; Scarus ghobban, FJ449707/NC011599; Scarus rubroviolaceus, FJ227899/NC011343; Scarus schlegeli, FJ595020/NC011936) were also included in the analysis.

Genomic DNA was extracted from either a small piece of flesh or a pelvic fin clip taken from the right side of the specimen using the E.Z.N.A.® Tissue DNA Kit (OMEGA, Beijing, China) and following the manufacturer's instructions. High-quality DNA samples were randomly broken into fragments with a length of 300~500 bp. Then complete genomic libraries were established using the Illumina TruSeq™ Nano DNA Sample Prep Kit (Illumina, San Diego, CA, USA) following the manufacturer's recommendation. The 150-bp paired-end sequencing was performed on the Illumina HiSeq2500 platform. Library construction and sequencing were performed by the Biozeron Corporation (Shanghai, China).

2.2. Sequence Assembly, Annotation, and Analyses

Prior to assembly, raw reads were filtered by Trimmomatic v0.39 [33] in order to remove the reads with adaptors, the reads showing a quality score below 20 (Q < 20), the reads containing a percentage of uncalled bases ("N" characters) equal to or greater than 10%, and the duplicated sequences. GetOrganelle 1.7.5 was used to assemble the mitogenomes [34]. The newly generated mitogenome sequences were deposited in Genbank under the accession numbers OQ349180–OQ349191. Annotation of the mitogenomes (PCGs, tRNAs, rRNAs, and CR) was performed using MITOS [35] and Mitoannotator v3.83 [36]. Transfer RNA (tRNA) genes and their secondary structures were determined by the MITOS webserver [35]. The base composition and codon distributions were analyzed in MEGA 7.0 [37], and the nucleotide composition skewness was calculated using the formulas (A − T)/(A + T) for AT skew and (G − C)/(G + C) for GC skew. Relative synonymous codon usage (RSCU) was calculated using DAMBE 7 [38]. The conserved sequence block domains (CSBs) were determined by comparing them with those of other species [1].

2.3. Phylogenetic Analyses

Prior to the phylogenetic analysis, the method of Xia et al. [39] was used to access substitution saturation of the sequences by comparing the information entropy-based index (I_{SS}) with critical values ($I_{SS.c}$) in DAMBE 7 [38]. If I_{SS} is significantly lower than $I_{SS.c}$, then sequences have not experienced substitution saturation. The sequence of the control region showed significant substitution saturation (I_{SS} = 1.1897 > $I_{SS.c}$ = 0.7851, $p < 0.001$) and was thus excluded from further analysis. The phylogenetic relationships were reconstructed using the 13 PCGs and 2 rRNAs of the 19 parrotfish mitogenomes. Three *Cheilinus* species (*C. fasciatus*, NC037707; *C. oxycephalus*, NC061045; *C. undulatus*, NC013842) were used as outgroup taxa. Multiple sequence alignment was performed using MAFFT [40] implemented in PhyloSuite [41] under default parameters and subsequently checked by eye in SeaView [42]. Our dataset was partitioned by gene and codon position, and then the best-fit nucleotide substitution model for each partition was determined using Modelfinder [43]. Phylogenetic relationships were reconstructed using Bayesian inference (BI) and maximum likelihood (ML) approaches. BI was carried out in Mr. Bayes 3.2.7 [44]. Two independent Markov chains were run with 1×10^6 iterations, and 10,000 trees were retained, with the first 25% of the samples discarded as burn-in. ML analysis was conducted in IQTREE [45] under 10,000 ultrafast bootstrap replicates. DNAsp 6 [46] was used to calculate non-synonymous substitution rates (dN), synonymous substitution rates (dS), and the ratio of dN/dS (ω).

MCMCTree, implemented in the PAML4.9i software package [47], was used to estimate the divergence time among the parrotfishes. The tree topology generated from BI was calibrated with fossil dates. The information of branch lengths, gradients, and hessian were first estimated with a maximum likelihood method in BsaeML of the PAML package. Then the MCMC approximation was performed with a burn-in period of 50,000 cycles, and a total of 10,000 samples were generated every 50 iterations. Two independent runs were performed. Tracer 1.7 [48] was used to check for effective sample sizes (ESS) of parameters. The ESS larger than 200 were considered to reach convergence.

Two fossil calibration points were used in the divergence time estimation. *Calotomus preisli* was known from the middle Miocene (~14 Ma) in Austria [49]. We calibrated the minimum age of the split between the sea grass clade and the coral reef clade using this fossil. The fossil elements belonging to the genus *Bolbometopon* were known from the late Miocene (~5.3 Ma) [49]. These fossils were used to set the minimum age of the separation between *Bolbometopon* and *Cetoscarus*. The root age of our phylogeny was set to be lower than 50 Ma, for the oldest known labrid fossil was dated back to 50 Ma from the Monte Bolca in Italy [50].

3. Results

3.1. General Features of Mitochondrial Genomes

The total length of the 12 newly sequenced complete mitogenomes ranged from 16,657 bp in *Scarus niger* to 16,816 bp in *Scarus globiceps*. The typical set of 37 genes, including 13 PCGs, two rRNAs, and 22 tRNAs, and a control region, were detected in all the mitogenomes (Table 1, Figure 1, Supplementary File S1: Table S1). All PCGs were encoded on the Heavy (H) strand except for NADH dehydrogenase subunit 6 (*ND6*), which was located on the Light (L) strand. Eight tRNAs (*tRNA*^{Gln}, *tRNA*^{Ala}, *tRNA*^{Asn}, *tRNA*^{Cys}, *tRNA*^{Tyr}, *tRNA*$^{Ser\,(UGA)}$, *tRNA*^{Glu} and *tRNA*^{Pro}) were located on the L-strand, and the remaining 14 tRNAs were on the H-strand (Figure 1, Table 1). This coding pattern on the H and L-strand was identical among the 19 parrotfish species (Additional File 1: Table S1) and was consistent with most vertebrates [51].

Table 1. Features of the mitochondrial genome of the parrotfishes. *Calotomus carolinus* was taken as an example.

Features	Start	Stop	Length/bp	Intergenic Nucleotide	Start Codon	Stop Codon	Anti-Codon	Strand
tRNA^{Phe}	1	69	69	0			GAA	+ *
12S-rRNA	70	1020	951	0				+
tRNA^{Val}	1021	1093	73	0			TAC	+
16S-rRNA	1094	2782	1689	0				+
tRNA$^{Leu(UAA)}$	2783	2855	73	0			TAA	+
ND1	2856	3830	975	7	ATG	TAA		+
tRNA^{Ile}	3838	3907	70	10			GAT	+
tRNA^{Met}	3918	3986	69	6			TTG	+
tRNA^{Gln}	3993	4063	71	68			CAT	−
ND2	4132	5176	1045	0	ATG	TAG		+
tRNA^{Trp}	5177	5247	71	4			TCA	+
tRNA^{Ala}	5252	5322	71	5			TGC	−
tRNA^{Asn}	5328	5400	73	41			GTT	−
tRNA^{Cys}	5442	5507	66	9			GCA	−
tRNA^{Tyr}	5517	5586	70	1			GTA	−
COI	5588	7138	1551	0	GTG	TAA		+
tRNA$^{Ser\,(UGA)}$	7139	7209	71	3			TGA	−
tRNA^{Asp}	7213	7283	71	4			GTC	+
COII	7288	7978	691	0	ATG	T		+
tRNA^{Lys}	7979	8052	74	1			TTT	+

Table 1. *Cont.*

Features	Start	Stop	Length/bp	Intergenic Nucleotide	Start Codon	Stop Codon	Anti-Codon	Strand
ATPase 8	8054	8221	168	−16	ATG	TAG		+
ATPase 6	8206	8894	689	0	CTG	TA		+
COIII	8895	9679	785	0	ATG	TAA		+
tRNA^{Gly}	9680	9750	71	1			TCC	+
ND3	9752	10,103	352	0	ATA	TAG		+
tRNA^{Arg}	10,104	10,172	69	0			TCG	+
ND4L	10,173	10,469	297	−7	ATG	TAA		+
ND4	10,463	11,843	1381	0	ATG	T		+
tRNA^{His}	11,844	11,912	69	2			GTG	+
tRNA$^{Ser\,(GCU)}$	11,915	11,980	66	32			GCT	+
tRNA$^{Leu\,(UAG)}$	12,013	12,084	72	4			TAG	+
ND5	12,089	13,930	1842	−4	ATG	TAA		+
ND6	13,927	14,448	522	1	ATG	TAA		−
tRNA^{Glu}	14,450	14,522	73	64			TTC	−
Cyt b	14,587	15,727	1141	0	ATG	T		+
tRNA^{Thr}	15,728	15,799	72	0			TGT	+
tRNA^{Pro}	15,800	15,872	73	0			TGG	−
D-loop	15,873	17,114	1242					+

* +/− indicated H strand and L strand, respectively; a negative value indicated overlapping nucleotides.

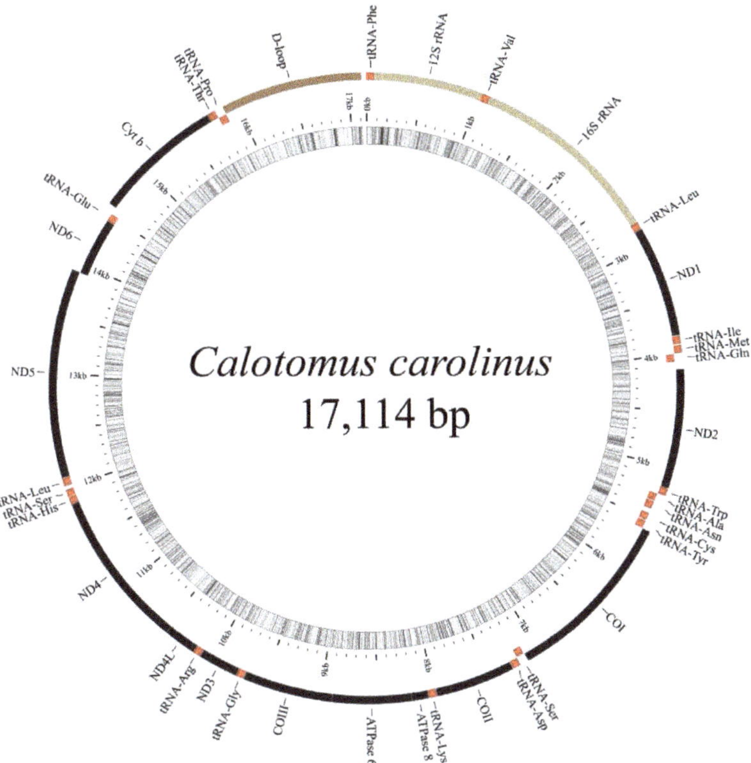

Figure 1. Organization of the parrotfish mitogenome. *Calotomus carolinus* was taken as an example. The inner ring indicated GC content.

3.2. Nucleotide Composition of the Parrotfish Mitogenomes and Unusual AT Skew of Calotomus Species

The nucleotide composition was similar among all of the parrotfish species, with the overall A + T content ranging from 53.0% in *Scarus globiceps* to 56.4% in *Bolbometopon muricatum*, and the A + T content was the lowest in *ND4L* (51.7 ± 3.6%) and the highest in the control region (62.6 ± 4.1%) (Figure 2a, Additional File 1: Table S2). All of the parrotfish mitogenomes exhibited AT bias, with the largest and most positive value observed in the rRNAs and the smallest and most negative value found in *ND6* (Figure 2b, Additional File 1: Table S3). Compared with other parrotfishes, species of the genus *Calotomus* exhibited an unusual AT skew for the whole mitogenomes with a slightly negative value (−0.04 to −0.02), while other species all displayed a positive value (Figure 2b, Additional File 1: Table S3). These results indicated that species of the genus *Calotomus* displayed an excess of T over A in the whole mitogenome.

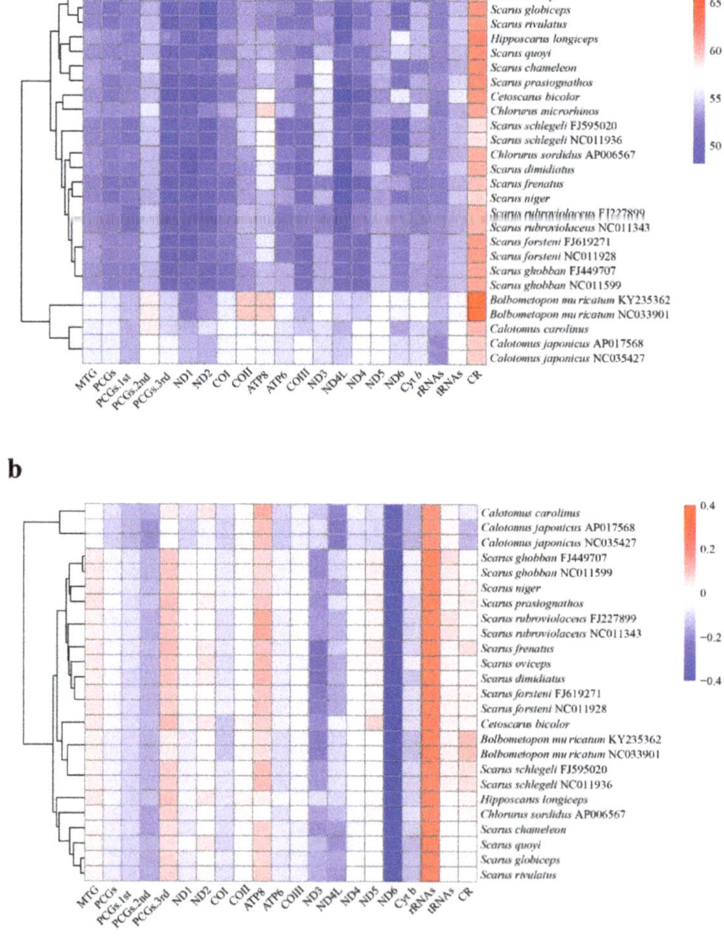

Figure 2. Base composition of various datasets among parrotfish mitogenomes, with hierarchical clustering of parrotfish species (y-axis) based on (**a**) AT content and (**b**) AT skew.

3.3. Protein-Coding Genes

The total length of PCGs ranged from 11,391 bp to 11,415 bp, with *ATP8* being the shortest (168 bp) and *ND5* being the longest (1839 bp to 1848 bp). Most genes exhibited the typical start codon ATN. However, *COI* initiated with GTG in all species, and *ATP6* started with GTG in *Calotomus japonicus* and *Scarus oviceps* or CTG in *Calotomus carolinus* (Additional File 1: Table S1). Four types of stop codons were detected, including two canonical (TAA and TAG) and two truncated codons (T– and TA-) (Additional File 1: Table S1). The incomplete stop codons were commonly observed in fish mitogenomes [1] and might be completed by post-transcriptional polyadenylation [52].

For all the parrotfish mitogenomes, Leu$^{(CUN)}$, Ala, and Thr were the three most frequently translated amnio acids, while Cys was the least used amnio acid (Figure 3a). Moreover, the most frequently used codon was CGA for arginine in all the parrotfish mitogenomes (Figure 3b). The RSCU revealed that degenerate codons were biased to use more A and T than G and C in the third codon position, which resulted in higher A + T content than G + C content in the third codon position of parrotfish mitogenomes (Figures 2a and 3b, Supplementary File S2: Figure S1).

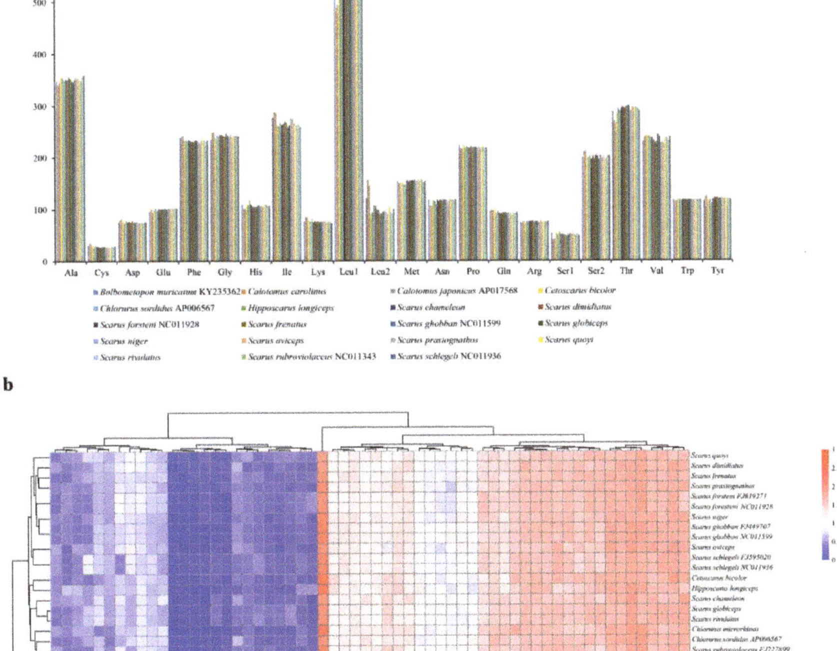

Figure 3. (**a**) Amino acid frequency in the parrotfish mitogenomes. (**b**) Heatmap based on the relative synonymous codon usage (RSCU) in the parrotfish mitogenomes.

3.4. Gene Rearrangement and Secondary Structure of tRNAs

All 22 tRNAs typical of the mitogenomes of vertebrates were found in the parrotfish mitogenomes (Figure 4a). Most tRNAs could be folded into the canonical clover-leaf secondary structure. The secondary structure of tRNAs generally consisted of four domains and a short variable loop: the amino acid acceptor (AA) stem, the dihydrouridine (D) arm (D stem and loop), the anticodon (AC) arm (AC stem and loop), the thymidine (T) arm (T stem and loop), and the variable (V) loop (Figure 4a). However, $tRNA^{Ser(AGN)}$ in *Bolbometopon muricatum* and *Calotomus japonicus* possessed only small loop(s) in their D arms (Figure 4b), thus not forming the typical clover-leaf structure. A gene rearrangement of the tRNA gene cluster between *ND1* and *ND2* was detected, with the $tRNA^{Met}$ inserted between the $tRNA^{Ile}$ and $tRNA^{Gln}$, and the $tRNA^{Gln}$ was followed by a putative $tRNA^{Met}$ pseudogene.

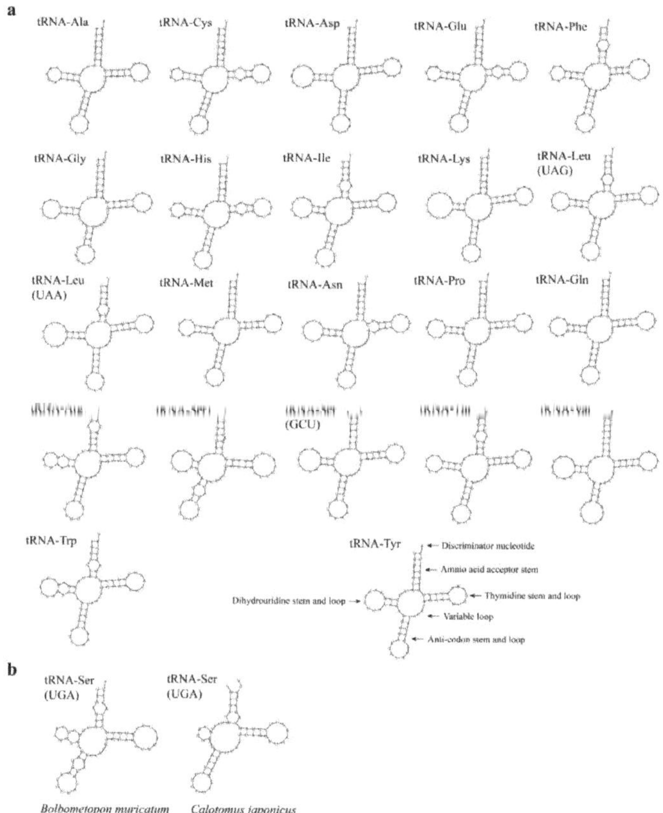

Figure 4. (**a**) Putative secondary structure of tRNAs in parrotfish mitogenomes. (**b**) Putative secondary structure of $tRNA^{Ser(AGN)}$ in *Bolbometopon muricatum* and *Calotomus japonicus*.

3.5. Overlaps and Non-Coding Intergenic Spacers

A total of four gene overlaps were detected in the mitogenome of *Calotomus carolinus* and five were observed in the mitogenomes of other parrotfishes (Table 1, Additional File 1: Table S1). The longest overlap was found between *ATP8* and *ATP6*, with highly conserved 10-bp motifs of "ATGGCACTAA" or "ATGACACTAA" detected in most parrotfish mitogenomes except for that of the genus *Calotomus*. The latter genus showed 16-bp overlaps of "CTGACCTTGGCACTAG" or "GTGGCCCTGGCACTAG". Apart from that, a 7-bp overlap was observed between *ND4L* and *ND4* in all parrotfish mitogenomes with highly conserved sequences of "ATGCTAA" or "ATGTTAA".

Two long intergenic spacers (IGS; $tRNA^{Gln}$-*ND2* and O_L) were found in all the parrotfish mitogenomes. Moreover, another long IGS between $tRNA^{Glu}$ and *Cyt b* was also found in the mitogenomes of the genus *Calotomus*. As mentioned above, the IGS between

$tRNA^{Gln}$ and *ND2* was assumed to be a pseudogene of $tRNA^{Met}$. O_L is located within the five tRNA gene cluster (WANCY), and its secondary structure showed a stable stem-loop hairpin, which is strengthened by 9 to 10 G-C base pairs (Figure 5). The G-C base pairs on the stem were highly conserved, while the loop varied in its base composition, with T being scarce.

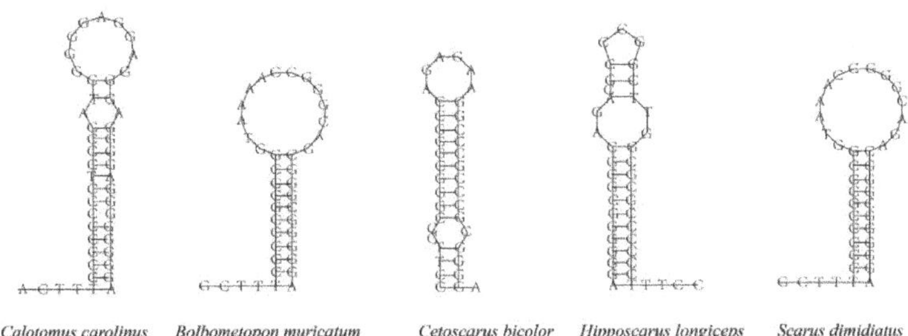

Figure 5. Putative secondary structure of the origin of L strand replication (O_L) in five parrotfish species.

The control region, located between $tRNA^{Pro}$ and $tRNA^{Phe}$, was the most variable region and constituted the majority of the length variation of the parrotfish mitogenomes (Additional File 1: Table S1). Only three conserved sequence blocks (CSB-D, CSB-I, and CSB-II) were detected (Figure 6), with CSB-III completely missing in all the parrotfish mitogenomes. The base composition was extremely unique to each CSB, with CSB-D being T rich, CSB-I being AT rich, and CSB-II being C rich (Table 2).

	CSB-D	CSB-I	CSB-II
Calotomus carolinus	TTCCTGGCATTTGGTTCC	ATATTATTCTTTCAGGTGC	GTATAAGTTACCCCCCCCTACCCCCC-------C
Cetoscarus bicolor	TTTCTGGCATTTGGCTCC	AACCTTGTTTTTCAAGAGC	ATCCATATT--CCCCCCAATCCCCCCAACAAAGC
Hipposcarus longiceps	TTACTGGCATCTGGTTCC	ATTAATAGTCATCAAGGAC	-------TTCCCCCTCCCCACTCCGC-------C
Scarus chameleon	TTACTGACATTTGGTTCC	ATTAGTGGATGTCATGAGC	AT-----TTTGCCCCCCCTCCCCCCA-------T
Scarus dimidiatus	TTACTGGCATTTGGTTCC	ATTATTGGATATCAAGAGC	-------TCTTCCCCCCTACCCCCCG-------C
Scarus frenatus	TTACTGGCATTTGGTTCC	ATTAAGGGATATCAAGAGC	-------TTTTCCCCCCTACCCCCCA-------C
Scarus globiceps	TTACTGGCATTTGGTTCC	ATTAATAGATATCATGAGC	-------TGTCCCCCCCTCCCCCCCA-------C
Scarus niger	TTACTGACATCTGGTTCC	ATTACTGGTTATCAGGAGC	GT-----TTTTCCCCCCCACCCCCCA-------C
Scarus oviceps	TTACTGGCATTTGGTTCC	ATTTTTGAATATCAAGAGC	TT-----TTTCCCCCCCAACCCCCCA-------C
Scarus prasiognathos	TTACTGGCATTTGGTTCC	ATTATCGTATATCAAGGAC	TT------TTACCCCCCCACCCCCCA-------C
Scarus quoyi	TTACTGGCATTTGGTTCC	ATTAATGGATATCATGAGC	-------TTTTCTCCCCCCCTCCCCCG-------C
Scarus rivulatus	TTACTGGCATTTGGTTCC	ATTAGTGGATATCATGAGC	-------TTGTCCCCCCCTCCCCCCA-------C
Chlorurus sordidus AP006567	TTACTGGCATTTGGTTCC	ATTAATAGCTATCAAGAGC	C-------TTTGCCCCCCTTCCCCCCG-------C
Bolbometopon muricatum KY235362/NC033901	TTTCTGGCATTTGGTTCC	AATATCTTTTATCAAGGAC	----------CCCCCCTTCCCCCCATAGA---C
Calotomus japonicus AP017568/NC035427	TTCCTGGCATTTGGTTCC	ATATTACTATATCAGGTGC	GG-----CTATCCCCCCCTACCCCCC-------C
Scarus forsteni FJ619271/NC011928	TTACTGGCATTTGGTTCC	ATTATCGTTTATCAAGAGC	GA-----GTTTCCCCCCTACCCCCG-------C
Scarus ghobban FJ449707/NC011599	TTACTGGCATTTGGTTCC	ATTATTGGATGTCAAGAGC	T-----TTTCCCCCCCCTCCCCCCA-------C
Scarus rubroviolaceus FJ227899/NC011343	TTACTGGCATTTGGTTCC	ATTACTGGATATCAAGAGC	GT-----CTTCCCCCCCCCTCCCCCCCC-------C
Scarus schlegeli FJ595020/NC011936	TTACTGGCATTTGGTTCC	ATTAGTGACTATCAAGAGC	-------TTGTCCCCCCCTCCCCCCA-------C

Figure 6. Conserved sequence blocks (CSBs) of the control region in the parrotfish mitogenomes.

Table 2. Base composition of the CSBs of parrotfish mitogenomes.

Base Composition (%)	CSB-D	CSB-I	CSB-II
A	10.5	34.2	8.8
T	44.4	32.2	20.1
G	21.6	19.6	5.2
C	23.5	14.0	65.9

3.6. Non-Synonymous and Synonymous Substitutions

To better understand the role of selective pressure and the evolutionary patterns of the protein coding genes, the dN/dS value (ω) of each PCG was calculated (Figure 7). All of the PCGs were subject to purifying selection, with a dN/dS value lower than 1 ($\omega < 1$). Among which, *ATP8* and *COI* presented the highest and lowest ω values ($\omega = 0.300$ and 0.016), respectively.

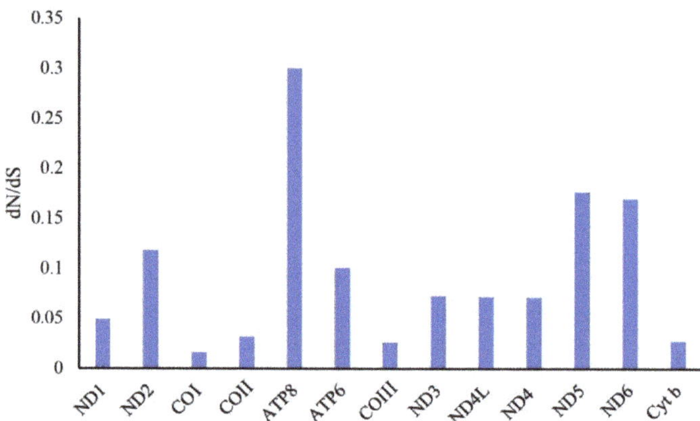

Figure 7. Non-synonymous/synonymous substitution ratios (ω) of the 13 PCGs in the parrotfish mitogenomes.

3.7. Phylogenetic Relationships of the Parrotfishes

The ML and BI trees based on the thirteen PCGs and two rRNAs yielded identical gene tree topologies (Figure 8), which congruently revealed two main clades. The first clade (clade A), located at the basal part of the tree, includes species of the genus *Calotomus*. The second clade (clade B) was comprised of the genera *Cetoscarus, Bolbometopon, Hipposcarus, Chlorurus*, and *Scarus*. The genus *Cetoscarus* was sister to *Bolbometopon*, positioned at the basal part of this clade. *Scarus* formed the sister genus to *Chlorurus*, then clustered with *Hipposcarus*. The monophyly of *Scarus* and *Chlorurus* was confirmed with strong support. Among the sampled *Scarus* species, *S. globiceps* showed a close relationship with *S. rivulatus* and exhibited little genetic difference (0.009 between the whole mitogenome). The nodes with high ML bootstrap support values and Bayesian posterior probabilities (BS > 70 and PP > 0.95) were shown.

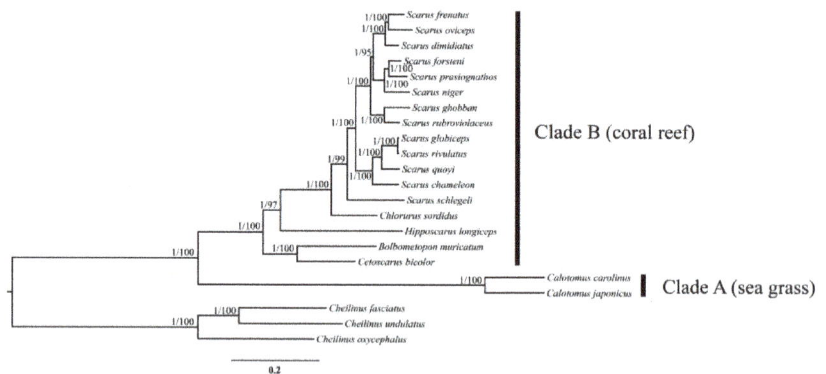

Figure 8. Phylogenetic relationships of the parrotfishes based on 13 PCGs and 2 rRNAs using Bayesian inference (BI) and maximum likelihood (ML). Numbers at nodes indicate Bayesian posterior probabilities (PP) and ultrafast bootstrap supports (UFBoot) from maximum likelihood analysis, respectively. Only well-supported numbers (PP > 0.95, UFBoot > 95) are shown.

3.8. Divergence Time Estimation

The estimated divergence time and 95% credible intervals (CIs) are shown in Figure 9. The split between clade A and clade B occurred at 26.9 Ma (95% CI 16.0~36.0 Ma) during the late Oligocene. The *Bolbometopon-Cetoscarus* clade differentiated at 15.9 Ma (95% CI 9.3~21.2

Ma) during the middle Miocene. *Hipposcarus* diverged at 14.3 Ma (95% CI 8.3~18.8 Ma). The split between *Chlorurus* and *Scarus* was dated back to 8.6 Ma (95% CI 5.2~11.6 Ma). The *Scarus* species diverged relatively recently, ranging from 0.2 Ma to 7.0 Ma.

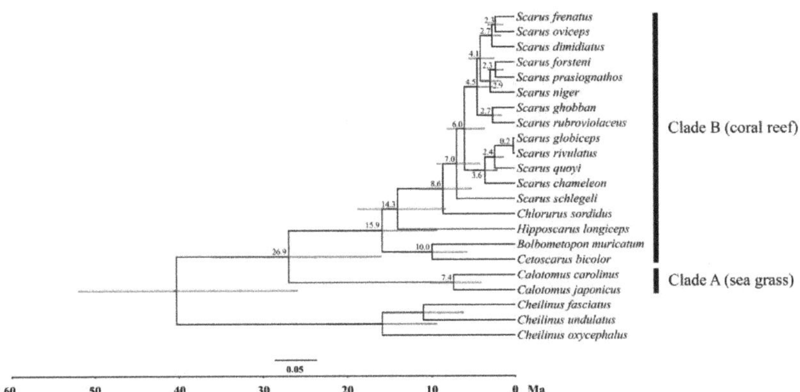

Figure 9. Divergence time estimation of the parrotfishes derived from MCMCTree analysis. Numbers at nodes indicate estimated age. Blue bars represent 95% credible age intervals for each node.

4. Discussion

The comparative analysis revealed that the mitogenome structure, organization, codon usage, and putative secondary structures of tRNAs were highly similar among all the analyzed parrotfish species. The gene rearrangement of the tRNA gene cluster between *ND1* and *ND2* was detected, which is typical of parrotfish [27–29]. In parrotfish mitogenomes, the $tRNA^{Met}$ was located between the $tRNA^{Ile}$ and $tRNA^{Gln}$, then a putative $tRNA^{Met}$ pseudogene was located after the $tRNA^{Gln}$. The gene rearrangements had been proposed to occur with tandem duplication of gene regions as a consequence of slipped-strand mispairing, followed by deletions of redundant genes [53]. The $tRNA^{Met}$ pseudogene was believed to function as punctuation marks for mitochondrial *ND2* mRNA processing [27].

Previous studies on insects suggested that the intergenic spacers were important for transcription and might be associated with gene rearrangement [54–56]. Our results showed significant variance in IGS among the parrotfish mitogenomes, especially for the longest IGS, the control region. Despite the great length variations found in the CR of the parrotfish mitogenomes, three conserved sequence blocks could still be detected. Compared with most fish species [1], the CSB-III cannot be observed in the CR of the parrotfish mitogenomes. The lack of CSB-III was also reported in other vertebrates [57]. Up until now, the functions of the CSBs were still not clear, however, the common existence of CSB-D and CSB-I in vertebrate mitogenomes suggested that they were vital in the replication and transcription of the genome [1].

CR was commonly used as genetic markers in phylogenetic and population genetic analysis due to its high variability among populations and closely related species [58,59]. However, our analysis suggested that the CR of parrotfishes experienced significant substitution saturation. Substitution saturation reduces the amount of phylogenetic signals to the point that sequence similarities could probably be the consequence of chance alone rather than homology. Therefore, phylogenetic signals are lost, and the sequences are no longer informative about the underlaying evolutionary processes that generate them if substitution saturation is reached [60]. For example, the mitochondrial markers *COI* and *ND3* that are commonly used in phylogenetic studies and DNA barcoding were proven to be subjected to significant substitution saturation in Caryophyllidean cestodes. Therefore, arbitrary application of these markers to the phylogenetic inference of this group of cestodes would jeopardize the well-supported phylogenetic estimates and evolutionary relationships [61]. In our case, the CR sequences have never been employed to infer the phylogenetic relationships among parrotfishes so far [30–32]. Future studies should avoid using the CR sequences when it comes to phylogenetic relationship inference or identification via DNA barcoding of the parrotfishes.

The RSCU revealed that degenerate codons were prone to use more A and T than G and C in the third codon position, therefore higher A + T content than G + C content was observed in the third codon position of parrotfish mitogenomes (Figures 2a and 3b, Additional File 2: Figure S1). This phenomenon was frequently observed in other teleosts [1] and might be related to genome bias, optimal selection of tRNA, or DNA repair efficiency [62].

Compared with other parrotfish species, the *Calotomus* species displayed an unusual AT skew for the whole mitogenome with a slightly negative skewness, while other species all showed positive values. Nucleotide skewness might be related to the balance between mutational and selective pressures during replication [63–65]. Some previous studies had indicated that the preference for certain nucleotides might be associated with selection rather than mutation [66]. For example, *Sinorhodeus microlepis*, a bitterling species with highly specialized ecological and behavioral preferences [67], also exhibited an unusual AT skew in its mitogenome [68] and this unique AT skewness was believed to be associated with unique selective forces [68]. Compared with other parrotfish species, the *Calotomus* species possessed some unique ecological aspects, such as the browsing feeding behavior and the lack of breeding territories [30]. It is suspected that distinct selective pressures or processes might lead to the preference of T in their mitogenomes. However, what and how the selective processes account for the unusual AT skew in the *Calotomus* species needs further investigation.

All of the PCGs were evolved under the purifying selection ($\omega < 1$). The lower ω value on the whole suggested a prevalent signature of strong functional restrictions across the mitogenome, which was largely in agreement with the functional importance of mitochondria as a respiration chain necessary for OXPHOS and electron transport [69]. Furthermore, the lower ω value indicated fewer variations in the amino acids; therefore, *COI* and *Cyt b* could serve as potential barcoding markers for the identification of parrotfish.

The phylogenetic relationships among the parrotfish genus based on thirteen PCGs and two rRNAs of the mitogenome indicated two distinct clades (A and B), which were identical with previous studies based on concatenated data of both mitochondrial and nuclear markers [30–32]. These two clades recovered by the phylogenetic analysis correspond to two distinct groups with different aspects of ecological adaptation, which had been defined as the sea grass clade and the coral reef clade [30]. The sea grass clade, as represented by *Calotomus* in this study, exhibited some less modified morphological characteristics (e.g., discrete teeth without cementation) [70] and showed some distinct ecological and behavioral aspects (e.g., browsing, no breeding territories, and no harem) [30]. These features differed greatly from the coral reef clades. In addition to the phylogenetic analysis, our results also revealed some unique features of the mitogenome composition and organization in the *Calotomus* mitogenomes (e.g., unusual AT skewness in the mitogenome and additional IGSs), indicating the evolutionary distinctiveness of the sea grass clade.

The first split of the parrotfish was estimated to have occurred in the late Oligocene (26.9 Ma, 95% CI 16.0~36.0 Ma), separating the sea grass clade from the coral reef clade. Geological evidence suggests that tectonic movements in the Indo-West Pacific region during the late Oligocene and early Miocene resulted in the formation of vast areas of shallow-water habitat between Australia and Indonesia [71], facilitating the expansion of sea grass habitat [72]. Our divergence time estimation was largely in congruence with the timing of the large-scale development of the sea grass habitat. This result probably indicated that the ecological differences between these two habitats acted as the major driving force in the early diversification of the parrotfishes. The differentiation within the coral reef clade had been initiated since the middle Miocene (about 15.9 Ma), which is well consistent with the closure of the Tethys Ocean [73]. Alterations in geomorphologies such as sea levels, sea surface temperatures, and ocean circulations exerted a great impact on coral reefs [74–76], likely functioning as the driving forces behind the rapid radiation of coral reef species. Previous studies indicated that the extensive diversification of coral reef taxa occurred during this period and was likely associated with the geomorphological reconfiguration of the marine realm [77]. In addition, natural and sexual selections might have also contributed to the diversification of parrotfishes. Some studies suggested that the protogynous mating system of parrotfishes might function as a possible driving force of speciation [30]. Though some research has suggested that ecological and selection may operate in tandem in the speciation processes [31], the function mechanisms and their relative roles still require further investigation.

5. Conclusions

In the present study, comparative analysis revealed both the conserved and unique characteristics of parrotfish mitogenomes. The mitogenome structure, organization, gene overlaps, putative secondary structures of tRNAs, and codon usage were relatively conserved among all the analyzed species. However, the base composition and the intergenic spacers varied largely among species. All of the PCGs were under purifying selection. Phylogenetic analysis revealed that the parrotfishes could be divided into two clades with distinct ecological adaptations. Early divergence of the sea grass and coral reef clades occurred in the late Oligocene, probably related to the expansion of sea grass habitat. Later diversification within the coral reef clade could be dated back to the Miocene, likely associated with the geomorphology alternation since the closing of the Tethys Ocean. This study offered fundamental molecular materials for further studies on the evolution and diversification of the parrotfishes and would contribute to their identification and conservation.

Supplementary Materials: The following supporting information can be downloaded at: https://www.mdpi.com/article/10.3390/biology12030410/s1, Table S1: General features of the parrotfish mitogenomes; Table S2: AT content of the parrotfish mitogenomes; Table S3: AT skew of the parrotfish mitogenomes; Figure S1: Relative synonymous codon usage (RSCU) of the 12 newly determined parrotfish species.

Author Contributions: Funding acquisition, T.W. and Y.L.; Methodology, J.G.; Resources, C.L., T.W., L.L., Y.X., P.W. and Y.L.; Software, J.G., C.L. and D.Y.; Supervision, T.W. and Y.L.; Visualization, J.G., D.Y., L.L., Y.X. and P.W.; Writing—original draft, J.G., C.L. and D.Y.; Writing—review and editing, T.W. and Y.L. All authors have read and agreed to the published version of the manuscript.

Funding: The study was funded by the Fundamental and Applied Fundamental Research Major Program of Guangdong Province (2019B030302004-05); Hainan Provincial Natural Science Foundation (322CXTD530); Hainan Provincial Natural Science Foundation (322MS153); Science and Technology Planning Project of Guangdong Province (2019B121201001); Central Public-interest Scientific Institution Basal Research Fund, CAFS (2020TD16); Financial Fund of the Ministry of Agriculture and Rural Affairs, P. R. of China (NFZX2021).

Institutional Review Board Statement: The animal study protocol was approved by the Institutional Animal Care and Use Committee of the Institute of Hydrobiology, Chinese Academic of Sciences (protocol code: 2022/LL/036; date of approval: 25 September 2022).

Informed Consent Statement: Not applicable.

Data Availability Statement: The data presented in this study are available in NCBI GenBank (Accession number: OQ349180-OQ349191).

Conflicts of Interest: The authors declare no conflict of interest.

References

1. Satoh, T.P.; Miya, M.; Mabuchi, K.; Nishida, M. Structure and variation of the mitochondrial genome of fishes. *BMC Genom.* **2016**, *17*, 719. [CrossRef] [PubMed]
2. Avise, J.C.; Arnold, J.; Ball, R.M.; Bermingham, E.; Lamb, T.; Neigel, J.E.; Reeb, C.A.; Saunders, N.C. Intraspecific Phylogeography: The Mitochondrial DNA Bridge between Population Genetics and Systematics. *Annu. Rev. Ecol. Syst.* **1987**, *18*, 489–522. [CrossRef]

3.   Avise, J.C. *Phylogeography: The History and Formation of Species*; Harvard University Press: Harvard, MA, USA, 2000.

4.   Avise, J.C. Phylogeography: Retrospect and prospect. *J. Biogeogr.* **2009**, *36*, 3–15. [CrossRef]

5.   Miya, M.; Takeshima, H.; Endo, H.; Ishiguro, N.B.; Inoue, J.G.; Mukai, T.; Satoh, T.P.; Yamaguchi, M.; Kawaguchi, A.; Mabuchi, K.; et al. Major patterns of higher teleostean phylogenies: A new perspective based on 100 complete mitochondrial DNA sequences. *Mol. Phylogenetics Evol.* **2003**, *26*, 121–138. [CrossRef] [PubMed]

6.   Inoue, J.G.; Miya, M.; Lam, K.; Tay, B.-H.; Danks, J.A.; Bell, J.; I Walker, T.I.; Venkatesh, B. Evolutionary Origin and Phylogeny of the Modern Holocephalans (Chondrichthyes: Chimaeriformes): A Mitogenomic Perspective. *Mol. Biol. Evol.* **2010**, *27*, 2576–2586. [CrossRef]

7.   Cole, T.L.; Ksepka, D.T.; Mitchell, K.J.; Tennyson, A.J.D.; Thomas, D.B.; Pan, H.; Zhang, G.; Rawlence, N.J.; Wood, J.R.; Bover, P.; et al. Mitogenomes Uncover Extinct Penguin Taxa and Reveal Island Formation as a Key Driver of Speciation. *Mol. Biol. Evol.* **2019**, *36*, 784–797. [CrossRef]

8.   Nie, R.-E.; Breeschoten, T.; Timmermans, M.J.T.N.; Nadein, K.; Xue, H.-J.; Bai, M.; Huang, Y.; Yang, X.-K.; Vogler, A.P. The phylogeny of *Galerucinae* (Coleoptera: Chrysomelidae) and the performance of mitochondrial genomes in phylogenetic inference compared to nuclear rRNA genes. *Cladistics* **2018**, *34*, 113–130. [CrossRef]

9.   Telford, M.J.; Herniou, E.A.; Russell, R.B.; Littlewood, D.T.J. Changes in mitochondrial genetic codes as phylogenetic characters: Two examples from the flatworms. *Proc. Natl. Acad. Sci. USA* **2000**, *97*, 11359–11364. [CrossRef]

10.   Shi, W.; Dong, X.-L.; Wang, Z.-M.; Miao, X.-G.; Wang, S.-Y.; Kong, X.-Y. Complete mitogenome sequences of four flatfishes (Pleuronectiformes) reveal a novel gene arrangement of L-strand coding genes. *BMC Evol. Biol.* **2013**, *13*, 173. [CrossRef]

11.   Zhang, D.; Zou, H.; Hua, C.-J.; Li, W.-X.; Mahboob, S.; Al-Ghanim, K.A.; Al-Misned, F.; Jakovlić, I.; Wang, G.-T. Mitochondrial Architecture Rearrangements Produce Asymmetrical Nonadaptive Mutational Pressures That Subvert the Phylogenetic Reconstruction in Isopoda. *Genome Biol. Evol.* **2019**, *11*, 1797–1812. [CrossRef]

12.   Sale, P.F. *The Ecology of Fishes on Coral Reefs*; Academic Press: San Diego, CA, USA, 1991.

13.   Parenti, P.; Randall, J.E. Checklist of the species of the families *Labridae* and *Scaridae*: An update. *Smithiana Bull.* **2011**, *13*, 29–44.

14.   Gobalet, K.W. Cranial Specializations of Parrotfishes, Genus *Scarus* (Scarinae, Labridae) for Scraping Reef Surfaces. In *Biology of Parrotfishes*; CRC Press: Boca Raton, FL, USA, 2018; pp. 1–25.

15.   Bellwood, D.R.; Choat, J.H. A functional analysis of grazing in parrotfishes (family *Scaridae*): The ecological im-plications. *Environ. Biol. Fishes* **1989**, *20*, 109–214. [CrossRef]

16.   Bellwood, D.R. Direct estimate of bioerosion by two parrotfish species, *Chlorurus gibbus* and *C. sordidus*, on the Great Barrier Reef, Australia. *Mar. Biol.* **1995**, *121*, 419–429. [CrossRef]

17.   Lewis, S.M.; Wainwright, P.C. Herbivore abundance and grazing intensity on a Caribbean coral reef. *J. Exp. Mar. Biol. Ecol.* **1985**, *87*, 215–228. [CrossRef]

18.   Bellwood, D.R.; Hoey, A.S.; Hughes, T.P. Human activity selectively impacts the ecosystem roles of parrotfishes on coral reefs. *Proc. R. Soc. B Boil. Sci.* **2012**, *279*, 1621–1629. [CrossRef]

19.   Thurber, R.V.; Burkepile, D.E.; Correa, A.M.S.; Thurber, A.R.; Shantz, A.A.; Welsh, R.; Pritchard, C.; Rosales, S. Macroalgae Decrease Growth and Alter Microbial Community Structure of the Reef-Building Coral, *Porites astreoides*. *PLoS ONE* **2012**, *7*, e44246. [CrossRef]

20.   Roos, N.C.; Pennino, M.G.; Lopes, P.F.D.M.; Carvalho, A.R. Multiple management strategies to control selectivity on parrotfishes harvesting. *Ocean Coast. Manag.* **2016**, *134*, 20–29. [CrossRef]

21.   Adam, T.; Burkepile, D.; Ruttenberg, B.; Paddack, M. Herbivory and the resilience of Caribbean coral reefs: Knowledge gaps and implications for management. *Mar. Ecol. Prog. Ser.* **2015**, *520*, 1–20. [CrossRef]

22.   Quan, Q.; Liu, Y.; Wang, T.; Li, C. Geographic Variation in the Species Composition of *Parrotfish* (Labridae: Scarini) in the South China Sea. *Sustainability* **2022**, *14*, 11524. [CrossRef]

23.   Morgan, K.M.; Kench, P.S. Parrotfish erosion underpins reef growth, sand talus development and island building in the Maldives. *Sediment. Geol.* **2016**, *341*, 50–57. [CrossRef]

24.   Eggertsen, L.; Goodell, W.; Cordeiro, C.A.M.M.; Mendes, T.C.; Longo, G.O.; Ferreira, C.E.L.; Berkström, C. Seascape Configuration Leads to Spatially Uneven Delivery of Parrotfish Herbivory across a Western Indian Ocean Seascape. *Diversity* **2020**, *12*, 434. [CrossRef]

25.   Knaus, B.J.; Cronn, R.; Liston, A.; Pilgrim, K.; Schwartz, M.K. Mitochondrial genome sequences illuminate maternal lineages of conservation concern in a rare carnivore. *BMC Ecol.* **2011**, *11*, 10. [CrossRef] [PubMed]

26.   Johri, S.; Fellows, S.R.; Solanki, J.; Busch, A.; Livingston, I.; Mora, M.F.; Tiwari, A.; Cantu, V.A.; Goodman, A.; Morris, M.; et al. Mitochondrial genome to aid species delimitation and effective conservation of the Sharpnose Guitarfish (*Glaucostegus granulatus*). *Meta Gene* **2020**, *24*, 100648. [CrossRef]

27.   Mabuchi, K.; Miya, M.; Satoh, T.P.; Westneat, M.W.; Nishida, M. Gene Rearrangements and Evolution of tRNA Pseudogenes in the Mitochondrial Genome of the Parrotfish (Teleostei: Perciformes: Scaridae). *J. Mol. Evol.* **2004**, *59*, 287–297. [CrossRef]

28.   Mabuchi, K. Complete mitochondrial genome of the parrotfish *Calotomus japonicus* (Osteichthyes: Scaridae) with implications based on the phylogenetic position. *Mitochondrial DNA Part B* **2016**, *1*, 643–645. [CrossRef]

29.   Chiang, W.-C.; Chang, C.-H.; Hsu, H.-H.; Jang-Liaw, N.-H. Complete mitochondrial genome sequence for the green humphead parrotfish *Bolbometopon muricatum*. *Conserv. Genet. Resour.* **2017**, *9*, 393–396. [CrossRef]

30. Streelman, J.T.; Alfaro, M.; Westneat, M.W.; Bellwood, D.R.; Karl, S.A. Evolutionary History of the Parrotfishes: Biogeography, Ecomorphology, and Comparative Diversity. *Evolution* **2002**, *56*, 961–971. [CrossRef]
31. Smith, L.L.; Fessler, J.L.; Alfaro, M.E.; Streelman, J.T.; Westneat, M.W. Phylogenetic relationships and the evolution of regulatory gene sequences in the parrotfishes. *Mol. Phylogenetics Evol.* **2008**, *49*, 136–152. [CrossRef]
32. Choat, J.H.; Klanten, O.S.; Van Herwerden, L.; Robertson, D.R.; Clements, K.D. Patterns and processes in the evolutionary history of parrotfishes (Family Labridae). *Biol. J. Linn. Soc.* **2012**, *107*, 529–557. [CrossRef]
33. Bolger, A.M.; Lohse, M.; Usadel, B. Trimmomatic: A flexible trimmer for Illumina sequence data. *Bioinformatics* **2014**, *30*, 2114–2120. [CrossRef]
34. Jin, J.-J.; Yu, W.-B.; Yang, J.-B.; Song, Y.; Depamphilis, C.W.; Yi, T.-S.; Li, D.-Z. GetOrganelle: A fast and versatile toolkit for accurate de novo assembly of organelle genomes. *Genome Biol.* **2020**, *21*, 241. [CrossRef]
35. Bernt, M.; Donath, A.; Jühling, F.; Externbrink, F.; Florentz, C.; Fritzsch, G.; Pütz, J.; Middendorf, M.; Stadler, P.F. MITOS: Improved de novo metazoan mitochondrial genome annotation. *Mol. Phylogenetics Evol.* **2013**, *69*, 313–319. [CrossRef]
36. Iwasaki, W.; Fukunaga, T.; Isagozawa, R.; Yamada, K.; Maeda, Y.; Satoh, T.P.; Sado, T.; Mabuchi, K.; Takeshima, H.; Miya, M.; et al. MitoFish and MitoAnnotator: A Mitochondrial Genome Database of Fish with an Accurate and Automatic Annotation Pipeline. *Mol. Biol. Evol.* **2013**, *30*, 2531–2540. [CrossRef]
37. Kumar, S.; Stecher, G.; Tamura, K. MEGA7: Molecular Evolutionary Genetics Analysis Version 7.0 for Bigger Datasets. *Mol. Biol. Evol.* **2016**, *33*, 1870–1874. [CrossRef]
38. Xia, X. DAMBE7: New and Improved Tools for Data Analysis in Molecular Biology and Evolution. *Mol. Biol. Evol.* **2018**, *35*, 1550–1552. [CrossRef]
39. Xia, X.; Xie, Z.; Salemi, M.; Chen, L.; Wang, Y. An index of substitution saturation and its application. *Mol. Phylogenetics Evol.* **2003**, *26*, 1–7. [CrossRef]
40. Katoh, K.; Standley, D.M. MAFFT Multiple Sequence Alignment Software Version 7: Improvements in Performance and Usability. *Mol. Biol. Evol.* **2013**, *30*, 772–780. [CrossRef]
41. Zhang, D.; Gao, F.; Jakovlić, I.; Zhou, H.; Zhang, J.; Li, W.X.; Wang, G.T. PhyloSuite: An integrated and scalable desktop platform for streamlined molecular sequence data management and evolutionary phylogenetics studies. *Mol. Ecol. Resour.* **2020**, *20*, 348–355. [CrossRef]
42. Gouy, M.; Guindon, S.; Gascuel, O. SeaView Version 4: A Multiplatform Graphical User Interface for Sequence Alignment and Phylogenetic Tree Building. *Mol. Biol. Evol.* **2010**, *27*, 221–224. [CrossRef]
43. Kalyaanamoorthy, S.; Minh, B.Q.; Wong, T.K.F.; Von Haeseler, A.; Jermiin, L.S. ModelFinder: Fast model selection for accurate phylogenetic estimates. *Nat. Methods* **2017**, *14*, 587–589. [CrossRef]
44. Ronquist, F.; Teslenko, M.; van der Mark, P.; Ayres, D.L.; Darling, A.; Höhna, S.; Larget, B.; Liu, L.; Suchard, M.A.; Huelsenbeck, J.P. MrBayes 3.2: Efficient Bayesian Phylogenetic Inference and Model Choice across a Large Model Space. *Syst. Biol.* **2012**, *61*, 539–542. [CrossRef] [PubMed]
45. Nguyen, L.-T.; Schmidt, H.A.; Von Haeseler, A.; Minh, B.Q. IQ-TREE: A Fast and Effective Stochastic Algorithm for Estimating Maximum-Likelihood Phylogenies. *Mol. Biol. Evol.* **2015**, *32*, 268–274. [CrossRef] [PubMed]
46. Rozas, J.; Ferrer-Mata, A.; Sánchez-DelBarrio, J.C.; Guirao-Rico, S.; Librado, P.; Ramos-Onsins, S.E.; Sánchez-Gracia, A. DnaSP 6: DNA Sequence Polymorphism Analysis of Large Data Sets. *Mol. Biol. Evol.* **2017**, *34*, 3299–3302. [CrossRef] [PubMed]
47. Yang, Z. PAML 4: Phylogenetic Analysis by Maximum Likelihood. *Mol. Biol. Evol.* **2007**, *24*, 1586–1591. [CrossRef]
48. Rambaut, A.; Suchard, M.A.; Xie, D.; Drummond, A.J. Tracer v1.7. 2014. Available online: http://beast.bio.ed.ac.uk/Tracer (accessed on 30 October 2022).
49. Bellwood, D.R.; Schultz, O. A Review of the Fossil Record of the Parrotfishes (Labroidei: Scaridae) with a Description of a New *Calotomus* Species from the Middle Miocene (Badenian) of Austria. *Ann. Nat. Mus. Wien* **1990**, *92*, 55–71.
50. Bellwood, D.R. A new fossil fish *Phyllopharyngodon longipinnis* gen. et sp. nov. (family labridae) from the Eocene, Monte Bolca, Italy. *Studi Ric. Sui Giacimenti Terziari Bolca* **1990**, *6*, 149–160.
51. Boore, J.L. Animal mitochondrial genomes. *Nucleic Acids Res.* **1999**, *27*, 1767–1780. [CrossRef]
52. Ojala, D.; Montoya, J.; Attardi, G. tRNA punctuation model of RNA processing in human mitochondria. *Nature* **1981**, *290*, 470–474. [CrossRef]
53. Levinson, G.; Gutman, G.A. Slipped-strand mispairing: A major mechanism for DNA sequence evolution. *Mol. Biol. Evol.* **1987**, *4*, 203–221. [CrossRef]
54. Taanman, J.-W. The mitochondrial genome: Structure, transcription, translation and replication. *Biochim. Et Biophys. Acta BBA-Bioenergies* **1999**, *1410*, 103–123. [CrossRef]
55. Mao, M.; Valerio, A.; Austin, A.D.; Dowton, M.; Johnson, N.F. The first mitochondrial genome for the wasp superfamily Platygastroidea: The egg parasitoid *Trissolcus basalis*. *Genome* **2012**, *55*, 194–204. [CrossRef]
56. Rodovalho, C.D.M.; Lyra, M.L.; Ferro, M.; Bacci, M., Jr. The Mitochondrial Genome of the Leaf-Cutter Ant Atta laevigata: A Mitogenome with a Large Number of Intergenic Spacers. *PLoS ONE* **2014**, *9*, e97117. [CrossRef]
57. Sbisà, E.; Tanzariello, F.; Reyes, A.; Pesole, G.; Saccone, C. Mammalian mitochondrial D-loop region structural analysis: Identification of new conserved sequences and their functional and evolutionary implications. *Gene* **1997**, *205*, 125–140. [CrossRef]
58. Chen, I.-S.; Wu, J.-H.; Huang, S.-P. The taxonomy and phylogeny of the cyprinid genus *Opsariichthys* Bleeker (Teleostei: Cyprinidae) from Taiwan, with description of a new species. *Environ. Biol. Fishes* **2009**, *86*, 165–183. [CrossRef]

59. Kang, B.; Hsu, K.; Wu, J.; Chiu, Y.; Lin, H.; Ju, Y. Population genetic diversity and structure of *Rhinogobius candidianus* (Gobiidae) in Taiwan: Translocation and release. *Ecol. Evol.* **2022**, *12*, e9154. [CrossRef]

60. Salemi, M. Nucleotide substitution models. Practice: The Phylip and Tree-Puzzle software packages. In *The Phylogenetic Handbook a Practical Approach to Phylogenetic Analysis and Hypothesis Testing*; Salemi, M., Vandamme, A.-M., Eds.; Cambridge University Press: Cambridge, UK, 2003; pp. 88–97.

61. Brabec, J.; Scholz, T.; Králová-Hromadová, I.; Bazsalovicsová, E.; Olson, P.D. Substitution saturation and nuclear paralogs of commonly employed phylogenetic markers in the *Caryophyllidea*, an unusual group of non-segmented tapeworms (Platyhelminthes). *Int. J. Parasitol.* **2012**, *42*, 259–267. [CrossRef]

62. Lv, W.; Jiang, H.; Bo, J.; Wang, C.; Yang, L.; He, S. Comparative mitochondrial genome analysis of *Neodontobutis hainanensis* and *Perccottus glenii* reveals conserved genome organization and phylogeny. *Genomics* **2020**, *112*, 3862–3870. [CrossRef]

63. Francino, M.P.; Ochman, H. Strand asymmetries in DNA evolution. *Trends Genet.* **1997**, *13*, 240–245. [CrossRef]

64. Frank, A.; Lobry, J. Asymmetric substitution patterns: A review of possible underlying mutational or selective mechanisms. *Gene* **1999**, *238*, 65–77. [CrossRef]

65. Nikolaou, C. A study on the correlation of nucleotide skews and the positioning of the origin of replication: Different modes of replication in bacterial species. *Nucleic Acids Res.* **2005**, *33*, 6816–6822. [CrossRef]

66. Charneski, C.A.; Honti, F.; Bryant, J.M.; Hurst, L.D.; Feil, E.J. Atypical AT Skew in Firmicute Genomes Results from Selection and Not from Mutation. *PLOS Genet.* **2011**, *7*, e1002283. [CrossRef] [PubMed]

67. Li, F.; Liao, T.-Y.; Arai, R.; Zhao, L. *Sinorhodeus microlepis*, a new genus and species of bitterling from China (Teleostei: Cyprinidae: Acheilognathinae). *Zootaxa* **2017**, *4353*, 69. [CrossRef] [PubMed]

68. Yu, P.; Zhou, L.; Zhou, X.-Y.; Yang, W.-T.; Zhang, J.; Zhang, X.-J.; Wang, Y.; Gui, J.-F. Unusual AT-skew of *Sinorhodeus microlepis* mitogenome provides new insights into mitogenome features and phylogenetic implications of bitterling fishes. *Int. J. Biol. Macromol.* **2019**, *129*, 339–350. [CrossRef] [PubMed]

69. Meiklejohn, C.D.; Montooth, K.L.; Rand, D.M. Positive and negative selection on the mitochondrial genome. *Trends Genet.* **2007**, *23*, 259–263. [CrossRef] [PubMed]

70. Bellwood, D.R. A phylogenetic study of the parrotfish family Scaridae (Pisces: Labroidea), with a revision of genera. *Rec. Aust. Museum, Suppl.* **1994**, *20*, 1–86. [CrossRef]

71. Wilson, M.E.J., Rosen, D.R. Implications of paucity of corals in the Paleogene of SE Asia: Plate tectonics or Centre of Origin? In *Biogeography and Geological Evolution of SE Asia*; Backhuys Publishers: Laiden, The Netherlands, 1998; pp. 165–195.

72. Braisier, M.D. An outline of sea grass communities. *Paleontology* **1975**, *18*, 681–702.

73. Sun, J.; Sheykh, M.; Ahmadi, N.; Cao, M.; Zhang, Z.; Tian, S.; Sha, J.; Jian, Z.; Windley, B.F.; Talebian, M. Permanent closure of the Tethyan Seaway in the northwestern Iranian Plateau driven by cyclic sea-level fluctuations in the late Middle Miocene. *Palaeogeogr. Palaeoclim. Palaeoecol.* **2021**, *564*, 110172. [CrossRef]

74. Pomar, L.; Hallock, P. Changes in coral-reef structure through the Miocene in the Mediterranean province: Adaptive versus environmental influence. *Geology* **2007**, *35*, 899. [CrossRef]

75. Coletti, G.; Balmer, E.M.; Bialik, O.M.; Cannings, T.; Kroon, D.; Robertson, A.H.; Basso, D. Microfacies evidence for the evolution of Miocene coral-reef environments in Cyprus. *Palaeogeogr. Palaeoclim. Palaeoecol.* **2021**, *584*, 110670. [CrossRef]

76. Riera, R.; Bourget, J.; Håkansson, E.; Paumard, V.; Wilson, M.E. Middle Miocene tropical oligotrophic lagoon deposit sheds light on the origin of the Western Australian coral reef province. *Palaeogeogr. Palaeoclim. Palaeoecol.* **2021**, *576*, 110501. [CrossRef]

77. Bellwood, D.R.; Goatley, C.H.R.; Bellwood, O. The evolution of fishes and corals on reefs: Form, function and interdependence. *Biol. Rev.* **2017**, *92*, 878–901. [CrossRef]

Article

Genetic Diversity and Connectivity of *Ocypode ceratophthalmus* in the East and South China Seas and Its Implications for Conservation

Feng Zhao, Yue Liu, Zihan Wang, Jiaying Lu, Ling Cao and Cong Zeng *

School of Oceanography, Shanghai Jiao Tong University, Shanghai 200030, China
* Correspondence: congzeng@sjtu.edu.cn

Simple Summary: This study investigated the genetic diversity and connectivity of 15 *Ocypode ceratophthalmus* populations in the East and South China Seas based on two genetic markers. The results showed that *O. ceratophthalmus* had a high genetic diversity among all collected populations, and an insignificant population structure was observed by a hierarchical analysis of molecular variance and fixation index. Additionally, Migrate-n revealed high historical gene flow and migration rates among populations. The results of this study could inform the construction and management of marine protected areas in the East and South China Seas.

Abstract: The East and South China Seas are rich in marine resources, but they are also under great pressure from climate change and human activities. Maintaining diversity and connectivity between communities is thought to be effective in mitigating these pressures. To assess the diversity and connectivity among the populations of *Ocypode ceratophthalmus* in the East and South China Seas, 15 populations from or near 15 marine protected areas in the two seas were studied using *COI* and *D-Loop* as genetic markers. The results showed that *O. ceratophthalmus* populations had high diversity, and the results of a hierarchical analysis of molecular variance and fixation index found that there were no significant genetic structures among these populations. High historical gene flow and high migration rates were further observed among populations by Migrate-n. Furthermore, the *COI* sequences further showed the asymmetric migration rate with a higher migration rate from south to north than from north to south. This information could provide recommendations for the management of marine protected areas in the East and South China Seas.

Keywords: population structure; East China Sea; South China Sea; horn-eyed ghost crab; conservation

Citation: Zhao, F.; Liu, Y.; Wang, Z.; Lu, J.; Cao, L.; Zeng, C. Genetic Diversity and Connectivity of *Ocypode ceratophthalmus* in the East and South China Seas and Its Implications for Conservation. *Biology* 2023, 12, 437. https://doi.org/10.3390/biology12030437

Academic Editor: José María Conde-Porcuna

Received: 17 January 2023
Revised: 7 March 2023
Accepted: 10 March 2023
Published: 12 March 2023

1. Introduction

The East and South China Seas are important areas for the utilization of marine resources and marine conservation. Currently, 12,933 tropical and subtropical species in the East China Sea (ECS) have been recorded, of which half (48%) were endemic [1], and the ECS was also an important fishery for the world, contributing about 6 million tons of catches annually [2]. The South China Sea (SCS), adjacent to the ECS, is regarded as the hotspot of tropical biodiversity in shallow seas, with 450 species of coral (about 7% of the world's total coral reef areas), 2 million hectares of mangrove (12% of the world's total mangrove areas), 1027 species of fish, 91 species of shrimp and 73 species of cephalopods [3]. Furthermore, the SCS provides 6 billion tons of catch annually, accounting for 10% of the world's total catch [3]. Nevertheless, with the rapid economic development in the East and South China Seas, the regions were also facing threats, such as overfishing, biodiversity decline and habitat degradation, which urgently need to be strengthened for conservation [3]. The marine protected area (MPA) was considered the core initiative of marine biodiversity conservation [4], and MPAs were established to protect the ECS and SCS 20 years ago by the Chinese government. So far, 22 national-level MPAs and more than 50 local-level MPAs have been established in the ECS, and more than 20 national-level MPAs have been set up in the SCS [5]. Most of these protected areas were usually designated by local governments, and these bottom-up designations may lack systematic planning [6].

Protecting biodiversity as a whole required not only the design of protected areas with reasonable functions, but also the consideration of species migration among different habitats on a larger scale, such as the establishment of several protected areas with the same functions to form a MPA network. Previous studies have also revealed that the MPA network could effectively prevent or mitigate the negative impacts of biodiversity [7], which required connectivity to promote population persistence in marine ecosystems [8].The Convention on Biological Diversity has recognized connectivity as a fundamental principle in the planning of the MPA network. Additionally, it was requisite to determine the boundaries of ecosystems based on connectivity in the ecosystem-based marine reserve management system [9]. Therefore, it is important to understand the connectivity in these two seas and that connectivity barriers may not be present in these two seas.

The ECS belongs to the East China Sea Ecoregion in the Warm Temperate Northern Pacific Marine Biogeographic province, and the SCS belongs to the Gulf of Tonkin Ecoregion, Southern China Ecoregion and South China Sea Oceanic Islands Ecoregion in the South China Sea Marine Biogeographic province [10]. Whether the transition in different ecoregions will have an impact on population connectivity remains to be researched, and this question relates to whether the protected areas between the two seas are constructed as a network of protected areas or separately. At present, studies on this issue do not have the same results. On the one hand, high connectivity reports were increasing in coastal species, which span different ecoregions. For example, *Atrina pectinate* [11], *Nibea albiflora* [12], *Octopus ovulum* [13], *Parasesarma affinis* [14], *Periophthalmus modestus* [15], *Siphonaria japonica* [16] and *Thamnaconus hypargyreus* [17] had high connectivity among populations in both ECS and SCS. On the other hand, it has been reported that there was limited connectivity between populations in the ECS and populations in the SCS, such as *Chelon haematocheilus* [18], *Eleutheronema tetradactylum* [19], *Epinephelus akaara* [20], *Pagrus major* [21] and *Scomber japonicus* [22]. These inconsistent findings, which could be caused by limited spatial coverage and insufficient sampling (usually less than 10 populations, see summary in Table S1), require further research on the connectivity of the two seas.

Horn-eyed ghost crab (*Ocypode ceratophthalmus*), a species widely distributed in the East and South China Seas, has a planktonic larval stage and a settled adult stage like most marine invertebrates. Furthermore, *O. ceratophthalmus* is expected to suffer less human disturbance due to its low economic value, and it is therefore considered to be better to reflect the diversity and connectivity patterns among protected areas [23]. Although the previous study did not find any genetic differences in the SCS population, it was tested in only three populations and with a limited sample [24]. To reveal its population structures and provide suggestions for the marine protection in the East and South China Seas, two molecular markers, *COI* and *D-Loop*, were employed in this study to assess the genetic connectivity among horn-eyed ghost crab populations in or near marine protected areas in the two seas. The results from this study were expected to inform the construction and management of marine protected areas in China.

2. Materials and Methods

2.1. Sample Collection and DNA Extraction

From September 2021 to April 2022, 8 populations of horn-eyed ghost crab were collected from sandy beach in 8 marine protected areas. In addition, because not every MPA has the ghost crab habitat, 7 populations were collected in the adjoining sandy beach of 7 marine protected areas (Figure 1; Table 1). The morphological details of each group are presented in Table S2.

Table 1. Location and sampling details for each population.

Region	Ecoregion	MPA *	Population	Number of Samples	Weight (g)	Sampling Time
East China Sea	East China Sea Ecoregion	Dongtou National Marine Park (in)	DT	20	18.06 ± 6.17	27 April 2022
		Fuyao Archipelago National Marine Park (in)	FYLD	15	18.63 ± 5.70	30 March 2022
		Haitan Bay National Marine Park (in)	PT	18	13.68 ± 4.54	6 March 2022
		Hua'ao Island National Marine Park (near)	HA	15	15.89 ± 4.15	8 April 2022
		Ma'an Archipelago Special Marine Protected Area (in)	MA	15	5.41 ± 2.25	10 December 2021
		Meizhou Island National Marine Park (in)	MZD	16	13.38 ± 2.66	5 March 2022
		Nanji Archipelago National Nature Reserve (in)	NJLD	10	16.15 ± 6.95	18 April 2022
		Zhongjieshan Archipelago Special Marine Protected Area (near)	ZJS	14	22.82 ± 7.11	18 April 2022
	Southern China Ecoregion	Chongwu National Marine Park (in)	CW	24	16.59 ± 5.40	8 January 2022
		Xiamen National Marine Park (in)	XM	18	16.65 ± 5.72	2 January 2022
South China Sea	Southern China Ecoregion	Qing'ao Bay National Marine Park (near)	QA	22	15.01 ± 3.54	8 January 2022
		Zhelang Peninsula National Marine Park (near)	DH	15	14.43 ± 4.56	22 March 2022
	Gulf of Tonkin Ecoregion	Dongzhai Bay National Nature Reserve (near)	DZG	21	5.28 ± 2.51	18 September 2021
		Jinhai Bay Mangrove Protected Area (near)	JH	15	18.89 ± 6.09	1 April 2022
		Sanya River Mangrove Nature Reserve (near)	SYH	15	5.54 ± 2.50	18 September 2021

* In this column, the contents in parentheses indicate that the sampling site is in or near the marine protected areas, "in" indicates that the sampling site is in the reserve, and "near" indicates that the sampling site is near the marine protected areas.

The obtained samples were morphologically identified according to the previous study [25], samples were transported back to the laboratory, measurements of carapace width, carapace length, abdomen width, abdomen length, weight and gender were conducted [26]. Then, samples were stored in 95% ethanol at $-20\ ^{\circ}$C until DNA extraction. About 20 mg of muscle tissue from the ambulatory leg was obtained to extract the total DNA by using MolPure® Cell/Tissue DNA Kit (Yeason, Shanghai, Chian), and the DNA was stored at $-20\ ^{\circ}$C and used as a template in the PCR reactions. PCR was implemented to amplify *COI* and *D-Loop* using the following primers, LCO1490l (5′-GGTCAACAAATCATAAAGATATTGG-3′) and HCO2198 (5′-TAAACTTCAGGGTGAC CAAAAAATCA-3′) for *COI*, 13,323F (5′-GCGAATGCTGGCACAAACAT-3′) and 14,378R (5′-AGGGAGTGGTGCAATTCCAT-3′) for *D-loop*. The PCR amplification reaction included 25 μL 2× Hieff® PCR Master Mix, 2 μL of upstream and downstream primers (10 μM), about 55 ng of template DNA, and double-distilled water was added to the total volume of 50 μL. Then, the PCR thermal cycling program was set as follows: denaturation at 94 $^{\circ}$C for 5 min, followed by 35 cycles of denaturation for 94 $^{\circ}$C 30 s, annealing at 55 $^{\circ}$C for 30 s, elongation at 72 $^{\circ}$C for 1 min; and the final extension step for 10 min at 72 $^{\circ}$C. All PCR products were detected by 1.5% agarose gel electrophoresis. Finally, the PCR products were purified and sequenced by Tsingke Biotech Company (Shanghai, China).

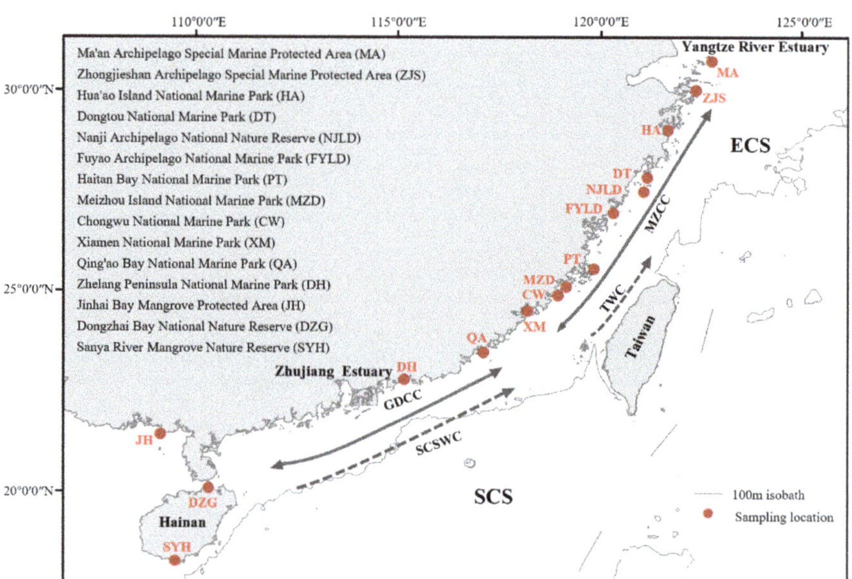

Figure 1. Sampling locations and ocean currents schematic in the studied area. TWC, Taiwan Warm Current; SCSWC, South China Sea Warm Current; MZCC, the coastal currents of Zhejiang and Fujian; GDCC, Guangdong coastal current. The direction of MZCC and GDCC varies seasonally, from south to north in summer and vice versa in winter [27], which is indicated by the solid line with bidirectional arrows in the figure.

2.2. Statistical Analyses of Genetic Data

All sequences were checked and edited using Geneious. Multiple sequences were aligned using the Geneious Alignment (Identity 1.0/0.0), and all alignments were checked visually. MEGA11 was used to evaluate the composition of the nucleotides. DnaSP v.4.0 was used to calculate the genetic diversity indices such as the number of haplotypes (h), haplotype diversity (Hd) and nucleotide diversity (π).

A map of haplotypes distribution was constructed using ArcGIS, R 4.2 and Arlequin 3.5, and minimum spanning networks were constructed using PopART 1.7. Hierarchical analysis of molecular variance (AMOVA) of the different levels was used to further examine the population structure. The AMOVA based on three levels included population level (MA, ZJS, HA, DT, NJLD, FYLD, PT, MZD, CW, XM, QA, DH, JH, DZG, SYH), region level (the East China Sea: MA, ZJS, HA, DT, NJLD, FYLD, PT, MZD, CW, XM; the South China Sea: QA, DH, JH, DZG, SYH) and ecoregion level (East China Sea Ecoregion: MZD, PT, FYLD, NJLD, DT, HA, ZJS, MA; Southern China Ecoregion: DH, QA, XM, CW; Gulf of Tonkin Ecoregion: JH, DZG, SYH). The significance of AMOVA was analyzed by 10,000 permutations. The fixation index (F_{ST}) was employed to assess the pairwise genetic divergence between different populations and the significance was obtained by 10,000 permutations.

Migrate-n 3.6 was conducted to estimate the mutation-scaled migration rate M (M = m / μ, where m was the historical migration rate among populations and μ was mutation per generation) and mutation-scaled population size θ (θ = Ne × μ, where Ne was historical effective population size). The runs were recorded every 100 steps for a total of 1,000,000 long-chain Markov chain Monte Carlo (MCMC) steps and a burn-in of 1000. To improve the efficiency of the MCMC search, a static heating scheme with four different temperatures (1.0, 1.5, 3.0 and 1,000,000.0) was used. We inspected histograms of estimated θ and M posterior values to assess convergence. We calculated historical gene flow (Nm) by using the equation Nm = θ × M (when using mitochondrial gene).

3. Results

3.1. Population Genetic Diversity

A total of 253 *COI* sequences (538 bp, OP989704-956) and 95 *D-Loop* sequences (611 bp-612 bp, ON504951-995) were obtained from 15 populations of horn-eyed ghost crab. For *COI* sequences, the average contents of A, T, G and C were 26.6%, 34.6%, 16.8% and 22.0%, respectively. A total of 50 polymorphic sites were detected, including 24 singleton variable sites and 26 parsimony informative sites. Haplotype diversity ranged from 0.80 to 0.98, and nucleotide diversity ranged from 0.0022 to 0.0054 (Table 2). When the 15 populations were considered as a metapopulation, the haplotype diversity was 0.87, and the nucleotide diversity was 0.0036 (Table 2).

Table 2. Genetic diversity for 15 populations of horn-eyed ghost crab based on *COI/D-Loop* sequences. N, number of specimens for each population; h, the number of haplotypes; Hd, haplotype diversity; π, nucleotide diversity.

Region	Population	COI				D-Loop			
		N	h	Hd	π	N	h	Hd	π
The East China Sea	CW	24	12	0.87 ± 0.05	0.0034 ± 0.0006	9	9	1.00 ± 0.05	0.0279 ± 0.0033
	XM	18	12	0.95 ± 0.03	0.0054 ± 0.0007	9	9	1.00 ± 0.05	0.0281 ± 0.0028
	MA	15	9	0.88 ± 0.07	0.0037 ± 0.0007	5	5	1.00 ± 0.13	0.0314 ± 0.0052
	PT	18	10	0.88 ± 0.06	0.0050 ± 0.0010	5	4	0.90 ± 0.16	0.0271 ± 0.0050
	MZD	16	8	0.88 ± 0.05	0.0030 ± 0.0005	6	6	1.00 ± 0.10	0.0230 ± 0.0033
	HA	15	9	0.88 ± 0.07	0.0034 ± 0.0006	6	6	1.00 ± 0.10	0.0192 ± 0.0035
	NJLD	10	9	0.98 ± 0.05	0.0043 ± 0.0008	5	5	1.00 ± 0.13	0.0298 ± 0.0051
	ZJS	14	9	0.90 ± 0.06	0.0036 ± 0.0007	10	10	1.00 ± 0.05	0.0306 ± 0.0028
	FYLD	15	7	0.82 ± 0.08	0.0022 ± 0.0004	7	6	0.95 ± 0.10	0.0233 ± 0.0035
	DT	20	11	0.86 ± 0.07	0.0037 ± 0.0007	5	5	1.00 ± 0.13	0.0257 ± 0.0048
The South China Sea	QA	22	14	0.93 ± 0.04	0.0036 ± 0.0005	5	5	1.00 ± 0.13	0.0331 ± 0.0063
	JH	15	7	0.81 ± 0.08	0.0027 ± 0.0006	5	5	1.00 ± 0.13	0.0379 ± 0.0061
	DZG	21	9	0.86 ± 0.05	0.0027 ± 0.0004	5	5	1.00 ± 0.13	0.0268 ± 0.0064
	SYH	15	7	0.80 ± 0.08	0.0032 ± 0.0009	7	6	0.95 ± 0.10	0.0278 ± 0.0038
	DH	15	12	0.94 ± 0.05	0.0052 ± 0.0010	6	6	1.00 ± 0.10	0.0313 ± 0.0052

The average contents of A, T, G and C were 42.7%, 32.6%, 9.6% and 15.0%, respectively, according to the results based on *D-Loop* sequences. A total of 116 polymorphic sites were found, including 89 parsimony informative sites and 27 singleton variable sites. The genetic diversity parameters showed that nucleotide diversity ranged from 0.0192 to 0.0379, and haplotype diversity ranged from 0.90 to 1.00 (Table 2). When the 15 populations were considered as a whole, the nucleotide diversity was 0.0279, and the haplotype diversity was 1.00. In conclusion, both genetic markers demonstrated high genetic diversity in horn-eyed ghost crab populations.

3.2. Haplotype Analysis

The *COI* sequences identified 62 haplotypes (Hap 1-62), including 24 shared haplotypes and 38 exclusive haplotypes. The frequencies of 62 haplotypes varied widely; the combined frequencies of Hap 1 and Hap 2 were as high as 49.40%, with the highest frequency of Hap 2 (26.48%), followed by Hap 1 (22.92%), while the remaining haplotypes were less frequent. Two haplotypes (Hap 1-2) were shared by 15 populations (Figure 2), while common haplotypes (Hap 1-2, Hap 7, Hap 10 and Hap 13) also existed in populations that were far apart (up to 1000 km), according to the geographic distribution of haplotypes (Figure 2). The haplotype network also showed that Hap 1 and Hap 2 were shared by 15 populations, Hap 13 was shared by 11 populations. In general, the haplotype network showed a shallow double star-shaped structure (Figure S1).

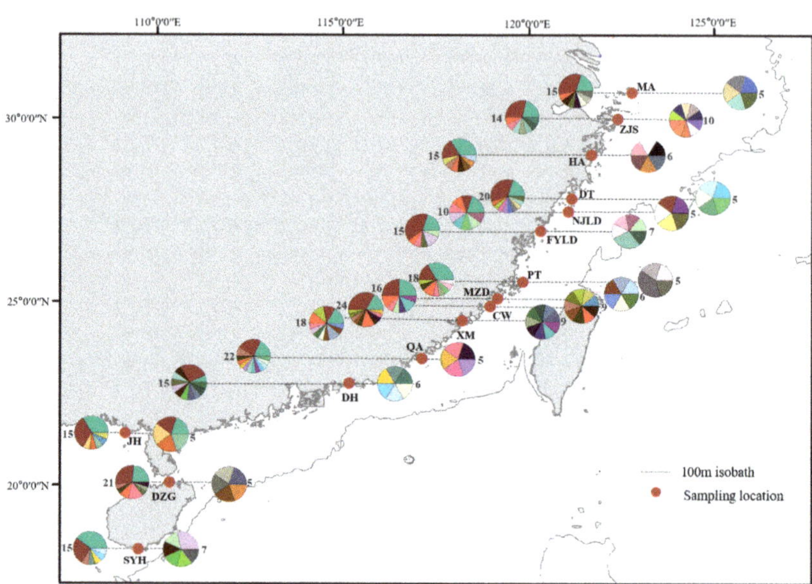

Figure 2. Haplotype map for horn-eyed ghost crab *COI* (**left**) and *D-loop* (**right**). Numbers represented the number of sequences used for analysis. Different colors represented different haplotypes.

A total of 91 haplotypes were detected in 95 *D-Loop* sequences from the 15 populations, which were defined as Hap 1-91, consisting of one shared haplotype and 90 exclusive haplotypes. The frequencies of Hap 25, Hap 29, Hap 47 and Hap 72 were all 2.11%, and the rest of haplotypes were found only once. According to the geographic distribution of haplotypes, Hap 29 was shared by two populations (XM and DZG). There were few shared haplotypes between diverse populations and no dominant haplotypes (Figure 2). The haplotype network also showed that only Hap 29 was shared by two populations. Additionally, the haplotype network presented a bush-like shape (Figure S2).

3.3. Population Genetic Structure

Based on *COI* sequences, at the region level, -0.22% of the variation was found between regions, with -0.18% of the variance found within regions among populations and 100.40% within populations. At the ecoregion level, most of the total variation (100.04%) was accounted for by differentiation within populations, with a further -0.80% accounting for variation within ecoregions among populations, and the remainder (0.76%) partitioned among ecoregions (Table 3). At the population level, -0.29% of the total variance was found among populations, whilst 100.29% of the variance was found within populations. The AMOVA statistical tests were not significant in all three levels ($p > 0.05$). The pairwise population F_{ST} showed that the genetic differences between populations ranged from -0.0456 to 0.0602. All pairwise F_{ST} values were not statistically significant ($p > 0.05$; Table S3).

Based on the *D-Loop* sequences, at the region level, -0.86% of the variation was found between regions, with -0.44% of the variance found within regions among populations and 101.29% within populations. At the ecoregion level, most of the total variation (101.27%) was accounted for by differentiation within populations, with a further 0.04% accounting for variation within ecoregions among populations, and the remainder (-1.31%) partitioned among ecoregions. At the population level, -0.82% of the total variance was found among populations, whilst 100.82% of the variance was found within populations (Table 3). AMOVA statistical tests in all three levels were not significant (Table 3). Furthermore, the F_{ST} values between populations were distributed between -0.1128 and 0.2675, the genetic differentiation between the remaining populations showed insignificant differences ($p > 0.05$), except for the significant differences between MZD and FYLD, and between MZD and HA ($p < 0.05$; Table S4). Two genetic markers revealed a high level of genetic homogeneity among the 15 populations of horn-eyed ghost crab and there was no significant genetic structure.

Table 3. AMOVA results of horn-eyed ghost crab based on *COI* and *D-Loop* sequences.

Different Classifications	Genetic Markers	Source of Variation	df	Sum of Squares	Variance Components	Percentage of Variation
Region level (ECS and SCS)	*COI*	Between regions	1	0.708	−0.002	−0.22
		Within regions among populations	13	12.383	−0.002	−0.18
		within populations	238	233.873	0.983	100.40
		Total	252	246.964	0.979	
	D-Loop	Between regions	1	5.481	−0.072	−0.86
		Within regions among populations	13	108.307	−0.037	−0.44
		within populations	80	685.181	8.565	101.29
		Total	94	798.968	8.455	
Ecoregion level (East China Sea Ecoregion, Southern China Ecoregion, Gulf of Tonkin Ecoregion)	*COI*	Among ecoregions	2	2.857	0.007	0.76
		Within ecoregions among populations	12	10.234	−0.008	−0.80
		within populations	238	233.873	0.983	100.04
		Total	252	246.964	0.982	
	D-Loop	Among ecoregions	2	10.766	−0.111	−1.31
		Within ecoregions among populations	12	103.021	0.003	0.04
		within populations	80	685.181	8.565	101.27
		Total	94	798.968	8.457	
Population level	*COI*	Among populations	14	13.091	−0.003	−0.29
		within populations	238	233.873	0.983	100.29
		Total	252	246.964	0.980	
	D-Loop	Among populations	14	113.787	−0.070	−0.82
		within populations	80	685.181	8.565	100.82
		Total	94	798.968	8.495	

3.4. Migration and Connectivity

Based on *COI* sequences, the estimated historical gene flow ranged from 4.921 (MZD-FYLD) to 248.595 (QA-DH), while the estimated historical gene flow varied from 10.738 (SYH-DT) to 334.675 (CW-FYLD) based on *D-Loop* sequences. Two markers revealed the high historical gene flow among populations of horn-eyed ghost crab. The historical gene flow from MZD to most other populations was lower than that of most other populations to MZD based on two markers.

Based on *COI* sequences, the estimated migration rate ranged from 134.7 (QA-FYLD) to 390.1 (MZD-QA), while the estimated migration rate varied from 159.3 (PT-SYH) to 284.1 (SYH-NJLD) based on *D-Loop* sequences. The analysis consequences of two markers exhibited the high migration rates among populations. The *COI* sequences further showed the asymmetric migration rate with the higher migration rate from south to north than from north to south. Meanwhile, the migration rate from SYH to most other populations was higher than the migration rate from that of most other populations to SYH based on two markers (Figure 3). In addition, Migrate-n also revealed that the effective population sizes of each population based on both markers were relatively close (Table S5).

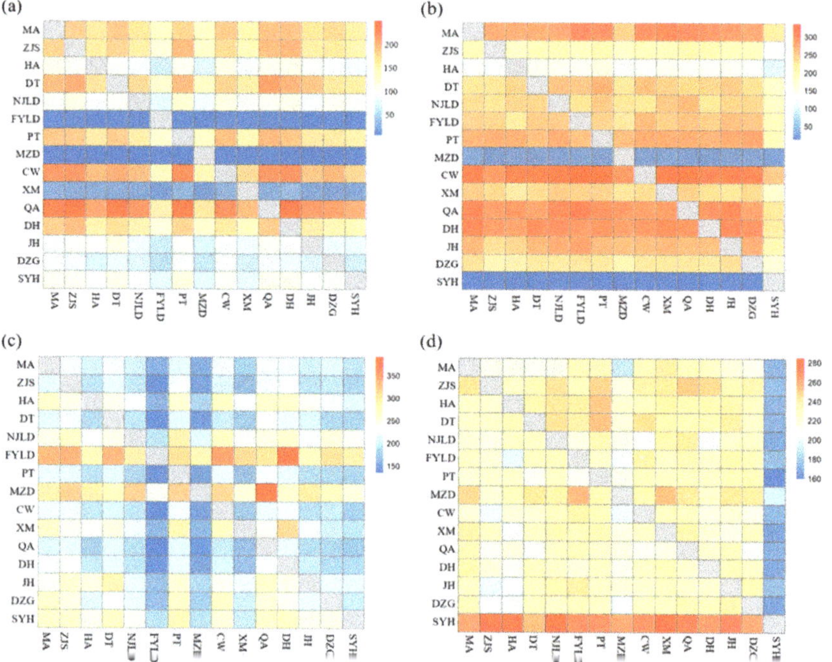

Figure 3. Historical gene flow (Nm) and migration rate (M) among 15 populations of horn-eyed ghost crab based on *COI* and *D-Loop* sequences. (**a**) Nm values based on *COI* sequences; (**b**) Nm values based on *D-Loop* sequences; (**c**) M values based on *COI* sequences; (**d**) M values based on *D-Loop* sequences.

4. Discussion

In this study, 15 populations of horn-eyed ghost crab were collected in or near 15 MPAs in the East and South China Seas to analyze connectivity among populations based on *COI* and *D-Loop*. The results revealed that the high genetic diversity and connectivity among populations, and a high historical gene flow and migration rate among 15 populations were supported by two markers. Additionally, the *COI* sequences further showed the asymmetric migration rate with a higher migration rate from south to north than from north to south. The results of population structure and connectivity could reflect the connectivities between the marine protected areas in the two seas. These results could offer information for the management of the marine protected areas in the study areas.

4.1. Differences of Genetic Diversity in Populations

High haplotype diversity (Hd > 0.5) and low nucleotide diversity (π < 0.005) were observed based on *COI* sequences when the 15 populations were considered as a whole. This low nucleotide diversity and high haplotype diversity were in line with the genetic diversity of *COI* sequences from other marine organisms, particularly marine crustacean species such as *Pachygrapsus crassipes* [28] and *Portunus trituberculatus* [29]. *D-Loop*, on the other hand, showed high haplotype diversity (Hd > 0.5) and high nucleotide diversity (π > 0.005). In contrast, the higher genetic diversity was revealed by *D-Loop* marker, which may be related to the mutation rate of the fragment [30]. *D-Loop* is the high mutation region, located on either side of the central conserved region, which evolves two to five times faster than mitochondrial protein-coding genes [31], and thus may be more sensitive in genetic diversity analysis than *COI* marker [32]. This disagreement was often seen in studies of multiple molecular markers [31].

Because the sampling sites of horn-eyed ghost crab were located in or near MPAs, the results of genetic diversity may provide some reference for the management effectiveness of MPAs. Species and populations are adversely affected by anthropogenic activities, such as habitat modification, which are a global phenomenon [33]. These negative impacts eventually would reduce the genetic diversity [34]. The management effectiveness of a protected area is generally considered to be closely related to its conservation effectiveness, and a good protection management is also expected to have a higher level of biodiversity [35]. In this study, we also found that the marine protected areas with high genetic diversity also had a higher management effectiveness. For example, the high genetic diversity in NJLD and MA was observed, for which a high management effectiveness in both protected areas was reported [36]. FYLD had the low genetic diversity based on two markers, and it also had a lower score of management effectiveness in this protected area [37]. Therefore, it might support that the level of population genetic diversity is related to the management effectiveness of the marine protected areas. However, more tests are required to explore the relationship between genetic diversity and MPA management effectiveness.

4.2. Connectivity Differences among Populations

The spatial distribution of haplotypes and the haplotype network based on *COI* sequences showed that the most common Hap 1 and Hap 2, accounting for almost 49.40% of samples, were found in all 15 populations. Shared haplotypes may be related with the large population sizes and high rates of migration [38]. Among the *COI* haplotypes, shared haplotypes existed between distant populations (up to 1000 km), which indicated that dispersal of this species may occur over long distances. In contrast, the existence of lower-frequency haplotypes was related with large population sizes, and this is because these haplotypes were not eliminated by the selection when in a large stable population [39]. Whereas the spatial distribution of haplotypes and the haplotype network from *D-Loop* revealed few shared haplotypes among populations, it may not represent low connectivity among populations, but be due to DNA superdiversity [40,41]. This result may be explained by the combination of small samples and high mutation rates in *D-loop* sequences [41]. Therefore, results from *COI* might provide more reliable results on the spatial distribution and connectivity.

In this research, the high level of connectivity among 15 populations, between ECS populations and SCS populations and among three ecoregion populations of horn-eyed ghost crab was observed based on the results of AMOVA and F_{ST} for two markers, which was also consistent with the findings of *Collichthys lucidus* [42], *Larimichthys crocea* [43], *Pampus chinensis* [44] and *Thamnaconus hypargyreus* [17]. The high connectivity among populations may be due to the long planktonic larval period of 34–42 days for horn-eyed ghost crab [45], which could allow distant geographical distances between populations to be overcome and produce genetic homogeneity [46]. On the other hand, ocean currents may promote connectivity among populations [42]. During the planktonic larval stage of horn-eyed ghost crab in summer, the current from the surface to the bottom is almost consistent with the northeast flow parallel to the coast in the coastal area from Xiangshan to Pearl River Estuary [47], which may result in high connectivity among populations. Meanwhile, this may account for the higher migration rate from south to north revealed by the *COI* marker in this study. It was also in line with the results of an analysis of seven mangrove species on the southern coast of China [48]. Nevertheless, the directional difference was not observed in the migration rate results based on *D-Loop* sequences, which may be due to the fact that the higher mutation rate of the *D-Loop* marker was not conducive to the prediction of gene flow or less number of *D-Loop* sequences [49]. In addition, Migrate-n also revealed a large population size in each population of horn-eyed ghost crab. Such population may be insensitive to the loss and reorganization of variation by genetic drift, and could maintain the ancestral genetic information [50], leading to high genetic connectivity among populations.

4.3. Conservation and Management Implications for MPA

Although it is not clear whether marine protected areas in the East and South China Seas could form a network, the results of this study may provide some suggestions for the construction of such a network due to the sampling sites of horn-eyed ghost crab being located in or near protected areas. The higher migration rate from south to north related to the current in summer was revealed based on *COI* sequences in this research. Consequently, the hydrodynamic characteristics of the East and South China Seas should be considered in constructing a MPA network. In the near-shore areas of the East and South China Seas, the Guangdong coastal current (GDCC) and the coastal currents of Zhejiang and Fujian (MZCC) are the main currents [27]. In summer, the direction of MZCC and GDCC is northward along the parallel shoreline, but in winter, the MZCC and GDCC flow in the southwesterly direction [47]. These season-changed currents should be included in the design of the MPA network, because organisms that reproduce in different seasons may have completely different connectivity patterns [51], especially those that reproduce in only one season.

Many protected species such as *Acropora solitarvensis*, *Epinephelus akaara* and *Epinephelus bruneus* spawn during summer [52]. The summer fishing moratorium implemented by the Chinese government in the East and South China Seas [53,54] contributes to the protection of larvae and may complement the MPA network [55]. This has also been shown in previous studies. A study of fish resources in Daya Bay showed that the stock density, species number, biodiversity and evenness index increased after the summer fishing moratorium period, indicating that the structure of the ecological community improved [56]. Additionally, in a study in the coastal ocean of Ningbo, the SCS showed that species richness, abundance and biomass of macrobenthos communities increased significantly during the summer fishing moratorium period, which implied this policy could facilitate the resilience of microbenthic communities [57]. However, focusing solely on the spawning patterns of these species for reserve design will decrease protection of fall–winter spawners when the direction of the gyres and the location of upstream larval sources reverses [51]. For example, at least some important species such as *Anguilla japonica* and *Penaeus japonicus* also spawn exclusively during fall–winter [52]. So, flexible management strategies could be carried out during the winter for the conservation of these species. For example, seasonal closures or fishing intensity restrictions could be assigned to their spawning sites and core areas of activity [58]. Moreover, seasonal protection corridors could be set up in areas with high connectivity during the planktonic larval period, and these areas should be a priority for new protected areas in the future.

5. Conclusions

This study investigated the genetic connectivity among the populations of *Ocypode ceratophthalmus* in the East and South China Seas based on wide geographic sampling and two genetic markers. The results showed that the horn-eyed ghost crab had high genetic diversity. The AMOVA analysis revealed no significant genetic structure and high genetic connectivity among the populations. All the F_{ST} values obtained from *COI* sequences were inapparent. The F_{ST} values based on *D-Loop* sequences were only significant between MZD and FYLD as well as between MZD and HA. Migrate-n revealed the high historical gene flow and connectivity among 15 populations based on two markers. The *COI* sequences further showed asymmetric connectivity with higher connectivity from south to north than from north to south. The outcomes from this study provide suggestions for the construction of marine protected areas.

Supplementary Materials: The following supporting information can be downloaded at: https://www.mdpi.com/article/10.3390/biology12030437/s1, Table S1: Current status of genetic connectivity between populations in the East China Sea and populations in the South China Sea; Table S2: The information of carapace width, carapace length, abdomen width, abdomen length, weight and gender for the horn-eyed ghost crab; Table S3: Pairwise F_{ST} values between populations based on *COI* sequences; Table S4: Pairwise F_{ST} values between populations based on *D-Loop* sequences; Table S5: The effective population sizes for each population based on *COI* and *D-Loop* sequences; Figure S1: The minimum spanning network based on *COI* sequences; Figure S2: The minimum spanning network based on *D-Loop* sequences.

Author Contributions: Conceptualization, C.Z. and F.Z.; methodology, C.Z., F.Z. and J.L.; data analysis, F.Z., Y.L. and Z.W. writing, F.Z. and C.Z.; funding acquisition, C.Z. and L.C. All authors have read and agreed to the published version of the manuscript.

Funding: We would like to thank the National Natural Science Foundation of China (Grant No. 42206082 and 42142018), Ministry of Science and Technology of China (Grant No. 2022YFC3102400), New Faculty Start-Up Program of Shanghai Jiao Tong University (22X010503822), Shanghai Pilot Program for Basic Research at Shanghai Jiao Tong University (Grant No. 21TQ1400220), Oceanic Interdisciplinary Program of Shanghai Jiao Tong University (Grant No. SL2022PT110), Blue Planet Found (OPF & WWF)(Grant No. PORO001426), Key Laboratory of Marine Ecological Monitoring and Restoration Technologies (Grant No. MEMRT202112), and Shanghai Frontiers Science Center of Polar Science (SCOPS) for financial support. Any opinions, findings and conclusions or recommendations expressed in this material are those of the authors and do not necessarily reflect the views of the funders.

Institutional Review Board Statement: Not applicable.

Informed Consent Statement: Not applicable.

Data Availability Statement: DNA sequences have been deposited in GenBank.

Conflicts of Interest: The authors declare no conflict of interest.

References

1. Ding, H.; Xu, H.; Wu, J.; Le Quesne, W.; Sweeting, C.; Polunin, N. An overview of spatial management and marine protected areas in the East China Sea. *Coast. Manag.* **2008**, *36*, 443–457. [CrossRef]
2. Teh, L.S.; Cashion, T.; Cheung, W.W.; Sumaila, U.R. Taking stock: A Large Marine Ecosystem perspective of socio-economic and ecological trends in East China Sea fisheries. *Rev. Fish Biol. Fish.* **2020**, *30*, 269–292. [CrossRef]
3. Yao, Y. On the construction of South China Sea Mrine Protected Area from the perspective of a community of shared future of mankind: Realistic requirements, theoretical basisi and China's countermeasures. *J. Guangxi Univ.* **2019**, *41*, 96–106.
4. Asaad, I.; Lundquist, C.J.; Erdmann, M.V.; Costello, M.J. Delineating priority areas for marine biodiversity conservation in the Coral Triangle. *Biol. Conserv.* **2018**, *222*, 198–211. [CrossRef]
5. He, J. A Legal Analysis of the Construction of Marine Protected Areas in the South China Sea and Its Impact. Master's Thesis, Wuhan University, Wuhan, China, 2020.
6. Zeng, X.; Chen, M.Y.; Zeng, C.; Cheng, S.; Wang, Z.H.; Liu, S.R.; Zou, C.X.; Ye, S.F.; Zhu, Z.G.; Cao, L. Assessing the management effectiveness of China's marine protected areas: Challenges and recommendations. *Ocean Coast. Manag.* **2022**, *224*, 106172. [CrossRef]
7. McLeod, E.; Salm, R.; Green, A.; Almany, J. Designing marine protected area networks to address the impacts of climate change. *Front. Ecol. Environ.* **2009**, *7*, 362–370. [CrossRef]
8. Planes, S.; Jones, G.P.; Thorrold, S.R. Larval dispersal connects fish populations in a network of marine protected areas. *Proc. Natl. Acad. Sci. USA* **2009**, *106*, 5693–5697. [CrossRef] [PubMed]
9. Toonen, R.J.; Andrews, K.R.; Baums, I.B.; Bird, C.E.; Concepcion, G.T.; Daly-Engel, T.S.; Eble, J.A.; Faucci, A.; Gaither, M.R.; Iacchei, M. Defining boundaries for ecosystem-based management: A multispecies case study of marine connectivity across the Hawaiian Archipelago. *J. Mar. Biol.* **2011**, *2011*, 460173. [CrossRef] [PubMed]
10. Spalding, M.D.; Fox, H.E.; Allen, G.R.; Davidson, N.; Ferdaña, Z.A.; Finlayson, M.; Halpern, B.S.; Jorge, M.A.; Lombana, A.; Lourie, S.A. Marine ecoregions of the world: A bioregionalization of coastal and shelf areas. *BioScience* **2007**, *57*, 573–583. [CrossRef]
11. Hashimoto, K.; Yamada, K.; Sekino, M.; Kobayashi, M.; Sasaki, T.; Fujinami, Y.; Yamamoto, M.; Choi, K.S.; Henmi, Y. Population genetic structure of the pen shell *Atrina pectinata* sensu lato (Bivalvia: Pinnidae) throughout East Asia. *Reg. Stud. Mar. Sci.* **2021**, *48*, 102024. [CrossRef]

12. Xu, D.D.; Lou, B.; Shi, H.L.; Geng, Z.; Li, S.L.; Zhang, Y.R. Genetic diversity and population structure of *Nibea albiflora* in the China Sea revealed by mitochondrial *COI* sequences. *Biochem. Syst. Ecol.* **2012**, *45*, 158–165. [CrossRef]

13. Dou, C.F. Analysis of genetic diversity and structure of Octopus ovulum. *China Water Transp.* **2017**, *17*, 165–167.

14. Xu, Y. Studies on Molecular Phylogeny of Sesarmid Crabs from the Coast of China and Molecular Phylogeography of Two Mangrove Crabs. Master's Thesis, Guangxi University, Nanjing, China, 2020.

15. He, L.J.; Mukai, T.; Chu, K.H.; Ma, Q.; Zhang, J. Biogeographical role of the Kuroshio Current in the amphibious mudskipper *Periophthalmus modestus* indicated by mitochondrial DNA data. *Sci. Rep.* **2015**, *5*, 15645. [CrossRef]

16. Wang, J.; Tsang, L.M.; Dong, Y.W. Causations of phylogeographic barrier of some rocky shore species along the Chinese coastline. *Bmc Evol. Biol.* **2015**, *15*, 114. [CrossRef]

17. Wang, Z.Y.; Zhang, Y.; Zhao, L.L.; Song, N.; Han, Z.Q.; Gao, T.X. Shallow mitochondrial phylogeographical pattern and high levels of genetic connectivity of *Thamnaconus hypargyreus* in the South China Sea and the East China Sea. *Biochem. Syst. Ecol.* **2016**, *67*, 110–118. [CrossRef]

18. JinXian, L.I.U.; TianYiang, G.A.O.; ShiFang, W.U. Pleistocene Isolation in the Northwestern Pacific Marginal Seas and Limited Dispersal in a Marine Fish, *Chelon haematocheilus* (Temminck & Schlegel, 1845). *J. Ocean Univ. China* **2007**, *37*, 931–938.

19. Wang, J.J.; Sun, P.; Yin, F. Low mtDNA Cytb diversity and shallow population structure of *Eleutheronema tetradactylum* in the East China Sea and the South China Sea. *Biochem. Syst. Ecol.* **2014**, *55*, 268–274. [CrossRef]

20. Chen, S.P.; Liu, T.; Li, Z.F.; Gao, T.X. Genetic population structuring and demographic history of red spotted grouper (*Epinephelus akaara*) in South and East China Sea. *Afr. J. Biotechnol.* **2008**, *7*, 3554–3562.

21. Perez-Enriquez, R.; Taniguchi, N. Genetic structure of red sea bream (*Pagrus major*) population off Japan and the Southwest Pacific, using microsatellite DNA markers. *Fish. Sci.* **1999**, *65*, 23–30. [CrossRef]

22. Zeng, L.Y.; Cheng, Q.Q.; Chen, X.Y. Microsatellite analysis reveals the population structure and migration patterns of *Scomber japonicus* (Scombridae) with continuous distribution in the East and South China Seas. *Biochem. Syst. Ecol.* **2012**, *42*, 83–93. [CrossRef]

23. Ni, G. Phylogeography of Four Marine Bivalves Along China' Coastline, with Views into the Evolutionary Processes and Mechanisms. Ph.D. Thesis, Ocean University of China, Qingdao, China, 2013.

24. Ma, K.Y.; Chow, L.H.; Wong, K.J.H.; Chen, H.N.; Ip, B.H.Y.; Schubart, C.D.; Tsang, L.M.; Chan, B.K.K.; Chu, K.H. Speciation pattern of the horned ghost crab *Ocypode ceratophthalmus* (Pallas, 1772): An evaluation of the drivers of Indo-Pacific marine biodiversity using a widely distributed species. *J. Biogeogr.* **2019**, *46*, 830. [CrossRef]

25. Shen, R.J.; Dai, A.Y. Illustrated Fauna of China: Crustacea. In *Volume 2 Crabs*; Science Press: Beijing, China, 1964; 142p.

26. Jenkins, T.L.; Stevens, J.R. Assessing connectivity between MPAs: Selecting taxa and translating genetic data to inform policy. *Mar. Policy* **2018**, *94*, 165–173. [CrossRef]

27. Roberts, K.E.; Cook, C.N.; Beher, J.; Treml, E.A. Assessing the current state of ecological connectivity in a large marine protected area system. *Conserv. Biol.* **2021**, *35*, 699–710. [CrossRef] [PubMed]

28. Cassone, B.J.; Boulding, E.G. Genetic structure and phylogeography of the lined shore crab, *Pachygrapsus crassipes*, along the northeastern and western Pacific coasts. *Mar. Biol.* **2006**, *149*, 213–226. [CrossRef]

29. Ren, G.; Miao, G.; Ma, C.; Lu, J.; Yang, X.; Ma, H. Genetic structure and historical demography of the blue swimming crab (*Portunus pelagicus*) from southeastern sea of China based on mitochondrial COI gene. *Mitochondrial DNA Part A* **2018**, *29*, 192–198. [CrossRef] [PubMed]

30. Canales-Aguirre, C.B.; Ferrada-Fuentes, S.; Galleguillos, R.; Oyarzun, F.X.; Hernandez, C.E. Population genetic structure of Patagonian toothfish (*Dissostichus eleginoides*) in the Southeast Pacific and Southwest Atlantic Ocean. *PeerJ* **2018**, *6*, e4173. [CrossRef]

31. Xu, H.; Zhang, Y.; Xu, D.; Lou, B.; Guo, Y.; Sun, X.; Guo, B. Genetic population structure of miiuy croaker (*Miichthys miiuy*) in the Yellow and East China Seas base on mitochondrial COI sequences. *Biochem. Syst. Ecol.* **2014**, *54*, 240–246. [CrossRef]

32. Ma, Z.F.; Pan, Q.Z.; An, M.; Yu, K.; Huang, S.; Li, S.; He, X.K. Analysis of genetic diversity of *Siniperca scherzeri* population in Qinshui river based on Mitochondrial *Cytb* and D-Loop Sequences. *Mar. Fish.* **2022**, *44*, 657–669.

33. Ewers, R.M.; Didham, R.K. Confounding factors in the detection of species responses to habitat fragmentation. *Biol. Rev.* **2006**, *81*, 117–142. [CrossRef]

34. Pacioni, C.; Hunt, H.; Allentoft, M.E.; Vaughan, T.G.; Wayne, A.F.; Baynes, A.; Haouchar, D.; Dortch, J.; Bunce, M. Genetic diversity loss in a biodiversity hotspot: Ancient DNA quantifies genetic decline and former connectivity in a critically endangered marsupial. *Mol. Ecol.* **2015**, *24*, 5813–5828. [CrossRef]

35. Feng, B. Study on the Management Effectiveness of Nature Reserve System of Guangxi in the Context of Climate Change. Ph.D. Thesis, Chinese Academy of Forestry, Beijing, China, 2020.

36. Zhuang, Q. Study on the Gap Analysis and Establishment Management of Xiamen Marine Protect Area. Master's Thesis, Third Institute of Oceanography, Xiamen, China, 2020.

37. Chen, M.Y.; Zeng, C.; Zeng, X.; Liu, Y.; Wang, Z.H.; Shi, X.J.; Cao, L. Assessment of Marine Protected Areas in the East China Sea Using a Management Effectiveness Tracking Tool. *Front. Mar. Sci.* **2022**, *10*, 174. [CrossRef]

38. Eastwood, E.K.; Lopez, E.H.; Drew, J.A. Population Connectivity Measures of Fishery-Targeted Coral Reef Species to Inform Marine Reserve Network Design in Fiji. *Sci. Rep.* **2016**, *6*, 19318. [CrossRef] [PubMed]

39. Duran, S.; Palacin, C.; Becerro, M.A.; Turon, X.; Giribet, G. Genetic diversity and population structure of the commercially harvested sea urchin *Paracentrotus lividus* (Echinodermata, Echinoidea). *Mol. Ecol.* **2004**, *13*, 3317–3328. [CrossRef]
40. Robalo, J.I.; Francisco, S.M.; Vendrell, C.; Lima, C.S.; Pereira, A.; Brunner, B.P.; Dia, M.; Gordo, L.; Castilho, R. Against all odds: A tale of marine range expansion with maintenance of extremely high genetic diversity. *Sci. Rep.* **2020**, *10*, 12707. [CrossRef] [PubMed]
41. Fourdrilis, S.; Backeljau, T. Highly polymorphic mitochondrial DNA and deceiving haplotypic differentiation: Implications for assessing population genetic differentiation and connectivity. *BMC Evol. Biol.* **2019**, *19*, 92.
42. Song, N.; Ma, G.; Zhang, X.; Gao, T.; Sun, D. Genetic structure and historical demography of *Collichthys lucidus* inferred from mtDNA sequence analysis. *Environ. Biol. Fishes* **2014**, *97*, 69–77. [CrossRef]
43. Han, Z.; Xu, H.; Shui, B.; Zhou, Y.; Gao, T. Lack of genetic structure in endangered large yellow croaker *Larimichthys crocea* from China inferred from mitochondrial control region sequence data. *Biochem. Syst. Ecol.* **2015**, *61*, 1–7. [CrossRef]
44. Sun, P.; Tang, B.; Yin, F. Population genetic structure and genetic diversity of Chinese pomfret at the coast of the East China Sea and the South China Sea. *Mitochondrial DNA Part A* **2018**, *29*, 643–649. [CrossRef]
45. Hughes, R.N.; Hughes, D.J.; Smith, I.P. The ecology of ghost crabs. *Oceanogr. Mar. Biol. Annu. Rev.* **2014**, *52*, 201–256.
46. Otwoma, L.M.; Reuter, H.; Timm, J.; Meyer, A. Genetic connectivity in a herbivorous coral reef fish (*Acanthurus leucosternon* Bennet, 1833) in the Eastern African region. *Hydrobiologia* **2018**, *806*, 237–250. [CrossRef]
47. Zhang, Z.X. Observation and Analysis of the Coastal Current and Its Adjacent Current System in the China Offshore Waters. Ph.D. Thesis, Ocean University of China, Qingdao, China, 2014.
48. Geng, Q.F.; Wang, Z.S.; Tao, J.M.; Kimura, M.K.; Liu, H.; Hogetsu, T.; Lian, C.L. Ocean Currents Drove Genetic Structure of Seven Dominant Mangrove Species Along the Coastlines of Southern China. *Front. Genet.* **2021**, *12*, 615911. [CrossRef] [PubMed]
49. Musilova, Z.; Kalous, L.; Petrtýl, M.; Chaloupková, P. Cichlid fishes in the Angolan headwaters region: Molecular evidence of the ichthyofaunal contact between the Cuanza and Okavango-Zambezi systems. *PLoS ONE* **2013**, *8*, e65047. [CrossRef]
50. Gagnaire, P.A.; Broquet, T.; Aurelle, D.; Viard, F.; Souissi, A.; Bonhomme, F.; Arnaud-Haond, S.; Bierne, N. Using neutral, selected, and hitchhiker loci to assess connectivity of marine populations in the genomic era. *Evol. Appl.* **2015**, *8*, 769–786. [CrossRef] [PubMed]
51. Munguia-Vega, A.; Green, A.L.; Suarez-Castillo, A.N.; Espinosa-Romero, M.J.; Aburto-Oropeza, O.; Cisneros-Montemayor, A.M.; Cruz-Pinon, G.; Danemann, G.; Giron-Nava, A.; Gonzalez-Cuellar, O.; et al. Ecological guidelines for designing networks of marine reserves in the unique biophysical environment of the Gulf of California. *Rev. Fish Biol. Fish.* **2018**, *28*, 749–776. [CrossRef]
52. Lu, J.Y.; Chen, Y.J.; Wang, Z.H.; Zhao, F.; Zhong, Y.S.; Zeng, C.; Cao, L. Larval dispersal modeling reveals low connectivity among national marine protected areas in the Yellow and East China. *Biology* **2023**, *12*, 396. [CrossRef]
53. Chen, L.L.; Zhu, X.L.; Ni, S.F.; Yan, Z.Q.; Xu, F.C. Research on the policy instrument of the fishing ban in the autumn season in the East China Sea of China- based on the content analysis of policy texts since 2011. *China Fish.* **2022**, *12*, 60–63.
54. Huang, C.D. A study on the administrative law of Fishing Ban in the summer in the South China Sea: A case of Naozhou island. *Leg. Syst. Soc.* **2020**, *13*, 125–126.
55. Chen, R.L.; Wu, X.Q.; Liu, B.J.; Wang, Y.Q.; Gao, Z.Q. Mapping coastal fishing grounds and assessing the effectiveness of fishery regulation measures with AIS data: A case study of the sea area around the Bohai Strait, China. *Ocean Coast. Manag.* **2022**, *223*, 106136. [CrossRef]
56. Yu, J.; Hu, Q.; Yuan, H.; Tong, F.; Chen, P.; Mao, J. Effects assessment of summer fishing moratorium in Daya Bay in the Northern South China Sea. *J. Geosci. Environ. Prot.* **2017**, *5*, 96. [CrossRef]
57. Liu, X.; Wang, Y.N.; Jiao, H.F.; Chen, C.; Liu, D.; Shi, H.X.; You, Z.J. Temporal Dynamics of Fishing Affect the Biodiversity of Macrobenthic Epifaunal Communities in the Coastal Waters of Ningbo, East China Sea. *Thalassas* **2021**, *37*, 39–49. [CrossRef]
58. Meyer, C.G.; Holland, K.N.; Papastamatiou, Y.P. Seasonal and diel movements of giant trevally *Caranx ignobilis* at remote Hawaiian atolls: Implications for the design of Marine Protected Areas. *Mar. Ecol. Prog. Ser.* **2007**, *333*, 13–25. [CrossRef]

Article

Should the Identification Guidelines for Siamese Crocodiles Be Revised? Differing Post-Occipital Scute Scale Numbers Show Phenotypic Variation Does Not Result from Hybridization with Saltwater Crocodiles

Nattakan Ariyaraphong [1,2,3], Wongsathit Wongloet [1,2], Pish Wattanadilokchatkun [1], Thitipong Panthum [1,2], Worapong Singchat [1,2], Thanyapat Thong [1], Artem Lisachov [1], Syed Farhan Ahmad [1,2], Narongrit Muangmai [1,2,4], Kyudong Han [1,5,6], Prateep Duengkae [1,2], Yosapong Temsiripong [7] and Kornsorn Srikulnath [1,2,3,8,9,*]

[1] Animal Genomics and Bioresource Research Unit (AGB Research Unit), Faculty of Science, Kasetsart University, 50 Ngamwongwan, Chatuchak, Bangkok 10900, Thailand; nattakan.ari@ku.th (N.A.); gamewongsathit@gmail.com (W.W.); pish.wa@ku.th (P.W.); thitipong.pa@ku.th (T.P.); worapong.singc@ku.ac.th (W.S.); thongthanyapat@gmail.com (T.T.); lisachev@bionet.nsc.ru (A.L.); syedfarhan.ah@ku.ac.th (S.F.A.); ffisnrm@ku.ac.th (N.M.); kyudong.han@gmail.com (K.H.); prateep.du@ku.ac.th (P.D.)

[2] Special Research Unit for Wildlife Genomics (SRUWG), Department of Forest Biology, Faculty of Forestry, Kasetsart University, 50 Ngamwongwan, Chatuchak, Bangkok 10900, Thailand

[3] Laboratory of Animal Cytogenetics and Comparative Genomics (ACCG), Department of Genetics, Faculty of Science, Kasetsart University, 50 Ngamwongwan, Chatuchak, Bangkok 10900, Thailand

[4] Department of Fishery Biology, Faculty of Fisheries, Kasetsart University, Bangkok 10900, Thailand

[5] Department of Microbiology, Dankook University, Cheonan 31116, Republic of Korea

[6] Bio-Medical Engineering Core Facility Research Center, Dankook University, Cheonan 31116, Republic of Korea

[7] R&D Center, Sriracha Moda Co., Ltd., Sriracha 20230, Thailand; yosapong@srirachamoda.com

[8] Center of Excellence on Agricultural Biotechnology (AG-BIO/PERDO-CHE), Bangkok 10900, Thailand

[9] Center for Advanced Studies in Tropical Natural Resources, National Research University-Kasetsart University, Bangkok 10900, Thailand

* Correspondence: kornsorn.s@ku.ac.th

Citation: Ariyaraphong, N.; Wongloet, W.; Wattanadilokchatkun, P.; Panthum, T.; Singchat, W.; Thong, T.; Lisachov, A.; Ahmad, S.F.; Muangmai, N.; Han, K.; et al. Should the Identification Guidelines for Siamese Crocodiles Be Revised? Differing Post-Occipital Scute Scale Numbers Show Phenotypic Variation Does Not Result from Hybridization with Saltwater Crocodiles. *Biology* **2023**, *12*, 535. https://doi.org/10.3390/biology12040535

Academic Editor: Cong Zeng

Received: 8 March 2023
Revised: 26 March 2023
Accepted: 28 March 2023
Published: 31 March 2023

Simple Summary: Morphological divergence between Siamese and other crocodiles has been identified by size, number of scales, and patterns of cervical squamation with post-occipital scutes (P.O.). However, a large variation of P.O. has been observed in captive Siamese crocodiles in Thailand, leading to questions about possible crocodile hybrids. The genetic diversity and population structure of Siamese crocodiles were studied using mitochondrial DNA D-loop and microsatellite genotyping. The STRUCTURE plot revealed numerous distinct gene pools, indicating that the crocodiles in each farm descended from distinct lineages. Researchers also discovered evidence of introgression in several individual crocodiles, implying that Siamese and saltwater crocodiles may have hybridized. A schematic protocol for screening hybrids was proposed based on patterns observed in phenotypic and molecular data.

Abstract: Populations of Siamese crocodiles (*Crocodylus siamensis*) have severely declined because of hunting and habitat fragmentation, necessitating a reintroduction plan involving commercial captive-bred populations. However, hybridization between Siamese and saltwater crocodiles (*C. porosus*) has occurred in captivity. Siamese crocodiles commonly have post-occipital scutes (P.O.) with 4–6 scales, but 2–6 P.O. scales were found in captives on Thai farms. Here, the genetic diversity and population structure of Siamese crocodiles with large P.O. variations and saltwater crocodiles were analyzed using mitochondrial DNA D-loop and microsatellite genotyping. Possible crocodile hybrids or phenotypic variations were ascertained by comparison with our previous library from the Siam Crocodile Bioresource Project. Siamese crocodiles with <4 P.O. scales in a row exhibit normal species-level phenotypic variation. This evidence encourages the revised description of Siamese crocodiles. Moreover, the STRUCTURE plot revealed large distinct gene pools, suggesting crocodiles in each farm were derived from distinct lineages. However, combining both genetic approaches provides evidence of introgression for several individual crocodiles, suggesting possible hybridization between Siamese and saltwater crocodiles. We proposed a schematic protocol with patterns observed in phenotypic and molecular data to screen hybrids. Identifying non-hybrid and hybrid individuals is important for long-term in situ/ex situ conservation.

Keywords: Siamese crocodile; saltwater crocodile; introgression; hybridization; post-occipital scutes

1. Introduction

Siamese crocodile (*Crocodylus siamensis*, Schneider, 1801) [1] is a freshwater species found in a wide range of lowland freshwater habitats including slow-moving rivers, streams, lakes, seasonal oxbow lakes, marshes, and swamps in mainland Southeast Asia, including Cambodia, Lao PDR and Thailand and on some islands of Indonesia and Malaysia [2–6]. This medium-sized crocodylian has a total length of less than 3.5 m [7]. However, its historical population distribution has decreased by 20% globally, with only 11% of its habitat range in nationally protected areas [8]. The Siamese crocodile was listed as a critically endangered species on the International Union for Conservation of Nature (IUCN) Red List and in Appendix I of the Convention on International Trade in Endangered Species of Wild Fauna and Flora (CITES) in 1996 to aid conservation efforts. In Thailand, Siamese crocodiles are widely distributed in lowland regions; however, most populations have been extirpated because of hunting, habitat loss, and collection to stock commercial crocodile farms over the last 40 years [8,9]. The severe decline of Siamese crocodile populations has led to there being fewer than 20 wild individuals in Khao Ang Rue Nai Wildlife Sanctuary (13°13′2.13″ N, 101°42′37″ E), Kaeng Krachan National Park (12°54′1.9″ N, 99°38′13.98″ E), Namnao National Park (16°57′57.65″ N, 101°30′28.08″ E), Yod Dom Wildlife Sanctuary (14°26′6.06″ N, 105°6′ 1.2″ E), and Bueng Boraphet (15°41′2.67″ N, 100°14′59.07″ E) [6,10]. To restore Siamese crocodile wild populations, it is essential to reintroduce captive-bred individuals and implement in situ/ex situ management practices. [11]. By contrast, 1.3 million Siamese crocodiles were present on 1400 farms in 2020, and crocodile farming now accounts for approximately 1% of Thailand's agricultural income [12]. However, the occurrence of hybridization between Siamese and saltwater crocodiles (*C. porosus*, Schneider, 1801) [13] is bidirectional between males and females of parental species in captivity. Interspecific hybridization frequently occurs in Southeast Asia due to the keeping of both species together in captivity, rather than from the wild [11,14]. Both F_1 hybrid and backcross crocodiles are fertile and reportedly grow faster than either parental species [13]. The genetic integrity of the species is at risk, which could harm conservation management. Alien outbreeding depression hybrids must be identified from the parental species before the reintroduction program or to improve genetic diversity in the wild. Using an effective genetic diagnosis approach, a genetically diverse captive population of pure Siamese crocodiles was identified while hybrids were differentiated from the parental species. Siamese crocodile sources serve as critical genetic resources for reintroduction efforts [11,15].

The morphological divergence between Siamese and other crocodiles (*Crocodylus* spp.) has been identified by size, the number of scales, and patterns of cervical squamation with post-occipital scutes (P.O.) [16]. The P.O.s of Siamese crocodiles show one row with 4–6 scales and several small scales, but no P.O. is seen in saltwater crocodiles. However, a large variation of P.O., ranging from two to six scales, has been observed in captive Siamese crocodiles in Thailand. This leads us to question whether possible crocodile hybrids remain in captivity, or whether these are actual phenotypic variations of Siamese crocodiles. A genetic approach to identifying Siamese and saltwater crocodiles was developed together with morphological observations in our previous study [11,15] but the P.O. pattern of each crocodile was not photo-recorded as evidence in our library report. In this study, to test these hypotheses, the genetic diversity and structure of Siamese and saltwater crocodile populations were assessed by screening the gene pool using 22 microsatellite markers and mitochondrial (mtDNA) D-loop sequences coupled with each crocodile photo record. MtDNA and nuclear DNA microsatellites are molecular genetic markers that can identify population diversity, origins of individuals, and hybrids along with their parents, especially in crocodiles [11,15]. Results were compared with those of the large gene pool library under "the Siam Crocodile Bioresource Project" from our previous study [11]. These findings provide pivotal information for prospective reintroduction programs and in situ/ex situ management.

2. Materials and Methods

2.1. Specimen Collection and DNA Extraction

A total of 136 Siamese and 29 saltwater crocodile specimens were collected from 4 captive locations under the auspices of the Thai Crocodile Farm Association (TCFA) and in accordance with CITES regulations for the leather and food industries. Table S1 provides detailed information on the sampled individuals. Scale samples were collected from the tail of captive crocodiles registered at four crocodile farms in Chonburi (CB) (13°09′06.57″ N, 101°28′36.01″ E), Nakhon Ratchasima (NR) (14°57′21.19″ N, 101°28′36.01″ E), Chainat (CN) (15°15′10.44″ N, 100°02′38.27″ E) and Nakhon Pathom (NP) (13°43′20.25″ N, 100°15′20.83″ E) between January and August 2022. A piece of scale clipped from the tail of each specimen was collected as a DNA source. Permission was granted by the farm owners and the TCFA and also from unnamed crocodile farms. Individuals were classified as Siamese or saltwater crocodiles based on external morphological observation [17,18] and photographed. The dataset comprised photo images of 165 individual crocodiles, each captured in 30–50 different postures to minimize testing redundancy and bias. The DNA extraction and quality assessment were performed using the same methods as in previous studies [19] (Supplementary Data S1). All experimental procedures and animal care were carried out in compliance with the Regulations on Animal Experiments at Kasetsart University and approved by the Animal Experiment Committee under Approval No. ACKU64-SCI-011.

2.2. Microsatellite Genotyping and Microsatellite Data Analysis

Twenty-two microsatellite primer sets, developed originally from saltwater crocodiles (Table S2) [20,21], were used for the genotyping of all crocodile individuals. The genotypic data resulting from this study were deposited in the Dryad Digital Repository Dataset (https://datadryad.org/stash/share/s4zREYQ1AUUXTsaIpk0r1HSdkYljvu8yvJOVR143K7Y, accessed on 18 February 2023). We used the same methods as previous studies for PCR amplification to analyze genetic diversity and population structure of the crocodile populations [11,22–27] (Supplementary Data S1).

2.3. Mitochondrial DNA D-Loop Sequencing and Data Analysis

The mtDNA D-loop sequences of DNA fragments were amplified using the primers mtCytbf2 (5′-TGCCATGTTCGCATCCATCC-3) and mt12srRNAr2 (5′-CCAGAGGCTA GGCGTCGTGG-3) [11]. We used the same methods as previous studies for PCR amplification and analyze genetic diversity of the crocodile populations [11,22] (Supplementary Data S1).

3. Results

Genetic Variability of Captive Crocodile Population Based on Microsatellite Data

All captive individuals were genotyped, and 459 alleles were found across all loci, with the mean number of alleles per locus being 20.864 ± 2.351 (Table 1). All allelic frequencies in the captive population significantly deviated from what would be expected under the Hardy–Weinberg equilibrium, indicating the presence of linkage disequilibrium (Tables S3–S7). Null alleles were frequently observed for 13 loci (CpP3001, CpP501, CpP214, CpP2206, CpP3313, CpP2504, CpP203, CpP1308, CpP4004, CpP3008, CpP2904, CpP3004, CpP1409), and all markers listed were treated similarly. Siamese crocodiles from CB and NR populations exhibited negative F values, but Siamese crocodiles from CN and NP populations exhibited positive F values, similar to saltwater crocodiles from NR. The PIC of all captive populations ranged from 0.057 to 0.932 and I ranged from 0.153 to 3.031 (Table S8). The H_o values ranged from 0.059 to 1.000 (mean ± standard error (SE): 0.629 ± 0.033) and the H_e values ranged from 0.058 to 0.936 (mean ± SE: 0.718 ± 0.037) (Tables 1 and S8). Welch's t-test showed that H_o was not significantly different from H_e in Table 2. By comparing pairwise H_o values between populations, there were statistical differences between six pairs, while pairwise H_e values were different between five pairs (Table 3). The AR value of the population was 20.787 ± 2.332. The standard genetic diversity indices are summarized in Table 1 and Supplementary Table S8.

Table 1. Genetic diversity of 136 Siamese (*Crocodylus siamensis*, Schneider, 1801) [1] and 29 saltwater crocodiles (*C. porosus*, Schneider, 1801) [13] based on 22 microsatellite loci. Table S1 provides detailed information on the sampled individuals.

Population		N	N_a	AR	N_{ea}	I	H_o	H_e	PIC	F
CB [1]	Mean	30	7.409	7.409	3.847	1.329	0.653	0.612	0.570	−0.092
	S.E.	0	0.993	0.993	0.713	0.140	0.044	0.044	0.214	0.045
NR [2]	Mean	30	8.727	8.727	3.847	1.382	0.602	0.609	0.572	−0.031
	S.E.	0	1.405	1.405	0.728	0.149	0.043	0.043	0.213	0.066
CN [3]	Mean	34	9.500	9.500	4.603	1.396	0.545	0.582	0.553	0.033
	S.E.	0	1.364	1.364	1.000	0.180	0.057	0.058	0.268	0.055
NP [4]	Mean	42	11.182	11.182	4.880	1.658	0.642	0.693	0.658	0.057
	S.E.	0	1.307	1.307	0.730	0.147	0.044	0.040	0.204	0.068
CP [5]	Mean	29	9.818	9.798	4.951	1.716	0.696	0.733	0.705	0.039
	S.E.	0	1.169	1.150	0.530	0.125	0.041	0.036	0.173	0.046
All Population	Mean	165	20.864	20.787	6.745	1.921	0.629	0.718	0.695	0.110
	S.E.	0	2.351	2.332	1.604	0.164	0.033	0.037	0.180	0.038

Sample size (N); number of alleles (N_a); allelic richness (AR); number of effective alleles (N_{ea}); Shannon's information index (I); observed heterozygosity (H_o); expected heterozygosity (H_e); polymorphic information content (PIC); fixation index (F). [1] CB = Chonburi (*Crocodylus siamensis*). [2] NR = Nakhon Ratchasima (*Crocodylus siamensis*). [3] CN = Chainat (*Crocodylus siamensis*). [4] NP = Nakhon Pathom (*Crocodylus siamensis*). [5] CP = Nakhon Ratchasima (*Crocodylus porosus*).

Table 2. Welch's t-test of observed heterozygosity (H_o) and expected heterozygosity (H_e) of Siamese (*Crocodylus siamensis*, Schneider, 1801) [1] and saltwater crocodiles (*Crocodylus porosus*, Schneider, 1801) [13] based on 22 microsatellite loci.

Population	H_o	H_e	df	t-Test	p-Value
CB [1]	0.641 ± 0.044	0.612 ± 0.044	0.041	0.466	0.643
NR [2]	0.602 ± 0.043	0.609 ± 0.043	−0.037	−0.115	0.909
CN [3]	0.545 ± 0.057	0.582 ± 0.058	0.007	−0.455	0.651
NP [4]	0.642 ± 0.044	0.693 ± 0.040	−0.051	−0.858	0.394
CP [5]	0.696 ± 0.041	0.733 ± 0.036	−0.037	−0.678	0.501

[1] CB = Chonburi (*Crocodylus siamensis*). [2] NR = Nakhon Ratchasima (*Crocodylus siamensis*). [3] CN = Chainat (*Crocodylus siamensis*). [4] NP = Nakhon Pathom (*Crocodylus siamensis*). [5] CP = Nakhon Ratchasima (*Crocodylus porosus*).

We determined the degree of relatedness between individuals in the captive crocodile population by employing a pairwise test. The mean pairwise r values were calculated for a total of 13,530 combinations of crocodiles, which included all 165 sampled individuals, including Siamese and saltwater crocodiles, were −0.018 ± 0.033 (CB population = −0.018 ± 0.042, NR population = −0.019 ± 0.029, CN population = −0.017 ± 0.030, NP population = −0.019 ± 0.032 for Siamese crocodiles and NR population of saltwater crocodiles = −0.020 ± 0.032). No pairs showed $r < -0.25$. There were 13,525 pairs with $-0.25 < r < 0.25$ and 5 pairs with $0.25 > r$ (Tables 2 and S9–S13). Distribution of r values for the crocodiles exhibited a left skew, indicating lower pairwise r values than what would be expected under a null hypothesis of unrelated individuals by chance. The distributions of pairwise r differed significantly between the CB and NP populations, and the mean pairwise r values were also significantly different across all populations. (Figure 1, Table S14). The mean F_{IS} was −0.074 ± 0.076 (Table 3), with individual values of F_{IS} ranging from −0.191 to 0.085

(Tables S15–S19). However, distributions of F_{IS} from all populations differed significantly from each other (Figure 1, Table S14). The N_e of Siamese crocodiles for individuals that contributed genetically to the CB population was 41.3 (95% CI: 32.2.8–46.5), 202.2 (95% CI: 87.7–113.3) for the NR population, 115.5 (95% CI: 69.5–88.1) for the CN population, 45.2 (95% CI: 37.1–136.1) for the NP population, and 103.7 (95% CI: 66.7–74.8) for saltwater crocodiles in the NR population (Table 4).

Table 3. Comparison of genetic diversity parameters between Siamese (*Crocodylus siamensis*, Schneider, 1801) [1] and saltwater crocodiles (*Crocodylus porosus*, Schneider, 1801) [13] based on 22 microsatellite loci. Table S1 provides detailed information on the sampled individuals.

	Population 1	Population 2	df	SE	*t*-Test	*p*-Value
	CB [1]	NR [2]	0.051	0.011	4.540	<0.05
	CB	CN [3]	0.108	0.013	8.536	<0.05
	CB	NP [4]	0.011	0.099	0.111	0.912
	CB	CP [5]	−0.051	0.010	−4.914	<0.05
Heterozygosity (H_o)	NR	CN	0.057	0.013	4.546	<0.05
	NR	NP	−0.040	0.099	−0.403	0.689
	NR	CP	−0.102	0.010	−9.962	<0.05
	CN	NP	−0.097	0.100	−0.974	0.335
	CN	CP	−0.159	0.012	−13.498	<0.05
	NP	CP	−0.062	0.099	−0.624	0.536
	CB	NR	0.003	0.011	0.267	0.790
	CB	CN	0.030	0.013	2.346	<0.05
	CB	NP	−0.081	0.107	−0.755	0.454
	CB	CP	−0.135	0.010	12.062	<0.05
Heterozygosity (H_e)	NR	CN	0.027	0.013	2.131	<0.05
	NR	NP	−0.084	0.107	−0.783	0.438
	NR	CP	−0.138	0.010	−13.324	<0.05
	CN	NP	−0.111	0.107	−1.034	0.307
	CN	CP	−0.165	0.012	−13.723	<0.05
	NP	CP	−0.054	0.107	−0.504	0.617

[1] CB = Chonburi (*Crocodylus siamensis*). [2] NR = Nakhon Ratchasima (*Crocodylus siamensis*). [3] CN = Chainat (*Crocodylus siamensis*). [4] NP = Nakhon Pathom (*Crocodylus siamensis*). [5] CP = Nakhon Ratchasima (*Crocodylus porosus*).

Table 4. Inbreeding coefficients, relatedness, effective population size, and ratio of effective population size and census population (N_e/N) of 136 Siamese (*Crocodylus siamensis*, Schneider, 1801) [1] and 29 saltwater crocodiles (*C. porosus*, Schneider, 1801) [13].

Population	N	F_{IS}	Relatedness (r)	Estimated N_e	95% CIs for N_e	N_e/N
CB [1]	30	−0.113 ± 0.201	−0.018 ± 0.042	41.300	32.200–46.500	1.377
NR [2]	30	−0.086 ± 0.039	−0.019 ± 0.029	202.200	87.700–113.300	6.740
CN [3]	34	−0.063 ± 0.039	−0.017 ± 0.030	115.500	69.500–88.100	3.397
NP [4]	42	−0.045 ± 0.046	−0.019 ± 0.032	45.200	37.100–136.100	1.076
CP [5]	29	−0.065 ± 0.055	−0.020 ± 0.032	103.700	66.700–74.800	3.576

Estimates were calculated using GenAlEx version 6.5 [28]. NeEstimator version 2.1 [29], and COANCESTRY version 1.0.1.9 [30]. Detailed information for all elephant individuals is presented in Table S1. Sample size (N); inbreeding coefficient (F_{IS}); effective population size (N_e). [1] CB = Chonburi (*Crocodylus siamensis*). [2] NR = Nakhon Ratchasima (*Crocodylus siamensis*). [3] CN = Chainat (*Crocodylus siamensis*). [4] NP = Nakhon Pathom (*Crocodylus siamensis*). [5] CP = Nakhon Ratchasima (*Crocodylus porosus*).

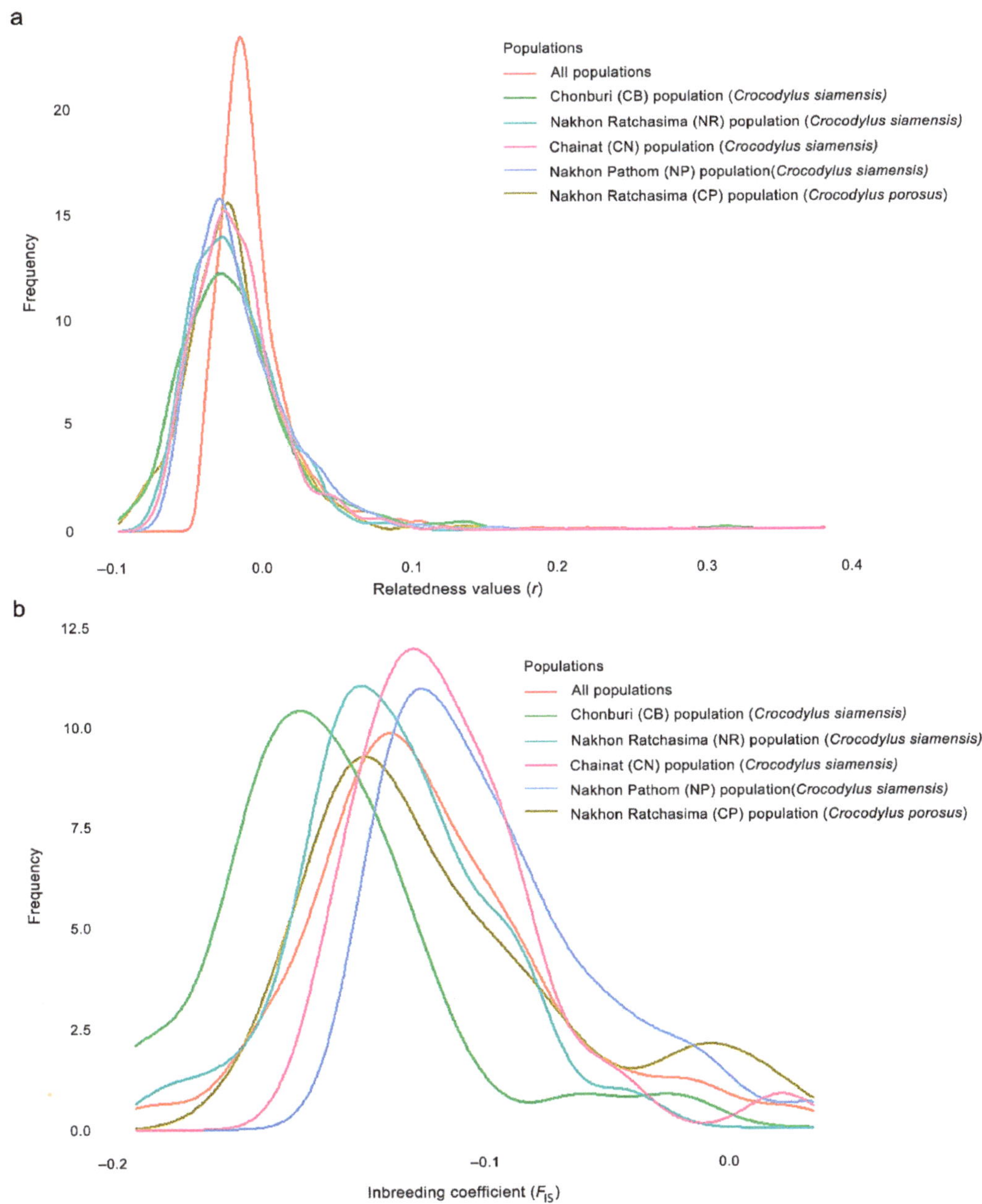

Figure 1. Observed distribution of (**a**) pairwise relatedness (*r*) and (**b**) inbreeding coefficients (F_{IS}) for 136 Siamese (*Crocodylus siamensis*, Schneider, 1801) [1] and 29 saltwater crocodiles (*C. porosus*, Schneider, 1801) [13] individuals, plotted against the expected distributions.

Significant differences ($p < 0.05$) were observed in the estimates of F_{ST} between captive populations after 110 permutations. The AMOVA showed that genetic variation was 84% among individuals crocodiles within a population and 11% between populations (Table S21). Nei's genetic distances and R_{ST} showed that the CN population was closer than the NP to the others (Tables S20 and S22). The distinction between the five crocodile groups and the three suspected individuals (CSI05, CSI06, and CPO09) from our previous study was supported by the first, second, and third principal components, which accounted for 10.48, 8.87, and 4.24% of the total variation, respectively, as revealed by PCoA [11] (Figure 2). Different population patterns were generated by the model-based Bayesian clustering algorithms implemented in STRUCTURE with increasing K values; however, the highest posterior probability with one peak ($K = 3$) based on Evanno's ΔK, while the mean ln $P(K)$ also revealed one peak ($K = 16$) (Figures 3 and S1).

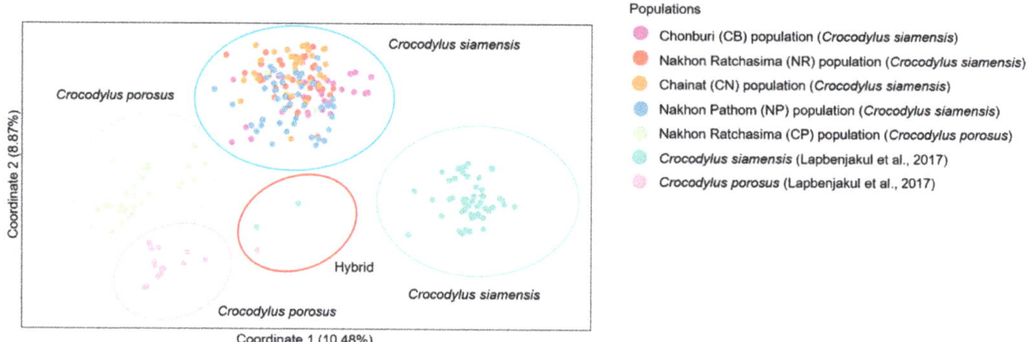

Figure 2. Principal component analysis of Siamese (*Crocodylus siamensis*, Schneider, 1801) [1] and saltwater crocodiles (*C. porosus*, Schneider, 1801) [13]. Table S1 provides detailed information on the sampled individuals and Lapbenjakul et al. [11].

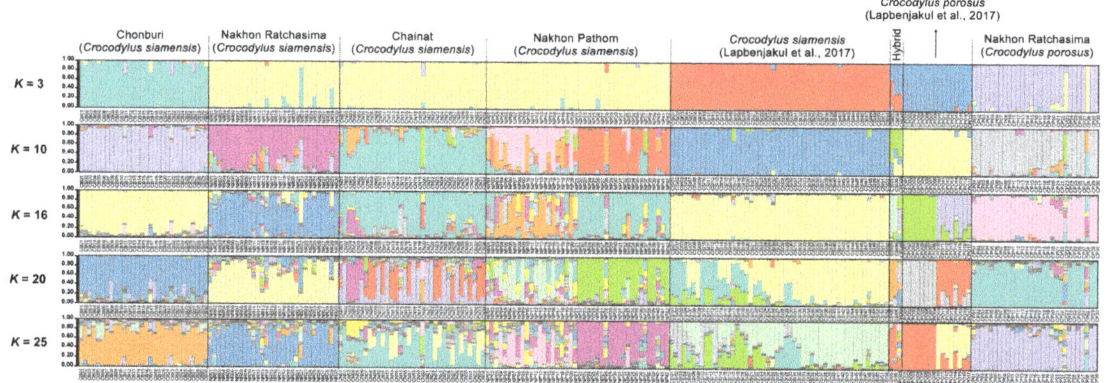

Figure 3. Population structure of 136 Siamese (*Crocodylus siamensis*, Schneider, 1801) [1] individuals and 29 saltwater crocodile (*C. porosus*, Schneider, 1801) [13] individuals. Each vertical bar on the *x*-axis represents an individual, while the *y*-axis represents the proportion of membership (posterior probability) in each genetic cluster. Crocodiles are superimposed on the plot, with black vertical lines indicating the boundaries. Detailed information for all crocodile individuals is presented in Table S1 and Lapbenjakul et al. [11].

The amplicon length and alignment length of the mtDNA D-loop sequences were 1500 and 1350 bp, respectively. The numbers of haplotypes were 46 and 28 for Siamese and saltwater crocodiles, respectively. Overall haplotype and nucleotide diversities were 0.846 ± 0.022 and 0.015 ± 0.003 for Siamese and 0.998 ± 0.055 and 0.069 ± 0.025 for saltwater crocodiles (Table 5). A complex haplotype network was constructed from the many detected polymorphic sites and haplotypes (Figure 4). The most common haplotype of all Siamese crocodile populations was haplotype CS36. Seven haplotypes (CS01, CS16, CS20, CS21, CS27, CS30, and CS36) were shared in the CB, NR, CN, and NP populations. Furthermore, the most common haplotype of all populations of saltwater crocodiles was haplotype CD04 and CD05. Phylogenetic analysis of a combined data set for the mtDNA D-loop sequences from both Siamese and saltwater crocodiles, together with those for 21 crocodile species obtained from the public repositories (GenBank/DDBJ/European Nucleotide Archive (ENA)), indicated that most Siamese and saltwater crocodile sequences each formed a monophyletic clade. However, NP04, NP06, NP07, NP09, NP13, NP14, NP16, and NP20, first assigned to Siamese crocodile, were grouped with Cuban crocodile (*C. rhombifer*) (GenBank accession number: NC_024513); whereas CP01, CP05, CP06, CP11, CP12, CP13, CP15, CP20, CP23, CP25, CP27, CP28, CP29, and CP30, categorized with the saltwater crocodile, were placed as a sister clade to Siamese crocodile (Figure S2). These results agreed with BLAST results of sequence identity (Table S1). To examine the genetic differentiation among the five populations, we calculated F_{ST}, G_{ST}, Φ_{ST}, D_{xy}, D_a, and N_m. The F_{ST} values ranged from -0.015 to 0.347, the G_{ST} values ranged from -0.008 to 0.071 and the Φ_{ST} values ranged from 0.013 to 0.330, The N_m values ranged from 0.947 to infinite, the D_{xy} values ranged from 0.002 to 0.065 and the D_a values ranged from 0.000 to 0.026 (Table 6).

Table 5. Mitochondrial DNA D-loop sequence diversity of Siamese (*Crocodylus siamensis*, Schneider, 1801) [1] and saltwater crocodiles (*Crocodylus porosus*, Schneider, 1801) [13].

Population	N	Number of Haplotypes (H)	Theta (per Site) from S	Average Number of Nucleotide Differences (k)	Overall Haplotype	Nucleotide Diversities (π)
CB [1]	30	20	0.016	6.786	0.959 ± 0.022	0.013 ± 0.006
NR [2]	30	18	0.007	3.471	0.940 ± 0.027	0.015 ± 0.008
CN [3]	34	20	0.010	4.783	0.934 ± 0.027	0.022 ± 0.011
NP [4]	42	24	0.036	40.539	0.922 ± 0.032	0.072 ± 0.035
CP [5]	29	28	0.052	56.924	0.998 ± 0.010	0.069 ± 0.025
All populations	165	54	0.043	18.024	0.725 ± 0.039	0.040 ± 0.020

[1] CB = Chonburi (*Crocodylus siamensis*). [2] NR = Nakhon Ratchasima (*Crocodylus siamensis*). [3] CN = Chainat (*Crocodylus siamensis*). [4] NP = Nakhon Pathom (*Crocodylus siamensis*). [5] CP = Nakhon Ratchasima (*Crocodylus porosus*).

Table 6. Genetic differentiation between the three populations of Siamese (*Crocodylus siamensis*, Schneider, 1801) [1] and saltwater crocodiles (*Crocodylus porosus*, Schneider, 1801) [13] for the mitochondrial DNA D-loop sequence. Genetic differentiation coefficient (G_{ST}), Wright's F-statistic for the subpopulations within the total population (F_{ST}), Φ_{ST}, gene flow (N_m) from the sequence data and the haplotype data, the average number of nucleotide substitutions per site between populations (D_{xy}), and the net nucleotide substitutions per site between populations (D_a).

Population 1	Population 2	G_{ST}	Φ_{ST}	F_{ST}	D_{xy}	D_a	N_m
CB [1]	NR [2]	-0.008	0.013	0.045 *	0.002	0.000	10.620
CB	CN [3]	0.000	0.017	0.027 [ns]	0.002	0.000	18.224
CB	NP [4]	0.005	0.083	0.127 **	0.021	0.003	3.432
CB	CP [5]	0.071	0.320	0.347 **	0.055	0.025	0.941
NR	CN	0.006	0.025	-0.015 [ns]	0.003	0.000	Infinite
NR	NP	0.009	0.081	0.094 **	0.021	0.003	4.840
NR	CP	0.066	0.318	0.345 **	0.055	0.025	0.949
CN	NP	0.000	0.087	0.090 *	0.021	0.003	5.064
CN	CP	0.067	0.330	0.345 **	0.055	0.026	0.950
NP	CP	0.050	0.198	0.261 *	0.065	0.020	1.418

ns = not significant, * $p < 0.05$, ** $p < 0.01$. [1] CB = Chonburi (*Crocodylus siamensis*). [2] NR = Nakhon Ratchasima (*Crocodylus siamensis*). [3] CN = Chainat (*Crocodylus siamensis*). [4] NP = Nakhon Pathom (*Crocodylus siamensis*). [5] CP = Nakhon Ratchasima (*Crocodylus porosus*).

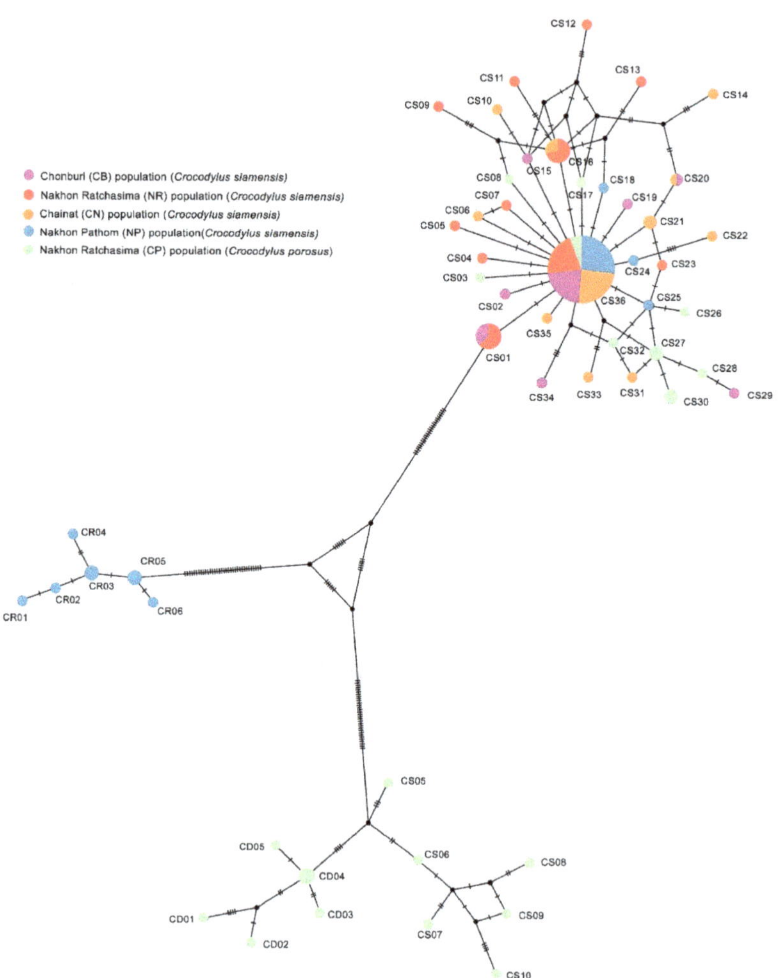

Figure 4. Haplotype network based on sequence data for the mitochondrial DNA D-loop region of Siamese (*Crocodylus siamensis*, Schneider, 1801) [1] and saltwater crocodiles (*Crocodylus porosus*, Schneider, 1801) [13].

4. Discussion

Siamese crocodile is well-represented in captivity, with possibly over 1.5 million individuals in farms in Thailand, Cambodia, and Vietnam [31–34], and smaller numbers in farms in China and zoos in Europe and North America. In Thailand, 1400 crocodile farms with 1,319,395 Siamese and 162,449 saltwater crocodiles were operating in 2020, while 47,367 skins were sold in international trade in 2020 [35–37]. However, the captive population in Thailand includes an unknown number of individuals hybridized with saltwater crocodiles [11,33,38–40], similar to captive crocodiles in Cambodia, Lao PDR, and Vietnam [33,41–43]. Observations of hybrids between Siamese and saltwater crocodiles have been reported as a consequence of anthropogenic impacts [11,39,44,45]. Most anthropogenic crocodile hybrids pose a serious problem for conservation management because hybrids a possess highly similar morphology to their parental species, which might lead to introgression if they are included in a reintroduction program [11].

Most crocodile farms are members of the TCFA, which aims to keep purely captive-bred individuals of both species to comply with the recommendations of CITES Appendix I captive breeding operation [11,15,46]. Most Thai crocodile farms have thus pledged not to produce hybrid offspring in the interest of conservation. However, we found discrepancies in two genetic markers, microsatellite genotyping and mtDNA D-loop in several crocodile individuals (both Siamese and saltwater crocodiles), suggesting the possibility of hybrids in the population examined. Crocodiles that were clearly identified as being pure specimens of a species were designated as either "identified Siamese crocodile" or "identified saltwater crocodile". However, if the results of the two genetic markers were not consistent, the crocodiles were designated "unidentified crocodile". To ensure compliance between genetic tools and phenotypic variation for Siamese and saltwater crocodile identification, we compared the results of genetic diversity and structure with phenotypic observation in each crocodile. The most frequently observed means of identifying hybrid characteristics between Siamese and saltwater crocodiles are the presence of P.O. This has led to the misunderstanding of many crocodile experts and non-governmental organizations (NGOs) who have visited crocodile farms in Southeast Asia as to whether Siamese crocodiles with fewer than four post-occipital scales in a row are hybrids [16,47].

4.1. Are Different Numbers of Post-Occipital Scutes Due to Phenotypic Variation within Siamese Crocodiles or the Consequence of Hybridization with Saltwater Crocodiles?

As shown in the STRUCTURE plot and PCoA, Siamese crocodiles from CB, CN, NR, and NP, and saltwater crocodiles from NR, were likely clustered into different groups. We analyzed the clustering order and gene pool pattern from $K = 2$–25. Identified pure Siamese crocodiles, which have two P.O. scales shared the same gene pool with Siamese crocodiles having three or four scales, whereas no P.O. scales were found in identified saltwater crocodiles (Figures 3, 5 and S3). For $K = 3$, Siamese crocodiles (both identified and unclear individuals) from CB, CN, NR, and NP were grouped in the same gene pool, while saltwater crocodiles were part of a new group, with a small part shared with Siamese crocodiles. At higher K levels, saltwater crocodiles (both identified and unclear individuals) became identifiable, whereas Siamese crocodiles from different farms were separated from each other. Gene pool structuring from both species or each farm showed admixture at higher K levels; however, P.O. scale number variation also appeared in Siamese crocodiles with mixtures of specific gene pools of Siamese crocodiles. This suggests that Siamese crocodiles with fewer than four scales in a row are part of normal phenotypic variations at the species level. Similarly, the first version of the species identification guideline was revised after DNA analysis and proved various characteristics under the same species such as in fighting fishes [48,49].

This misunderstanding of widespread hybridization in Thailand has probably resulted from personal communication among experts from IUCN/SSC/Crocodile Specialist Group and other NGOs who visited farms and followed the CITES Identification Guide based on morphological characteristics in Charette [16], leading to an erroneous judgment of crocodile hybridization events in Thai crocodile farms [46,47]. Revision of Siamese crocodile identification should be reconsidered for scientific taxonomic study, which is relevant to conservation management and economic value. However, the limited number of microsatellite markers located at regular intervals cannot cover species-specific genomic regions [50]. The 22 microsatellite marker loci in this study may have caused bias due to limited population history, the timing of selection, phasing error, and false LD resolution [51,52]. Therefore, larger sample sizes with higher numbers of microsatellite loci are required to extensively investigate the evidence. Genome-wide SNP are also needed to identify signature selection between species or specific phenotypic issues such as P.O.

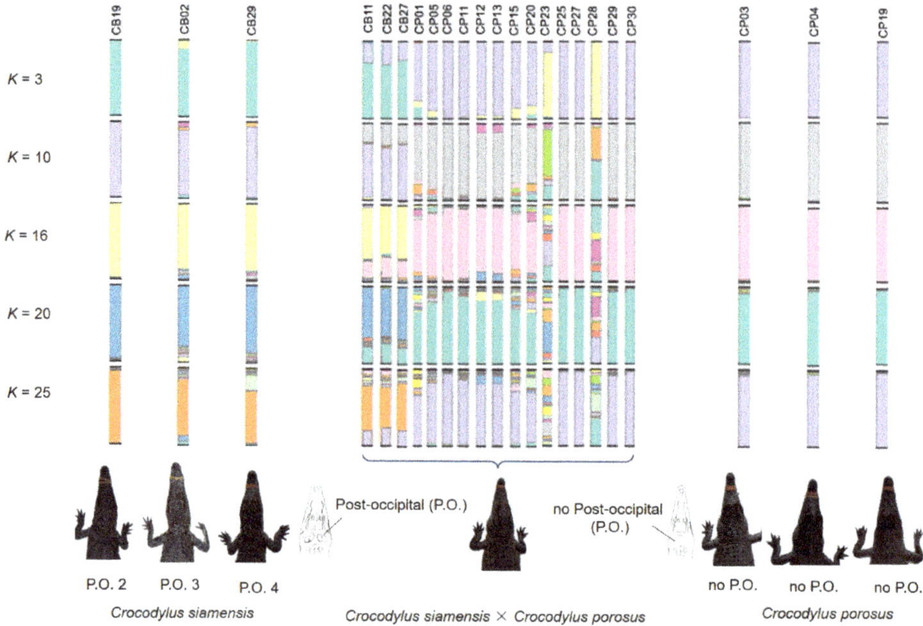

Figure 5. Representation of structure plot and post occipital scutes (P.O.) for hybrids between Siamese and saltwater crocodiles (*Crocodylus siamensis* × *Crocodylus porosus*) and pure Siamese and saltwater crocodiles. Detailed information for all hybrids between Siamese and saltwater crocodile individuals is presented in Figures 3 and S3.

4.2. Large Gene Pool Variation Reflects Different Historical Origins in the Wild Population

Using data generated from microsatellite loci derived from Siamese and saltwater crocodiles by Lapbenjakul et al. [11] and this study, we addressed the genetic structure and gene pool pattern between the two crocodile species. As shown by PCoA, Siamese and saltwater crocodiles were clustered into different groups. However, sharing of gene pools by Siamese crocodiles between the crocodile farms was observed at different K levels of the STRUCTURE plot. Although Siamese crocodile individuals from this study grouped in the same gene pool at $K = 3$, and Siamese and saltwater crocodile individuals from our previous study [11] were shown to be part of a new group with saltwater crocodiles, large differences in gene pools were observed among the four populations from $K = 5$–25. In the NP population, two subpopulations of different gene pools were found, consistent with the positive fixation index value. It can be inferred from this that Siamese crocodiles in each farm had different ancestral original lineages.

Historically, Siamese crocodiles were widespread in Central Thailand. The current captive Siamese crocodile populations might reflect differences in the original sources brought into farms in Thailand. This result agreed with the genetic diversity parameters showing high values of heterozygosity, AR, and N_e in Siamese crocodiles from the four captive sites in this study. The pairwise F_{ST} value was statistically significant among the four populations, implying genetic structure differentiation between the farms. The differentiation in genetic makeup reflects the accumulation of variations in allelic frequencies, providing crucial insights into the evolutionary history, genetic drift, and selection of distinct populations [53,54]. However, we have no evidence to identify gene pool associations and geographic origin as known ecotypes, as no capture records exist for Siamese crocodile individuals. Interestingly, many Siamese crocodiles from the four captivities showed a genetic admixture of different gene pools of Siamese crocodile from $K = 5$–25. This was also observed in the saltwater crocodile population, which might have resulted from the historical genetic exchange of parental stocks between farms. Regarding mtDNA D-loop sequences, positive N_m, low F_{ST}, and sharing haplotypes of crocodiles between farms also confirmed the presence of crocodile genetic exchange in the market farms. The exchange of Siamese crocodile parental stocks has been conducted to impede inbreeding within each captive site, which may provide a negative inbreeding level. These findings collectively suggest that large captive populations of Siamese crocodiles held on farms represent a good potential source for reintroduction programs. Crocodile farms under TCFA are willing to donate Siamese crocodiles for this purpose [55].

4.3. Siamese Crocodile Identification Protocol Based on Morphology and DNA Fingerprinting

Captive crocodiles must be genetically identified at the species level before release [9,56,57]. However, hybridization between Siamese and saltwater crocodiles is widespread among some captive populations in Southeast Asia [33,34,39,58]. Differentiating hybrids from parental species based on phenotype alone is very challenging; thus, genetic screening is necessary to confirm species identity [11,15,40,58]. To ensure the success of reintroduction programs, we must first address the complex hybrid issue. Cluster analysis using STRUCTURE can now determine the degree of hybridization and gene pool pattern by aggregating individuals into a single cluster relative to additional highly differentiated populations/species [11,59,60]. Our previous study indicated that three individuals (CPO09, CSI05, and CSI06) may have been the result of interspecific hybridization between Siamese and saltwater crocodiles [11]. In this study, after we added more Siamese and saltwater crocodiles to the library analysis, the three crocodile individuals were still identified as hybrids. However, high levels of genetic admixture were observed in many crocodile individuals, and this might result in misleading conclusions about the genetic admixture of gene pools under the STRUCTURE plot with the probability of identifying the state of hybrids alone. According to genetic diversity parameters and the STRUCTURE plot, great genetic diversity and large gene pools of both species likely remain in the population, while both species are very closely related lineages [15]. The two species may share some alleles of microsatellite repeats at the same genomic locus, which is often observed in many closely related species in vertebrates [61,62]. We, therefore, proposed criteria to screen hybrid crocodiles between Siamese and saltwater crocodiles as follows: (*i*) Consideration of genetic admixture at different K levels to examine the trend of clustering, separation of allelic signals and the majority of allelic pattern, although the best K level might be predicted from different algorithms; (*ii*) sharing a gene pool between the two species might be possible, but should not have more than a posterior probability of 0.05 at the K level, which shows the trend of separation between the two species; (*iii*) clustering by PCoA should be considered together with the STRUCTURE plot to test the group of crocodile specimens; (*iv*) determination of maternal lineage by mtDNA D-loop sequences may be added to confirm; and (*v*) external morphological observation with updated phenotypic variations in the P.O. should be scored together with genetic screening.

These five steps would provide evidence that can prove the hybridization status of each Siamese crocodile individual under reasonable time before they are used in the reintroduction program (Figure 6). Crocodiles that pass the five tests of characteristics would be key sources for release to the wild. However, if the crocodile fails on some aspects with unclear determination, the individual should undergo more experimental tests such as karyotyping. Siamese and saltwater crocodiles have different chromosome numbers, whereas the F_1 hybrid or backcross shows diverge chromosome constitution from the parental species [63,64]. However, karyotyping is time-consuming, expensive, and may not be practical to prove species purity for large numbers of individuals, whereas multiple types and generations of hybrid (both F_2, F_3 or backcross) might escape detection of chromosome number. More research utilizing genome-wide scans with single nucleotide polymorphisms (SNPs) is necessary to enhance our comprehension of selection signatures in diverse populations and species. However, genome-wide SNP analysis might not be reliable for multiple processes with small crocodile numbers in each reintroduction program. From this state, we found suspected hybrids with CB11 (P.O. 4), CB22 (P.O. 3), CB27 (P.O. 4), CP01 (P.O. 0), CP05 (P.O. 4), CP06 (P.O. 4), CP11 (P.O. 4), CP12 (P.O. 4), CP13 (P.O. 4), CP15 (P.O. 2), CP20 (P.O. 0), CP23 (P.O. 4), CP25 (P.O. 2), CP27 (P.O. 4), CP28 (P.O. 4), CP29 (P.O. 0), CP30 (P.O. 0), CSI05 (unidentified P.O.), CSI06 (unidentified P.O.), and CPO09 (unidentified P.O.) (Figure 5), which should be tested by karyotyping before release.

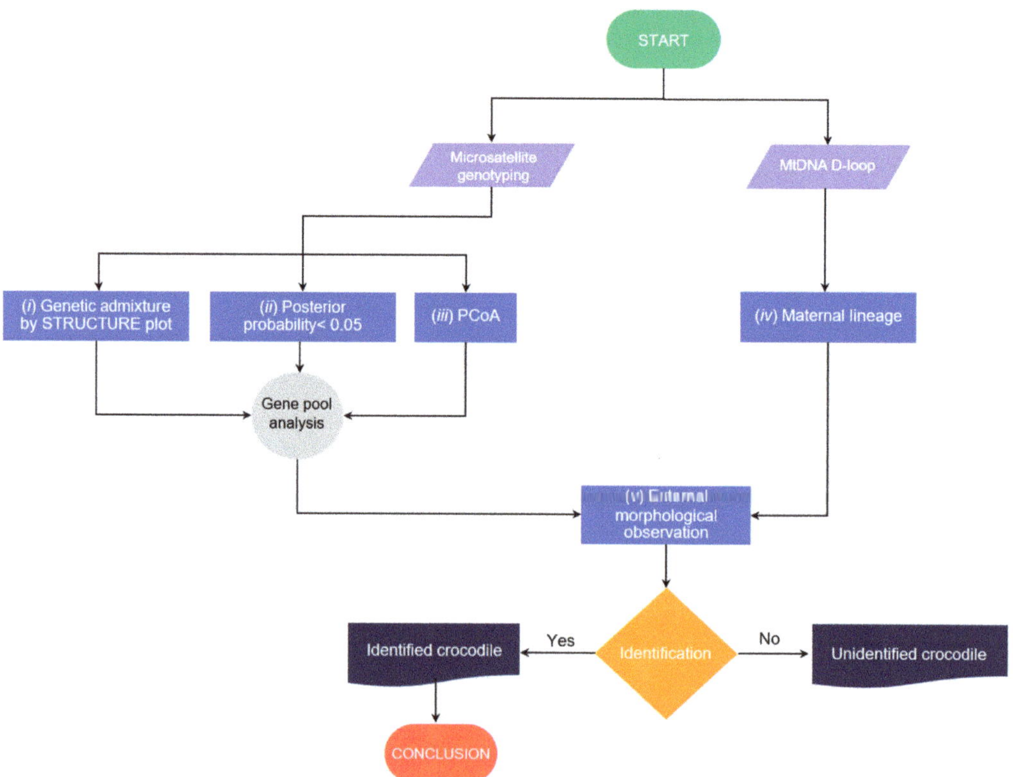

Figure 6. Schematic representation of criteria to screen hybrid crocodiles between Siamese and saltwater crocodiles to prove the hybridization level status in each Siamese crocodile individual before they can be used in the reintroduction program.

The Thai government and TCFA have never promoted hybridization in crocodile farms under Appendix I captive breeding operation [46]; however, the incidence of contamination by hybrids was observed to be 5–10% here and in our previous study [11]. Hybrid contamination may result from a long history of crocodile trade from 30 years ago, which had no concrete plan of protection by the Thai Government. Hybridization between Siamese and Cuban crocodiles (*C. rhombifer*) as a result of human-mediated movement has also been observed in Southeast Asia [11,39,44,45]. Current results suggest that NP04, NP06, NP07, NP09, NP13, NP14, NP16, and NP20 are hybrid crocodiles derived from the Cuban crocodile lineage. We also found signs of a unique gene pool from unidentified crocodiles (suspected hybrid with Cuban crocodiles). However, we could not identify genetic admixture and introgression of Cuban crocodiles using microsatellite genotyping with our library, as there were no pure Cuban crocodiles in our experimental genetic stock [11] or in this study. Mitochondrial DNA analysis could allow us to infer the maternal lineage of Cuban crocodiles by comparing the DNA sequences with nucleotide data repositories such as GenBank, but this might not be enough for the cut-off determination of crocodile species. Thus, collaboration with crocodile research groups, governments, and NGOs such as the Crocodile Specialist Group (CSG) will be necessary for documenting and monitoring the introgression of several crocodile species in Southeast Asia.

Identifying purebred individuals from captive populations is a significant challenge for restocking and reintroduction efforts. The CSG has proposed various recommendations to support the conservation of Siamese crocodiles, including legislative and regulatory measures, compliance with CITES obligations, appropriate management of captive populations, conducting surveys and conservation initiatives, controlling illegal trade, promoting regional conservation initiatives, and exploring restocking options. These recommendations are aimed at supporting the current efforts of Thai national agencies to conserve the species, including the Department of Fisheries under the Ministry of Agriculture and Cooperatives and the Department of National Parks, Wildlife and Plant Conservation (DNP) under the Ministry of Natural Resources and Environment. To restore, protect, and create habitats for the Siamese crocodile, initiatives for public–private partnerships and sustainable financing will be launched. Prioritizing the management of key threats and conducting large-scale assessments of Siamese crocodile conservation status will be crucial for addressing challenges related to populations, protected areas, and other conservation initiatives.

5. Conclusions

These results indicate that a P.O. variation of 2–4 is within the species-level variation of Siamese crocodiles. In short, this is a phenotypic variation and not the result of hybridization with saltwater crocodiles. This baseline information on the association between genetic status and phenotypic variation of Siamese crocodiles in captive populations in Thailand is important for future conservation. Large differences in gene pools were observed in Siamese crocodiles, suggesting different historical origins of Siamese crocodiles in the wild population before massive capture and collection. Recently, a call was made to redefine the role of admixture in species conservation. It emphasized that crocodiles that have undergone gene flow and introgression during their evolutionary history or have been impacted by anthropogenic issues require protection measures. Ultimately, hybridization presents a management problem for Siamese crocodiles and complicates species identification based on morphology alone [65]. Adequate protocols to identify introgression and hybridization are urgently needed. Here, the genetic approach we followed proved that combining information from genetic and phenotypic approaches yielded more robust results. Accurate data on captive populations is critical for ensuring the long-term survival of the species through reintroduction programs and in situ/ex situ management, which helps to maintain sustainable genetic diversity.

Supplementary Materials: The following supporting information can be downloaded at: https://www.mdpi.com/article/10.3390/biology12040535/s1, Figure S1. Population structure of 136 Siamese crocodiles (*Crocodylus siamensis*, Schneider, 1801) [1] and 29 saltwater crocodiles (*C. porosus*, Schneider, 1801) [13], (a) Plot of Evanno's ΔK and (b) plot of ln $P(K)$; Figure S2. Phylogenetic relationships among mitochondrial DNA D-loop region sequences were inferred using Bayesian inference analysis. Support values at each node denote the Bayesian posterior probability. Table S1 provides detailed information on the sampled individuals; Figure S3. Representation post-occipital scutes (P.O.) for hybrids between Siamese and saltwater crocodiles (*Crocodylus siamensis* × *Crocodylus porosus*). (a) CB11 (b) CB22 (c) CB27 (d) CP01 (e) CP05 (f) CP06 (g) CP11 (h) CP12 (i) CP13 (j) CP15 (k) CP20 (l) CP23 (m) CP25 (n) CP27 (o) CP28 (p) CP29 (q) CP30; Table S1. Specimen populations of Siamese (*Crocodylus siamensis*, Schneider, 1801) [1] and saltwater crocodiles (*C. porosus*, Schneider, 1801) [13] All sequences were deposited in the DNA Data Bank of Japan (DDBJ) and BLASTn (http://blast.ncbi.nlm.nih.gov/Blast.cgi, accessed on 8 February 2023) of sequence identity; Table S2. Microsatellite primers and sequences; Table S3. Pairwise differentiation of linkage disequilibrium of Siamese crocodiles (*Crocodylus siamensis*, Schneider, 1801) [1] at Chonburi (CB) based on 22 microsatellite loci. Numbers indicate *p*-values with 110 permutations; Table S4. Pairwise differentiation of linkage disequilibrium of Siamese crocodiles (*Crocodylus siamensis*, Schneider, 1801) [1] at Nakhon Ratchasima (NR) based on 22 microsatellite loci. Numbers indicate *p*-values with 110 permutations;

Table S5. Pairwise differentiation of linkage disequilibrium of Siamese crocodiles (*Crocodylus siamensis*, Schneider, 1801) [1] at Chainat (CN) based on 22 microsatellite loci. Numbers indicate *p*-values with 110 permutations; Table S6. Pairwise differentiation of linkage disequilibrium of Siamese crocodiles (*Crocodylus siamensis*, Schneider, 1801) [1] at Nakhon Pathom (NP) based on 22 microsatellite loci. Numbers indicate *p*-values with 110 permutations; Table S7. Pairwise differentiation of linkage disequilibrium of saltwater crocodiles (*Crocodylus porosus*, Schneider, 1801) [13] at Nakhon Ratchasima (CP) based on 22 microsatellite loci. Numbers indicate *p*-values with 110 permutations; Table S8. Genetic diversity of 136 Siamese crocodiles (*Crocodylus siamensis*, Schneider, 1801) [1] and 29 saltwater crocodiles (*C. porosus*, Schneider, 1801) [13] based on 22 microsatellite loci. Table S1 provides detailed information on the sampled individuals; Table S9. Pairwise genetic relatedness (*r*) for all 30 Siamese crocodiles (*Crocodylus siamensis*, Schneider, 1801) [1] in Chonburi (CB) Table S1 provides detailed information on the sampled individuals; Table S10. Pairwise genetic relatedness (*r*) for all 30 Siamese crocodiles (*Crocodylus siamensis*, Schneider, 1801) [1] in Nakhon Ratchasima (NR) Table S1 provides detailed information on the sampled individuals; Table S11. Pairwise genetic relatedness (*r*) for all 34 Siamese crocodiles (*Crocodylus siamensis*, Schneider, 1801) [1] in Chainat (CN). Table S1 provides detailed information on the sampled individuals; Table S12. Pairwise genetic relatedness (*r*) for all 42 Siamese crocodiles (*Crocodylus siamensis*, Schneider, 1801) [1] in Nakhon Pathom (NP). Table S1 provides detailed information on the sampled individuals; Table S13. Pairwise genetic relatedness (*r*) for all 29 saltwater crocodiles (*Crocodylus porosus*, Schneider, 1801) [3] in Nakhon Ratchasima (CP). Table S1 provides detailed information on the sampled individuals; Table S14. Distributions of *r* values and F_{IS} values for Siamese (*Crocodylus siamensis*, Schneider, 1801) [1] and saltwater crocodiles (*Crocodylus porosus*, Schneider, 1801) [1]; Table S15. Pairwise inbreeding coefficients (F_{IS}) for all 30 Siamese crocodiles (*Crocodylus siamensis*, Schneider, 1801) [1] in Chonburi (CB). Detailed information on all Siamese crocodile individuals is presented in Table S1; Table S16. Pairwise inbreeding coefficients (F_{IS}) for all 30 Siamese crocodiles (*Crocodylus siamensis*, Schneider, 1801) [1] in Nakhon Ratchasima (NR). Detailed information on all Siamese crocodile individuals is presented in Table S1; Table S17 Pairwise inbreeding coefficients (F_{IS}) for all 34 Siamese crocodiles (*Crocodylus siamensis*, Schneider, 1801) [1] in Chainat (CN). Detailed information on all Siamese crocodile individuals is presented in Table S1; Table S18. Pairwise inbreeding coefficients (F_{IS}) for all 42 Siamese crocodiles (*Crocodylus siamensis*, Schneider, 1801) [1] in Nakhon Pathom (NP). Detailed information on all Siamese crocodile individuals is presented in Table S1; Table S19. Pairwise inbreeding coefficients (F_{IS}) for all 29 saltwater crocodiles (*Crocodylus porosus*, Schneider, 1801) [1] in Nakhon Ratchasima (CP). Detailed information on all saltwater crocodile individuals is presented in Table S1; Table S20. Pairwise genetic differentiation (F_{ST}), pairwise F_{ST}^{ENA} values with ENA correction for null alleles and R_{ST} values using FSTAT version 2.9.3 [66] and of Siamese (*Crocodylus siamensis*, Schneider, 1801) [1] and saltwater crocodiles (*Crocodylus porosus*, Schneider, 1801) [13] between captive-bred individuals based on 22 microsatellite loci. The numbers indicate *p* values, with 110 permutations. Detailed information on all Siamese and saltwater crocodile individuals is presented in Table S1; Table S21. Analysis of molecular variance (AMOVA) results for Siamese crocodile (*Crocodylus siamensis*, Schneider, 1801) [1] and crocodiles (*Crocodylus porosus*, Schneider, 1801) [1] based on 22 microsatellite loci using Arlequin version 3.5.2.2 [67]. Detailed information on all Siamese and saltwater crocodiles is presented in Table S1; Table S22. Pairwise population Nei's genetic distance (*D*) values using GenAlEx version 6.5 [28] of 166 Siamese (*Crocodylus siamensis*, Schneider, 1801) [1] and saltwater crocodiles (*Crocodylus porosus*, Schneider, 1801) [13] based on 22 microsatellite loci. Detailed information on all Siamese and saltwater crocodiles is presented in Table S1; Supplementary Data S1 Materials and Methods [66–85].

Author Contributions: Conceptualization, N.A., Y.T. and K.S.; funding acquisition, K.S.; formal analysis, N.A., W.W., P.W., N.M. and K.S.; investigation, N.A., W.W., P.W., T.P., W.S., T.T., A.L., S.F.A., N.M., K.H., P.D., Y.T. and K.S.; methodology, N.A., W.W., P.W., T.P., W.S., T.T. and K.S.; project administration, K.S.; resources, N.A., W.W., P.W., W.S. and Y.T.; software, N.A., W.W., P.W., T.P., W.S. and N.M.; supervision, P.D. and K.S.; validation, N.A., W.S. and K.S.; visualization, N.A. and K.S.; writing—original draft, N.A. and K.S.; writing—review and editing, N.A., W.W., P.W., T.P., W.S., T.T., A.L., S.F.A., N.M., K.H., P.D., Y.T. and K.S. All authors have read and agreed to the published version of the manuscript.

Funding: This research was financially supported in part by the High-Quality Research Graduate Development Cooperation Project between Kasetsart University and the National Science and Technology Development Agency (NSTDA) (6417400247) awarded to T.P. and K.S.; the Thailand

Science Research and Innovation through the Kasetsart University Reinventing University Program 2021 (3/2564) awarded to T.P., A.L. and K.S.; the Higher Education for Industry Consortium (Hi-FI) (6414400777) awarded to N.A.; the e-ASIA Joint Research Program (no. P1851131) awarded to W.S. and K.S.; a grant from the National Science and Technology Development Agency (NSTDA) (NSTDA P-19-52238 and JRA-CO-2564-14003-TH) awarded to W.S. and K.S.; a grant from Betagro Group (no. 6501.0901.1/68) awarded to K.S.; a grant from Kasetsart University Research and Development Institute (FF(KU)25.64) awarded to W.S., S.F.A. and K.S.; and support from the Office of the Ministry of Higher Education, Science, Research, and Innovation. International SciKU Branding (ISB), Faculty of Science, Kasetsart University awarded funds to K.S.

Institutional Review Board Statement: The animal care and experimental procedures were conducted according to the Regulations on Animal Experiments at Kasetsart University and were approved by the Animal Experiment Committee under the code ACKU64-SCI-011.

Informed Consent Statement: Not applicable.

Data Availability Statement: All sequences were deposited in the DNA Data Bank of Japan (DDBJ). The online version contains Supplementary Materials available at https://datadryad.org/stash/share/s4zREYQ1AUUXTsaIpk0r1HSdkYljvu8yvJOVR143K7Y (accessed on 18 February 2018).

Acknowledgments: We would like to thank the Thai Crocodile Farm Association for assistance with specimen collection and the provision of useful information. This project was conducted under a Memorandum of Agreement between the Faculty of Science at Kasetsart University and the Thai Crocodile Farm Association. The authors are also grateful to Ekaphan Kraichak (Department of Botany, Faculty of Science, Kasetsart University, Thailand), Chalitra Saysuk, and Rasoarahona Rivo Niaina Hajatiana Kevin Ryan (AGB Research Unit, Kasetsart University, Thailand) for helpful discussion. We thank the Center for Agricultural Biotechnology (CAB) at the Kasetsart University Kamphaeng Saen Campus and the NSTDA Supercomputer Center (ThaiSC) for support with server 561 analysis services. We also thank the Faculty of Science for providing research facilities.

Conflicts of Interest: The authors declare no conflict of interest. No funding source was involved in the study design, collection, analysis, and interpretation of the data, nor in writing the report or the decision to submit the article for publication.

References

1. Schneider, J.G. *Crocodylus siamensis*. 1801. Available online: https://www.gbif.org/species/185108439 (accessed on 16 January 2023).
2. Smith, M.A. *The Fauna of British India including Ceylon and Burma; Reptilia and Amphibia V. 1. Loricata, Testudines*; Taylor and Francis: London, UK, 1931.
3. Daltry, J.C.; Chheang, D.; Em, P.; Poeung, M.; Sam, H.; Tan, T.; Simpson, B.K. Status of the Siamese crocodile in the central cardamom mountains, Cambodia. In *Fauna & Flora International: Cambodia Programme and Department of Forestry and Wildlife: Phnom Penh*; Fauna & Flora International: Phnom Penh, Cambodia, 2003.
4. Platt, S.G.; Lynam, A.J.; Temsiripong, Y.; Kampanakngarn, M. Occurrence of the Siamese crocodile (*Crocodylus siamensis*) in Kaeng Krachan National Park, Thailand. *Nat. Hist. Bull. Siam Soc.* **2002**, *50*, 7–14.
5. Platt, S.G.; Rainwater, T.R.; Finger, A.G.; Thorbjarnarson, J.B.; Anderson, T.A.; McMurry, S.T. Food habits, ontogenetic dietary partitioning and observations of foraging behaviour of Morelet's crocodile (*Crocodylus moreletii*) in northern Belize. *Herpetol. J.* **2006**, *16*, 281–290.
6. Simpson, B.K.; Bezuijen, M.R. Siamese crocodile *Crocodylus siamensis*. In *Crocodiles Status Survey and Conservation Action Plan*, 2nd ed.; Manolis, S.C., Stevenson, C., Eds.; Crocodile Specialist Group: Darwin, NT, Australia, 2010; pp. 120–126.
7. Smith, M.A. Crocodylus siamensis. *J. Nat. Hist. Soc. Siam* **1919**, *3*, 217–222.
8. Ilhow, F.; Bonke, R.; Hartman, T.; Geissler, P.; Behler, N.; Rödder, D. Habitat suitability, coverage by protected areas and population connectivity for the Siamese crocodile *Crocodylus siamensis* Schnider, 1801. *Aquatic. Conserv. Mar. Fresh. Ecosyst.* **2015**, *25*, 544–554. [CrossRef]
9. Platt, S.G.; Rainwater, T.R.; Nichols, S. A recent population assessment of the American crocodile (*Crocodylus acutus*) in Turneffe Atoll, Belize. *Herpetol. Bull.* **2004**, *89*, 26–32.
10. Thorbjarnarson, J.B.; Messel, H.; King, F.W.; Ross, J.P. *Crocodiles: An Action Plan for their Conservation*; IUCN: Gland, Switzerland, 1992.
11. Lapbenjakul, S.; Thanapa, W.; Twilprawat, P.; Muangmai, N.; Khancanaketu, T.; Temsiripong, Y.; Unajak, S.; Peyachoknagul, S.; Srikulnath, K. High genetic diversity and demographic history of captive Siamese and saltwater crocodiles suggest the first step toward the establishment of a breeding and reintroduction program in Thailand. *PLoS ONE* **2017**, *12*, e0184526. [CrossRef]
12. Suttimeechaikul, N.; Pakdeekong, M.; Tankitjanukit, S.; Keawjantavee, P.; Temsiripong, Y.; Phuangcharoen, W.; Kittiwanich, J.; Aiamsaart, P. *Croccodile aquaculture*. Available online: https://online.fliphtml5.com/vqklc/ljrd/?fbclid=IwAR0ZDRISD47Gh8t3#p=1 (accessed on 16 January 2023).

13. Schneider. *Crocodylus porosus*. 1801. Available online: https://www.gbif.org/species/144104483 (accessed on 16 January 2023).
14. Suvanakorn, P.; Youngprapakorn, C. Crocodile farming in Thailand. In *Wildlife Management: Crocodiles and Alligators*; Webb, G.J.W., Manolis, S.C., Whitehead, P.J., Eds.; Surrey Beatty and Sons, Pty. Ltd.: Chipping Norton, NSW, Australia, 1987; pp. 341–343.
15. Srikulnath, K.; Thongpan, A.; Suputtitada, S.; Apisitwanich, S. New haplotype of the complete mitochondrial genome of *Crocodylus siamensis* and its species-specific DNA markers: Distinguishing *C. siamensis* from *C. porosus* in Thailand. *Mol. Biol. Rep.* **2012**, *39*, 4709–4717. [CrossRef]
16. Charette, R. *CITES Identification Guide: Crocodilians*; Environment Canada: Ottawa, ON, Canada, 1995.
17. Ross, F.D.; Mayer, G.C. On the dorsal armor of the Crocodilia. In *Advances in Herpetology and Evolutionary Biology*; Museum of Comparative Zoology: Cambridge, MA, USA, 1983; pp. 305–331.
18. Dinets, V. Long-distance signaling in Crocodylia. *Copeia* **2013**, *2013*, 517–526. [CrossRef]
19. Supikamolseni, A.; Ngaoburanawit, N.; Sumontha, M.; Chanhome, L.; Suntrarachun, S.; Peyachoknagul, S.; Srikulnath, K. Molecular barcoding of venomous snakes and species-specific multiplex PCR assay to identify snake groups for which antivenom is available in Thailand. *Genet. Mol. Res.* **2015**, *14*, 13981–13997. [CrossRef]
20. Miles, L.G.; Isberg, S.R.; Glenn, T.C.; Lance, S.L.; Dalzell, P.; Thomson, P.C.; Moran, C. A genetic linkage map for the saltwater crocodile (*Crocodylus porosus*). *BMC Genom.* **2009**, *10*, 339. [CrossRef]
21. Miles, L.G.; Isberg, S.R.; Moran, C.; Hagen, C.; Glenn, T.C. 253 Novel polymorphic microsatellites for the saltwater crocodile (*Crocodylus porosus*). *Conserv. Genet.* **2009**, *10*, 963–980. [CrossRef]
22. Ariyaraphong, N.; Pansrikaew, T.; Jangtarwan, K.; Thintip, J.; Singchat, W.; Laopichienpong, N.; Pongsanarm, T.; Panthum, T.; Suntronpong, A.; Ahmad, S.F.; et al. Introduction of wild Chinese gorals into a captive population requires careful genetic breeding plan monitoring for successful long-term conservation. *Glob. Ecol. Conserv.* **2021**, *28*, e01675. [CrossRef]
23. Chailertrit, V.; Swatdipong, A.; Peyachoknagul, S.; Salaenoi, J.; Srikulnath, K. Isolation and characterization of novel microsatellite markers from Siamese fighting fish (*Betta splendens*, Osphronemidae, Anabantoidei) and their transferability to related species, *B. smaragdina* and *B. imbellis*. *Genet. Mol. Res.* **2014**, *13*, 7157–7162. [CrossRef]
24. Jangtarwan, K.; Koomgun, T.; Prasongmaneerut, T.; Thongchum, R.; Singchat, W.; Tawichasri, P.; Fukayama, T.; Sillapaprayoon, S.; Kraichak, E.; Muangmai, N.; et al. Take one step backward to move forward: Assessment of genetic diversity and population structure of captive Asian woolly-necked storks (*Ciconia episcopus*). *PLoS ONE* **2019**, *14*, e0223726. [CrossRef]
25. Jangtarwan, K.; Kamoongkram, P.; Subpayakom, N.; Sillapaprayoon, S.; Muangmai, N.; Kongphoomph, A.; Wongsodchuen, A.; Intapan, S.; Chamchumroon, W.; Safoowong, M.; et al. Predictive genetic plan for a captive population of the Chinese goral (*Naemorhedus griseus*) and prescriptive action for ex situ and in situ conservation management in Thailand. *PLoS ONE* **2020**, *15*, e0234064. [CrossRef]
26. Thintip, J.; Singchat, W.; Ahmad, S.F.; Ariyaraphong, N.; Muangmai, N.; Chamchumroon, W.; Pitiwong, K.; Suksavate, W.; Duangjai, S.; Duengkae, P.; et al. Reduced genetic variability in a captive-bred population of the endangered Hume's pheasant (*Syrmaticus humiae*, Hume 1881) revealed by microsatellite genotyping and D-loop sequencing. *PLoS ONE* **2021**, *16*, e0256573. [CrossRef]
27. Wongtienchai, P.; Lapbenjakul, S.; Jangtarwan, K.; Areesirisuk, P.; Mahaprom, R.; Subpayakom, N.; Singchat, W.; Sillapaprayoon, S.; Muangmai, N.; Songchan, R.; et al. Genetic management of a water monitor lizard (*Varanus salvator macromaculatus*) population at Bang Kachao Peninsula as a consequence of urbanization with Varanus Farm Kamphaeng Saen as the first captive research establishment. *J. Zool. Syst. Evol. Res.* **2021**, *59*, 484–497. [CrossRef]
28. Peakall, R.; Smouse, P.E. GenAlEx 6.5: Genetic analysis in Excel. Population genetic software for teaching and research—An update. *Bioinformatics* **2012**, *28*, 2537–2539. [CrossRef]
29. Do, C.; Waples, R.S.; Peel, D.; Macbeth, G.M.; Tillett, B.J.; Ovenden, J.R. NeEstimator v2: Re-implementation of software for the estimation of contemporary effective population size (N_e) from genetic data. *Mol. Ecol. Resour.* **2014**, *14*, 209–214. [CrossRef]
30. Wang, J. Coancestry: A program for simulating, estimating and analysing relatedness and inbreeding coefficients. *Mol. Ecol. Resour.* **2011**, *11*, 141–145. [CrossRef]
31. Temsiripong, Y.; Ratanakorn, P.; Kullavanijaya, B. Management of the Siamese crocodile in Thailand. In Proceedings of the 17th Working Meeting of the IUCN-SSC Crocodile Specialist Group, Darwin, NT, Australia, 24–29 May 2004; pp. 141–142.
32. Jelden, D.C.; Manolis, C.; Giam, H.; Thomson, J.; Lopez, A. *Crocodile Conservation and Management in Cambodia: A Review with Recommendations*; IUCN Crocodile Specialist Group: Darwin, NT, Australia, 2005.
33. Jelden, D.C.; Manolis, C.; Tsubouchi, T.; Nguyen Dao, N.V. *Crocodile Conservation and Farming in the Socialist Republic of Viet Nam: A Review with Recommendations*; Crocodile Specialist Group: Darwin, NT, Australia, 2008.
34. Manolis, C. 2nd Siamese crocodile meeting on husbandry and conservation and Siamese crocodile task force meeting (1–2 June 2017). *Crocodile Spec. Group Newsl.* **2017**, *36*, 5–9.
35. Caldwell, J. *World Trade in Crocodilian Skins 2008–2010*; UNEP-WCMC: Cambridge, CA, USA, 2010.
36. Getpech, Y. Crocodile conservation and captive breeding in Thailand. In Proceedings of the 1st Regional Species Meeting of the IUCN-SSC Crocodile Specialist Group, Bangkok, Thailand, 4–6 April 2011; pp. 27–28.
37. Caldwell, J. *World Trade in Crocodilian Skins 2018–2020*; UNEP-WCMC: Cambridge, CA, USA, 2022.
38. Chavananikul, V.; Wattanodorn, S.; Youngprapakorn, P. Karyotypes of 5 species of crocodile kept in Samutprakan Crocodile Farm and Zoo. In Proceedings of the 12th Working Meeting of the IUCN-SSC Crocodile Specialist Group, Pattaya, Thailand, 2–6 May 1994; pp. 58–62.

39. Thang, N.Q. The status of *Crocodylus rhombifer* in the Socialist Republic of Vietnam. In Proceedings of the 12th Working Meeting of the IUCN-SSC Crocodile Specialist Group, Pattaya, Thailand, 2–6 May 1994; pp. 141–142.

40. Fitzsimmons, N.N.; Buchan, J.C.; Lam, P.V.; Polet, G.; Hung, T.T.; Thang, N.Q.; Gratten, J. Identification of purebred *Crocodylus siamensis* for reintroduction in Vietnam. *J. Exp. Zool. A Ecol. Genet. Physiol.* **2002**, *294*, 373–381. [CrossRef]

41. Truyen, T. Country—Vietnam. In Proceedings of the 1st Regional Meeting of the IUCN-SSC Crocodile Specialist Group, Bangkok, Thailand, 4–6 April 2011; pp. 33–35.

42. Phothitay, C.; Phommachanh, B.; Bezuijen, M.R. Siamese crocodiles at Ban Kuen Zoo, Lao PDR. *Crocodile Spec. Group Newsl.* **2005**, *24*, 11–12.

43. Cox, J.H., Jr.; Phothitay, C. *Surveys of the Siamese Crocodile Crocodylus Siamensis in Savannakhet Province, Lao PDR, 6 May–4 June 2008*; OZ Minerals Ltd. & Wildlife Conservation Society: Vientiane, Laos, 2008.

44. Cohen, M.M.; Gans, C. The chromosomes of the order Crocodilia. *Cytogenet. Genome Res.* **1970**, *9*, 81–105. [CrossRef]

45. Chavananikul, V.; Suwattana, D.; Koykul, W.; Wattanodorn, S.; Sukkai, J. *A Research Report on Karyotypes of Freshwater Crocodiles (Crocodylus siamensis), Saltwater Crocodiles (Crocodylus porosus) and the Inter-Specific Hybrids by Conventional and Banding Techniques*; Chulalongkorn University: Bangkok, Thailand, 1998.

46. Doc, C. Convention on international trade in endangered species of wild fauna and flora. In Proceedings of the Nineteenth meeting of the Conference of the Parties, Panama City, Panama, 14–25 November 2022.

47. Wildlife Conservation Society. *Recommendations for CITES CoP19*; Wildlife Conservation Society: New York, NY, USA, 2022.

48. Kowasupat, C.; Panijpan, B.; Ruenwongsa, P.; Sriwattanarothai, N. *Betta mahachaiensis*, a new species of bubble-nesting fighting fish (Teleostei: Osphronemidae) from Samut Sakhon Province, Thailand. *Zootaxa* **2012**, *3522*, 49–60. [CrossRef]

49. Kowasupat, C.; Panijpan, B.; Ruenwongsa, P. *Betta siamorientalis*, a new species of bubble-nest building fighting fish (Teleostei: Osphronemidae) from eastern Thailand. *Vertebr. Zool.* **2012**, *62*, 387–397. [CrossRef]

50. Abdelkrim, J.; Robertson, B.C.; Stanton, J.A.L.; Gemmell, N.J. Fast, cost-effective development of species-specific microsatellite markers by genomic sequencing. *BioTechniques* **2009**, *46*, 185–192. [CrossRef]

51. Fischer, M.C.; Rellstab, C.; Leuzinger, M.; Roumet, M.; Gugerli, F.; Shimizu, K.K.; Holderegger, R.; Widmer, A. Estimating genomic diversity and population differentiation–an empirical comparison of microsatellite and SNP variation in *Arabidopsis halleri*. *BMC Genom.* **2017**, *18*, 69. [CrossRef]

52. Fola, A.A.; Kattenberg, E.; Razook, Z.; Lautu-Gumal, D.; Lee, S.; Mehra, S.; Bahlo, M.; Kazura, J.; Robinson, L.J.; Laman, M.; et al. SNP barcodes provide higher resolution than microsatellite markers to measure *Plasmodium vivax* population genetics. *Malar. J.* **2020**, *19*, 375. [CrossRef]

53. Egito, A.A.; Paiva, S.R.; Albuquerque, M.D.S.M.; Mariante, A.S.; Almeida, L.D.; Castro, S.R.; Grattapaglia, D. Microsatellite based genetic diversity and relationships among ten Creole and commercial cattle breeds raised in Brazil. *BMC Genet.* **2007**, *8*, 83. [CrossRef]

54. Cortes, O.; Cañon, J.; Gama, L.T. Applications of microsatellites and single nucleotide polymorphisms for the genetic characterization of cattle and small ruminants: An overview. *Ruminants* **2022**, *2*, 456–470. [CrossRef]

55. Department of Fisheries. Meeting to Discuss Action Guidelines for Crocodiles in Thailand. Available online: https://www4.fisheries.go.th/dof/news_local/131/138388 (accessed on 16 January 2023).

56. Kanwatanakid-Savini, C.; Pliosungnoen, M.; Pattanavibool, A.; Thorbjarnarson, J.B.; Limlikhitaksorn, C.; Platt, S.G. A survey to determine the conservation status of Siamese crocodiles in Kaeng Krachan National Park, Thailand. *Herpetol. Conserv. Biol.* **2012**, *7*, 157–168.

57. Platt, S.G. *Community-Based Crocodile Conservation in Lao PDR*; Wildlife Conservation Society: Bronx, NY, USA, 2012.

58. Starr, J.C.D.A. Development of a re-introduction and re-enforcement program for Siamese crocodiles in Cambodia. In *Global Re-Introduction Perspectives: Additional Case Studies from Around the Globe*; IUCN: Gland, Switzerland, 2010; p. 118.

59. Hata, A.; Nunome, M.; Suwanasopee, T.; Duengkae, P.; Chaiwatana, S.; Chamchumroon, W.; Suzuki, T.; Koonawootrittriron, S.; Matsuda, Y.; Srikulnath, K. Origin and evolutionary history of domestic chickens inferred from a large population study of Thai red junglefowl and indigenous chickens. *Sci. Rep.* **2021**, *11*, 2035. [CrossRef]

60. Singchat, W.; Chaiyes, A.; Wongloet, W.; Ariyaraphong, N.; Jaisamut, K.; Panthum, T.; Ahmed, S.F.; Chaleekarn, W.; Suksavate, W.; Inpota, M.; et al. Red junglefowl resource management guide: Bioresource reintroduction for sustainable food security in Thailand. *Sustainability* **2022**, *14*, 7895. [CrossRef]

61. Adams, R.H.; Blackmon, H.; Reyes-Velasco, J.; Schield, D.R.; Card, D.C.; Andrew, A.L.; Waynewood, N.; Castoe, T.A. Microsatellite landscape evolutionary dynamics across 450 million years of vertebrate genome evolution. *Genome* **2016**, *59*, 295–310. [CrossRef]

62. Wattanadilokchatkun, P.; Panthum, T.; Jaisamut, K.; Ahmad, S.F.; Dokkaew, S.; Muangmai, N.; Duengkae, P.; Singchat, W.; Srikulnath, K. Characterization of microsatellite distribution in Siamese fighting fish genome to promote conservation and genetic diversity. *Fishes* **2022**, *7*, 251. [CrossRef]

63. Srikulnath, K.; Thapana, W.; Muangmai, N. Role of chromosome changes in *Crocodylus* evolution and diversity. *Genom. Inform.* **2015**, *13*, 102–111. [CrossRef]

64. Kawagoshi, T.; Nishida, C.; Ota, H.; Kumazawa, Y.; Endo, H.; Matsuda, Y. Molecular structures of centromeric heterochromatin and karyotypic evolution in the Siamese crocodile (*Crocodylus siamensis*) (Crocodylidae, Crocodylia). *Chromosome Res.* **2008**, *16*, 1119–1132. [CrossRef]

65. Nguyen, T.T.; Ziegler, T.; Rauhaus, A.; Nguyen, T.Q.; Tran, D.T.A.; Wayakone, S.; Luu, V.Q.; Vences, M.; Le, M.D. Genetic screening of Siamese crocodiles (*Crocodylus siamensis*) in Laos and Vietnam: Identifying purebred individuals for conservation and release programs. *Crocodile Spec. Group Newsl.* **2018**, *37*, 8–14.
66. Goudet, J.F. FSTAT (version 1.2): A computer program to calculate F-statistics. *J. Hered.* **1995**, *86*, 485–486. [CrossRef]
67. Excoffier, L.; Lischer, H.E. Arlequin suite ver 3.5: A new series of programs to perform population genetics analyses under Linux and Windows. *Mol. Ecol. Resour.* **2010**, *10*, 564–567. [CrossRef] [PubMed]
68. Guo, S.W.; Thompson, E.A. Performing the exact test of Hardy-Weinberg proportion for multiple alleles. *Biometrics* **1992**, *48*, 361–372. [CrossRef] [PubMed]
69. Raymond, M.; Rousset, F. GENEPOP (version 1.2): Population genetics software for exact tests and ecumenicism. *J. Hered.* **1995**, *86*, 248–249. [CrossRef]
70. R Core Team. *R: A Language and Environment for Statistical Computing*; R Foundation for Statistical Computing: Vienna, Austria, 2022.
71. Welch, B.L. The generalization of student's' problem when several different population variances are involved. *Biometrika* **1947**, *34*, 28–35. [CrossRef]
72. Van, O.C.; Hutchinson, W.F.; Wills, D.P.; Shipley, P. MICRO-CHECKER: Software for identifying and correcting genotyping errors in microsatellite data. *Mol. Ecol. Notes.* **2004**, *4*, 535–538. [CrossRef]
73. Park, S.D.E. Trypanotolerance in West African Cattle and the Population Genetic Effects of Selection. Ph.D. Thesis, University of Dublin, Dublin, Ireland, 2001.
74. Præstgaard, J.T. Permutation and bootstrap Kolmogorov-Smirnov tests for the equality of two distributions. *Scand. J. Stat.* **1995**, *22*, 305–322.
75. Lynch, M.; Ritland, K. Estimation of pairwise relatedness with molecular markers. *Genetics* **1999**, *152*, 1753–1766. [CrossRef]
76. Chapuis, M.P.; Estoup, A. Microsatellite null alleles and estimation of population differentiation. *Mol. Biol. Evol.* **2007**, *24*, 621–631. [CrossRef]
77. Nei, M. Genetic distance between populations. *Am. Nat.* **1972**, *106*, 283–292. [CrossRef]
78. Pritchard, J.K.; Stephens, M.; Donnelly, P. Inference of population structure using multilocus genotype data. *Genetics* **2000**, *155*, 945–959. [CrossRef]
79. Earl, D.A. Structure harvester: A website and program for visualizing STRUCTURE output and implementing the Evanno method. *Conserv. Genet. Resour.* **2012**, *4*, 359–361. [CrossRef]
80. Tamura, K.; Stecher, G.; Kumar, S. MEGA11: Molecular evolutionary genetics analysis version 11. *Mol. Biol. Evol.* **2021**, *38*, 3022–3027. [CrossRef]
81. Rozas, J.; Ferrer-Mata, A.; Sánchez-DelBarrio, J.C.; Guirao-Rico, S.; Librado, P.; Ramos-Onsins, S.E.; Sanchez-Gracia, A. DnaSP6: DNA sequence polymorphism analysis of large data sets. *Mol. Biol. Evol.* **2017**, *34*, 3299–3302. [CrossRef]
82. Weir, B.; Cockerham, C. Estimating F-Statistics for the analysis of population-structure. *Evolution* **1984**, *38*, 1358–1370. [CrossRef]
83. Excoffier, L.; Smouse, P.E.; Quattro, J.M. Analysis of molecular variance inferred from metric distances among DNA haplotypes: Application to human mitochondrial DNA restriction data. *Genetics* **1992**, *131*, 479–491. [CrossRef]
84. Clement, M.; Snell, Q.; Walker, P.; Posada, D.; Crandall, K. TCS: Estimating gene genealogies. *Parallel Distrib. Process. Symp. Int. Proc.* **2002**, *2*, 184.
85. Huelsenbeck, J.P.; Ronquist, F. MRBAYES: Bayesian inference of phylogenetic trees. *Bioinformatics* **2001**, *17*, 754–755. [CrossRef]

Article

Molecular Characterization and Expression Pattern of *leptin* in Yellow Cheek Carp (*Elopichthys bambusa*) and Its Transcriptional Changes in Response to Fasting and Refeeding

Min Xie [1], Jinwei Gao [1], Hao Wu [1], Xiaofei Cheng [1], Zhou Zhang [1], Rui Song [1], Shaoming Li [1,*], Jie Zhou [2], Cheng Li [1,3] and Guoqing Zeng [1,*]

[1] Hunan Fisheries Science Institute, Changsha 410153, China; xieminhaha@126.com (M.X.); gaojinwei163@163.com (J.G.); wh17380133463@163.com (H.W.); chengxiaofei19@126.com (X.C.); zz19961022@hotmail.com (Z.Z.); ryain1983@163.com (R.S.); licheng1969001@sina.com (C.L.)

[2] College of Animal Science and Technology, Hunan Agricultural University, Changsha 410125, China; 17674042583@163.com

[3] Hunan Aquatic Foundation Seed Farm, Changsha 410153, China

* Correspondence: lishaoming1977@126.com (S.L.); zengguoqing001@163.com (G.Z.)

Simple Summary: This article is mainly about that molecular characterization and expression pattern of *leptin* in yellow cheek carp (*Elopichthys bambusa*) and its transcriptional changes in response to fasting and refeeding. In this study, the authors used PCR to clone the CDS of *leptin* in yellow cheek carp, and analyzed the sequence differences of the gene with other species, constructed the phylogenetic tree, used real-time PCR for analyzing the expression of *leptin* in different tissues, including the expression of *leptin* in the brain and liver after fasting–refeeding of yellow cheek carp. This paper found that the full-length cDNA sequence of *Eblep* was 1140 bp and the length of the open reading frame (ORF), which can encode a protein of 174 amino acids, was 525 bp. The Eblep mRNA transcript was detected in all tested tissues, with the highest expression in the liver and lowest expression in the spleen. It was found that the change in the mRNA expression of EbLep may be an adaptive strategy for different energy levels by studying the expression of EbLep mRNA in the brain and liver under fasting and refeeding.

Abstract: *Leptin*, a secretory protein encoded by obese genes, plays an important role in regulating feeding and energy metabolism in fish. To study the structure and function of the *Leptin* gene in yellow cheek carp (*Elopichthys bambusa*), the full-length cDNA sequence of *leptin* was cloned, named *EbLep*. The full-length cDNA of *Eblep* was 1140 bp, and the length of the open reading frame (ORF), which can encode a protein of 174 amino acids, was 525 bp. The signal peptide was predicted to contain 33 amino acids. Sequence alignment showed that the amino acid sequence of *Leptin* was conserved in cyprinid fish. Despite large differences between primary structures, the tertiary structure of the EbLep protein was similar to that of the human protein and had four α-helices. The *EbLep* mRNA transcript was detected in all tested tissues, with the highest expression in the liver and lowest expression in the spleen. In this study, short-term fasting significantly increased the mRNA expression of *EbLep* in the liver, which returned to a normal level after 6 days of refeeding and was significantly lower than the normal level after 28 days of refeeding. In the brain, the mRNA expression of *EbLep* significantly decreased during short-term fasting and significantly increased to a higher value than the control group after 1 h of refeeding. It then rapidly decreased to a lower value than the control group after 6 h of refeeding, returning to the normal level after 1 day of refeeding, and significantly decreasing to a lower value than the control group after 28 days of refeeding. To sum up, the change in the mRNA expression of *EbLep* in the brain and liver may be an adaptive strategy for different energy levels.

Keywords: yellow cheek carp; *leptin*; gene structure; tissue expression; fasting and refeeding

Citation: Xie, M.; Gao, J.; Wu, H.; Cheng, X.; Zhang, Z.; Song, R.; Li, S.; Zhou, J.; Li, C.; Zeng, G. Molecular Characterization and Expression Pattern of *leptin* in Yellow Cheek Carp (*Elopichthys bambusa*) and Its Transcriptional Changes in Response to Fasting and Refeeding. *Biology* **2022**, *11*, 758. https://doi.org/10.3390/biology12050758

Academic Editor: Nuria De Pedro

Received: 20 April 2023
Revised: 19 May 2023
Accepted: 19 May 2023
Published: 22 May 2023

1. Introduction

Leptin is a product of the obesity gene, which was identified by positional cloning technology [1]. *Leptin* is secreted by the adipose tissue in mammals and is considered an anorexia hormone [2–4]. The physiological role and regulatory mechanism of *leptin* have been extensively studied in many animals [5–9]. In mammals, *leptin* can reduce feeding and increase energy consumption through the regulation of feedback in the hypothalamic–pituitary axis [10]. In frogs (*Xenopus laevis*), *leptin* can influence limb growth and differentiation during early development [7]. In chickens, the expression of *leptin* and its receptor were detected in the brain and digestive tract [8], which suggests that *leptin* may be involved in brain and digestive tract-related functions. In *Anolis carolinensis*, *leptin* can ameliorate immunity [9]. In fish, *leptin* was first identified in *Takifugu rubripes* [11]. At present, *leptin* has been cloned from many species of fish, such as grass carp (*Ctenopharyngodon idella*), common carp (*Cyprinus carpio*), zebrafish (*Barchydanio rerio*), mandarin fish (*Siniperca chuatsi*), and orange-spotted grouper (*Epinephelus coioides*) [12–16]. Previous studies have shown that *leptin* plays a role in regulating food intake, glucose and lipid metabolism, reproduction, and immunity in fish [12,14,17–20]. Since teleost fishes exhibit a remarkable level of anatomical, ecological, behavioral, and genomic diversity, the structure and function of *leptin* may also vary considerably between fish species [21]. However, *leptin* from yellow cheek carp (*Elopichthys bambusa*) has not been cloned so far. Cloning of the *leptin* gene of yellow cheek carp is helpful to enrich the basic data of *leptin* in fishes, and provides reference material for further functional research on yellow cheek carp.

The liver is a central organ that controls metabolism in fish [22]. The liver, a main organ secreting *leptin* in fish, is first affected by different feeding states [18,23]. Several studies have shown that feeding status can affect the expression of *leptin* in fish [14,24] (Gambardella et al., 2012 and Gorissen et al., 2009). Gorissen et al., (2009) [14] confirmed that after fasting for a week, the mRNA level of *leptin-B* in the liver of zebrafish significantly decreased, while the mRNA level of *leptin-A* increased. In perch, the mRNA level of *leptin* in the liver significantly decreased after 3 weeks of starvation, and then increased after 3 weeks of feeding [25]. The regulation of food intake in fish was based on the integration of hypothalamic metabolic, endocrine, and circadian rhythm information [26], with mechanisms comparable but not identical [27] to those known in mammals [21]. The function of *leptin* in fish may be more complex than in mammals. There are few studies on the expression of *leptin* in the brain under different energy states in fish. More extensive and in-depth studies are needed to clarify the function and mechanism of *leptin* in the regulation of feeding status and energy metabolism.

Yellow cheek carp (*Elopichthys bambusa*), a member of the Cyprinidae and Leuciscinae, is a fierce carnivorous fish, which is mainly distributed in the plain area of the middle and lower Yangtze River. Known as the "freshwater tuna", it has fresh and tender meat and a beautiful taste, high protein content, and a low fat and cholesterol content [28]. In China, managers of traditional fisheries are controlling populations of fierce carnivorous fish to reduce the threat to other fish larvae [29]. As a result, the population of fierce carnivorous fish, including the yellow cheek carp, has dramatically declined in natural water bodies. With the adjustment of aquaculture industry structure and the implementation of the "ten-year fishing ban" policy in China, the price of yellow cheek carp has continued to rise in recent years. However, basic research on the yellow cheek carp is particularly scarce. In the present study, the *Leptin* gene of yellow cheek carp has been cloned and its distribution patterns have been identified by analyzing the tissue expression of *EbLep*. Additionally, the changes in *Eblep* mRNA expression in the liver and brain during short-term fasting and refeeding has been studied. This study aimed to enrich the research conducted on the yellow cheek carp thus far and provide information for studying the function of *leptin* and its role in food intake and energy metabolism regulation.

2. Materials and Methods

2.1. Animals and Samples

The yellow cheek carp were provided by the Hunan Fisheries Research Institute and were obtained from the same parent group. A total of 390 healthy and neatly sized *Elopichthys bambusa*, with an average body weight of (221.36 ± 6.75 g) and average body length of (33.09 ± 1.33 cm), were selected from the pond. They were randomly and equally assigned to 6 cement pools (10 m × 5 m × 1 m) to acclimatize for 1 week. Commercial feed (crude protein ≥ 48%, crude fat ≥ 5.0%, lysine ≥ 2.8%, moisture ≤ 10%, ash ≤ 18%) was used twice a day (8:00 and 18:00) in all groups during the experimental period.

Six fishes were randomly collected from the cement pools and dissected after anesthesia with MS222. The liver, intestine, spleen, kidney, heart, gill, brain, head kidney, skin, and muscle were collected for cloning and the tissue expression analysis of *leptin*. Samples were quickly frozen in liquid nitrogen and then stored at −80 °C.

In order to study the changes in the expression of *leptin* in the liver and brain under fasting and refeeding conditions, a control group and a treatment group were set up. There were three replicates in each group. The control group was fed continuously for 36 days, and the treatment group was fed for 28 days after 8 days of fasting. The control group and the treatment group were fed twice a day (8:00 and 18:00). Six *Elopichthys bambusa* were randomly collected from each group after fasting for 3 d (F3) and 8 d (F8), and were refed for 1 h (F8R1h), 6 h (F8R6h), 1 d (F8R1d), 6 d (F8R6d), and 28 d (F8R28d), respectively. Additionally, the liver and brain were sampled after anesthesia with MS-222. Samples were quickly frozen in liquid nitrogen and stored at −80 °C. During the experiment, the water temperature was (23.5 ± 3.4) °C, the pH was 6.7–7.2, and the dissolved oxygen was 5.7–6.5 mg/L.

2.2. Cloning of EbLep Gene

Liver RNA was extracted using Trizol, and first-strand cDNA was synthesized by reverse transcription using the RevertAid First Strand cDNA Synthesis Kit (Fermentas, Waltham, MA, USA). The quantity and quality of RNA were detected by Eppendorf (Hamburg, Germany) BioPhotometer Plus spectrophotometry. The ratio of absorbance at 260 and 280 nm (A260/A280) for samples was ranged from 1.8 to 2.0. According to the *leptin* sequence of grass carp (GenBank accession numbers FJ373293.1), the degenerate primer LEP-1 (Table 1) was designed by clustal X alignment software, version 2.1. The 50 µL reaction system contained PCR-grade water at 15.0 µL, 2X Ex taq buffer (Takara, Kusatsu, Japan) at 25.0 µL, dNTP mix (10 mM) at 1.0 µL, Ex taq (Takara) at 1.0 µL, cDNA first strand at 5.0 µL, primer F at 1.5 µL, and primer R at 1.5 µL. Reaction program: pre-denaturation at 94 °C for 2 min, denaturation at 94 °C for 30 s, annealing at 55 °C for 30 s, extension at 72 °C for 1 min, 35 cycles, and finally extension at 72 °C for 10 min. The PCR products were detected by 1% agarose gel electrophoresis, and the target fragments were purified using Gel Extraction Kit (OMEGA, Norcross, GA, USA). Then, the purified PCR products were cloned, and positive clones were screened for sequencing.

Table 1. Primer sequences for PCR.

Purpose	Name	Sequence (5′-3′)	Annealing Temperature	E-Values
Core Fragment Sequence Acquisition	LEP-1	F: ATGTATTYTCCAGYTCTTC R: GCATGAACTSTKTCAGTC	55 °C	
5′-RACE PCR	LEP-5′-RACE-1 LEP-5′-RACE-2 LEP-5′-RACE-3	ATGATGGTGTCTGCCT TGAGGCTATCCTGATG CCATCAATCAGACCAAGAAT		
3′-RACE PCR	LEP-3′-RACE-1 LEP-3′-RACE-2	CCAAGGGGCATGTGAGCCAGTTAC GCACACCAAAGGAGCCAGCCAATG		
Expression Examination	q-leptin	F: ACTGTCTCCAAAGATCCTCA R: AAAAGGGTGGACACATCATT	60 °C	100.3%
Reference Genes	β-actin	F: CCTGTATGCCTCTGGTCG R: CTCGGCTGTGGTGGTGAA	60 °C	98.6%

Note: Primer synthesis and sequencing in this experiment were performed by Sangon Biotech (Shanghai) Co., Ltd.

According to the core fragment sequence, the PCR primers for 5′-and 3′-RACE were designed (Table 1). To obtain the complete cDNA sequence of the *Eblep* gene, the 5′- and 3′-RACE PCRs were carried out using SMARTer RACE 5′/3′ Kit (Code No. 634858/59, Takara) by following the manufacturer's instructions. The target product was cut from gel for purification. The purified PCR product was ligated with pMD18T, and the positive clones were sequenced.

According to the intermediate fragment sequence 5′RACE and 3′RACE results, the full-length cDNA sequence of *leptin* gene was spliced out. The gene start and stop codon positions were predicted by NCBI alignment analysis.

2.3. Sequence Analysis and Phylogenetic Tree Construction

The software DNAMAN was used to translate the sequence of the *leptin* gene to obtain the amino acid sequence. The amino acid sequences of *leptin* of other fishes and vertebrates were obtained from NCBI, and the multiple sequence alignment was performed using the software DNAMAN. The GenBank accession numbers of the sequences used in the figure were as follows: *Ctenopharyngodon idella* LEP, ACI32423.1; *Culter alburnus* LEP, AHI13615.1; *Megalobrama amblycephala* LEP, XP_048024644.1; *Carassius gibelio* LEP-A, ULE27184.1; *Cyprinus carpio* LEP-A2, AGK24956.1; *Labeo rohita* LEP, KAI2653424.1; *Schizothorax prenanti* LEP, AIE45855.1; *Schizothorax richardsonii* LEP, UNO36740.1; *Danio rerio* LEP-A, CAP47064.1; *Tachysurus fulvidraco* LEP, AFO67938.1; *Tachysurus vachellii* LEP, UNW38763.1; *Pygocentrus nattereri* LEP-A, XP_037396125.1; *Myxocyprinus asiaticus* LEP-A, XP_051580568.1; *Oreochromis niloticus* LEP-A, AHL37667.1; *Silurus meridionalis* LEP-A, XP_046720980.1; *Acipenser ruthenus* LEP, RXM28554.1; *Acipenser dabryanus* LEP-A, QSZ40475.1; *Chanos chanos* LEP-A, QNG41929.1; *Alosa sapidissima* LEP-A, XP_041965310.1; *Salvelinus alpinus* LEP, BAH83535.1; *Oncorhynchus keta* LEP-A, XP_035653655.1; *Astyanax mexicanus* LEP-A, XP_022535129.1; *Colossoma macropomum* LEP-A, XP_036441558.1; *Xenopus laevis* LEP, XP_018108304.1; *Homo sapiens* LEP, AAH69452.1.

A protein phylogenetic tree was established based on amino acid sequences by the NJ method (neighbor joining) in MEGA 11.0. The number of bootstrap verifications was set to 1000. Signal peptides were predicted using signalP-5.0 (https://services.healthtech.dtu.dk/service.php?SignalP-5.0 accessed on 12 October 2022). The SWISS-MODEL (https://swissmodel.expasy.org/ accessed on 12 October 2022) modelling function was used to predict the secondary and tertiary structure of *leptin*.

2.4. Tissue Distribution and Fasting–Refeeding Expression

The primer *q-leptin* was designed according to the cloned *leptin* gene cDNA sequence, and *β-actin* was selected as the reference gene [30] (Table 1). The molality of *q-leptin* F and R are 5.26 and 4.83 nmol/OD, and the molality of *β-actin* F and R are 6.37 and 5.89 nmol/OD. cDNA used for *leptin* gene quantification was synthesized using the PrimeScript™ RT reagent Kit with the gDNA Eraser kit.

Real-time PCR assays were carried out to examine the distribution of the *leptin* gene in various tissues—including the liver, intestine, spleen, kidney, heart, gill, brain, head kidney, skin, and muscle, and the expressions of the brain and liver under different feeding conditions—with 12.5 µL reaction volume containing 6 µL of SYBR Premix Ex TaqTM II, 1 µL of cDNA obtained by reverse transcription, 4.5 µL of RNase-free water, 0.5 µL of upstream primer, and 0.5 µL of downstream primer. The reaction was pre-denatured at 95 °C for 3 min, denatured at 95 °C for 5 s, annealed at 60 °C, and extended for 20 s for 39 cycles. Three technical replicates were used for each set of reactions. All operations are carried out according to the kit instructions.

2.5. Data Analysis

The relative expression results were calculated using the formula $R = 2^{-\Delta\Delta Ct}$. All data were shown as mean $\pm$ standard deviation (SD) and were subjected to the one-way ANOVA using the software SPSS 18.0, with $p < 0.05$ as a significant difference. After the one-way ANOVA, the homogeneity-of-variance was conducted, and when the Sig value was greater than 0.05, the variance could be considered homogeneous.

3. Results

3.1. Characterization of EbLep

The full-length cDNA sequence of *leptin* gene of yellow cheek carp was obtained by RT-PCR and RACE (GenBank Accession No. MW794324). The full-length cDNA of *leptin* is 1140 bp, and the length of the open reading frame (ORF) is 525 bp, which can encode a protein of 174 amino acids. The signal peptide was predicted to contain 33 amino acids by NCBI BLAST (Figure 1).

```
      TATCATACAATCCTCACATCTCAAAGCAGGTATTTCACTATTGCCAAGAAAAAGACCCCCATATACAGGAACC
            10        20        30        40        50        60        70
1     ATGTATTCTCCAGTTCTTCTCTACACCTGCTTTTTGAGCATTCTTGGTCTGATTGATGGCCATTCAATCCCTGTT
1      M  Y  S  P  V  L  L  Y  T  C  F  L  S  I  L  G  L  I  D  G  H  S  I  P  V
                                Signal peptide
            85        95       105       115       125       135       145
76    CATCAGGATAGCCTCAAAAACTGCAAACTGCAGGCAGACACCATCATCCGCAGAATTAAGGAACACAATGAGAAG
26     H  Q  D  S  L  K  N  C  K  L  Q  A  D  T  I  I  R  R  I  K  E  H  N  E  K
           160       170       180       190       200       210       220
151   GTAAAGCTAAAACTGTCTCCAAAGATCCTCATTGGGGATTCAGAGCTTTACCCTGAGGTTCCTGCTGATAAACCT
51     V  K  L  K  L  S  P  K  I  L  I  G  D  S  E  L  Y  P  E  V  P  A  D  K  P
           235       245       255       265       275       285       295
226   ATCCAAGGGCTTGGGTCCATCATGGACACCCTAACTACCTTCCAGAAGGTTCTTCAAACACTGCCCAAGGGGCAT
76     I  Q  G  L  G  S  I  M  D  T  L  T  T  F  Q  K  V  L  Q  T  L  P  K  G  H
           310       320       330       340       350       360       370
301   GTGAGCCAGTTACACAATGATGTGTCCACCCTTTTGGACTCATTTAAAGATAGGATGACATTTATGCGTTGCACA
101    V  S  Q  L  H  N  D  V  S  T  L  L  D  S  F  K  D  R  M  T  F  M  R  C  T
           385       395       405       415       425       435       445
376   CCAAAGGAGCCAGCCAATGGGAAGTCACTGGACACTTTCATAGAGAAGAACGCCACCCACCACGTTACTTTTGGG
126    P  K  E  P  A  N  G  K  S  L  D  T  F  I  E  K  N  A  T  H  H  V  T  F  G
           460       470       480       490       500       510       520
451   TACATGGCTTTAGACAGACTGAAACAGTTCATGCAAAAGCTGATAGTTAGTCTTGACCAGTTGAAGATCTGCTAA
151    Y  M  A  L  D  R  L  K  Q  F  M  Q  K  L  I  V  S  L  D  Q  L  K  I  C  *
      TCTGCCACTATTATAGCATTATGGTATACTATTTATTATATTTATTTAAAACCCTGTATATTTATAGACCAAAGCAGCATTTTTTG
      GCACATTTTAATGTACAAACAGATTTAAAAATTATTCCCTGCTATTAAACAGAAAGCCATAAATGTTGCTGGCGACATCATGGTGC
      ACCAATCAGTCTGTCCTTACGATGTAAACAAACACTTGCACACATTGCCATGTCATACTTCATGCTGTTTGCAGGATGACATCAGC
      ACTTGCTGCGCTTGCTCACTGATGTCACTGTGTTTACCAGAAGAAGCGAGGTCAGACTCTGCTGGGAAGCTGGAATTATGTTCCCA
      TTTTACTTCTTGCTGCAACACCAGCAATGTAGGACACACACCAAAACATATATATTTTTTCCTACACATTGTAAATCTATGCACTT
      TAAATATCCTGTGATTGTAAATAGTTTTGTATTTTGTGTAATGATGCATTTTTCCATAAAAATAAAGCAATAAAATAAACGTATTG
      TAAAAAAAAAAAAAAAAAAAAAAAAAAA
```

Figure 1. Nucleotide sequence and deduced amino acid sequence of the *leptin* gene of Elopichthys bambusa. The sense strand is displayed from the 5′ to 3′ direction. Stop codons are marked with "*" in the nucleotide sequence. Signal peptides are marked in bold letters. The GenBank accession number is MW794324.

Sequence alignment shows that the amino acid sequence of *leptin* was conserved in cyprinid fish, such as yellow cheek carp (*Elopichthys bambusa*), grass carp (*Ctenopharyngodon idella*), *Megalobrama amblycephaloid*, common carp (*Cyprinus carpio*), and crucian carp (*Carassius gibelio*), but had a large difference from the sequences of non-cynic fishes and other vertebrates. However, the secondary structure of *leptin* proteins from various animals was highly conserved, with four α-helices (Helix A–D) (Figure 2).

The amino acid sequence conservation of *leptin* was low between the yellow cheek carp and human, with only 23.20% sequence identity. The amino acid sequence identity of *EbLep* and other fishes was Ctenopharyngodon idella (91.60%), *Culter alburnus* (91.60%), *Megalobrama amblycephala* (91.00%), *Carassius gibelio* (76.10%), *Cyprinus carpio* (76.80%), *Labeo rohita* (74.80%), *Schizothorax prenanti* (74.20%), *Schizothorax richardsonii* (74.20%), *Danio rerio* (60.00%), *Tachysurus fulvidraco* (40.00%), and *Tachysurus vachellii* (40.00%), respectively (Table 2).

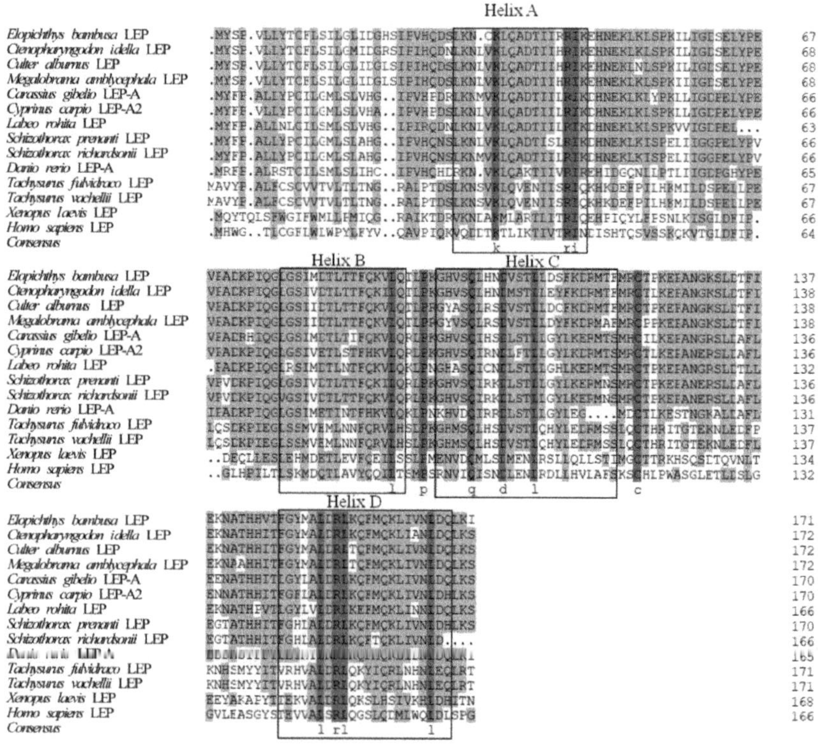

Figure 2. Amino acid sequence alignment among *EbLep* and *leptin* of other vertebrates. Omitted portions are indicated by dots, and shaded areas indicate residues that are 100% shared by all sequences. Four α-helices are framed.

The phylogenetic tree constructed based on the amino acid sequence of *leptin* is shown in Figure 3. All teleost fish LEPs are grouped into one clade, and two cartilaginous fishes (*Acipenser ruthenus* and *Acipenser dabryanus*) and other vertebrates (*Xenopus laevis* and *Homo sapiens*) LEPs are grouped into one clade. The closest to *EbLep* were *Culter alburnus* and *Megalobrama amblycephala*. Despite large differences between primary structures (Figure 2 and Table 2), the tertiary structure of EbLep protein was also predicted to be similar to that of humans (Figure 4).

Table 2. Amino acid sequence identities of EbLep compared with *leptin* of various vertebrates.

	Eb LEP	Ci LEP	Ca LEP	Ma LEP	Cg LEP-A	CcLEP-A2	Lr LEP	Sp LEP	Sr LEP	Dr LEP-A	Tf LEP	Tv LEP	Xl LEP	Hs LEP
Eb LEP	100.00%													
Ci LEP	91.60%	100.00%												
Ca LEP	91.60%	89.70%	100.00%											
Ma LEP	91.00%	89.70%	96.10%	100.00%										
Cg LEP-A	76.10%	75.50%	72.90%	73.50%	100.00%									
Cc LEP-A2	76.80%	76.80%	74.20%	74.20%	87.10%	100.00%								
Lr LEP	74.80%	73.40%	72.90%	71.60%	78.80%	77.40%	100.00%							
Sp LEP	74.20%	74.20%	72.30%	73.50%	83.20%	86.50%	75.50%	100.00%						
Sr LEP	74.20%	72.90%	72.30%	73.50%	83.20%	86.50%	75.50%	97.40%	100.00%					
Dr LEP-A	60.00%	59.40%	58.70%	59.40%	63.90%	65.20%	63.90%	61.90%	61.90%	100.00%				
Tf LEP	40.00%	40.00%	40.00%	40.60%	38.70%	41.30%	37.40%	40.60%	41.30%	36.80%	100.00%			
Tv LEP	40.00%	40.00%	40.00%	40.60%	39.40%	41.90%	38.10%	41.30%	41.90%	37.40%	99.40%	100.00%		
Xl LEP	27.70%	30.30%	28.40%	28.40%	28.40%	26.50%	27.10%	27.10%	26.50%	27.70%	29.00%	29.00%	100.00%	
Hs LEP	23.20%	22.60%	21.30%	21.30%	22.60%	25.20%	23.90%	25.20%	23.90%	23.20%	21.30%	21.30%	35.50%	100.00%

Note: The abbreviations of the species listed in this table are as follows: *Elopichthys bambusa* (Eb); *Ctenopharyngodon idella* (Ci); *Culter alburnus* (Ca); *Megalobrama amblycephala* (Ma); *Carassius gibelio* (Cg); *Cyprinus carpio* (Cc); *Labeo rohita* (Lr); *Schizothorax prenanti* (Sp); *Schizothorax richardsonii* (Sr); *Danio rerio* (Dr); *Tachysurus fulvidraco* (Tf); *Tachysurus vachellii* (Tv); *Xenopus laevis* (Xl); *Homo sapiens* (Hs).

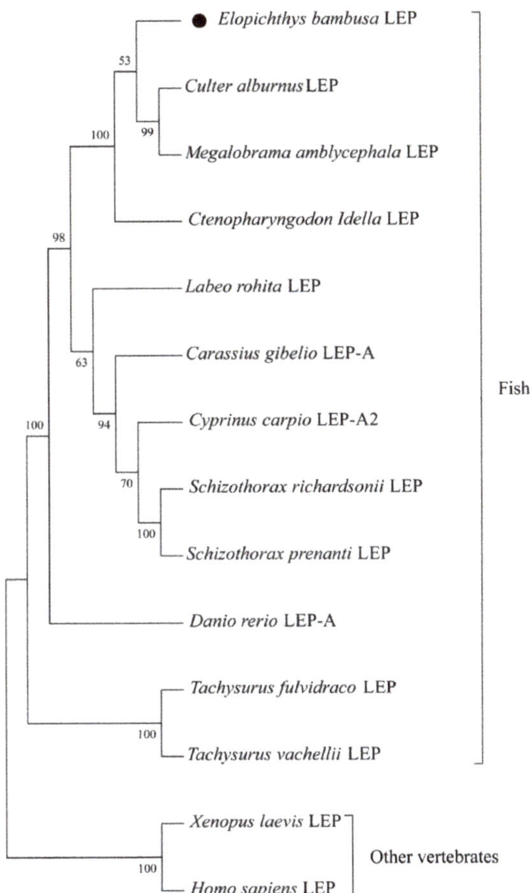

Figure 3. Phylogenetic tree based on amino acid sequence of *leptin*. The phylogenetic tree was established by the NJ method (neighbor-joining) in MEGA 11.0. The number of bootstraps verifications was set to 1000 times.

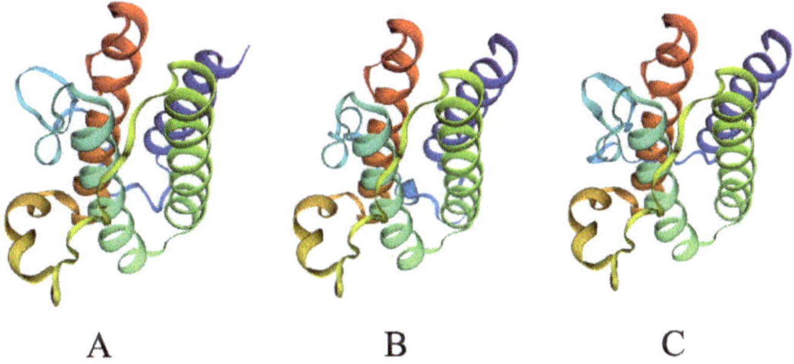

Figure 4. Tertiary structure model of *Leptin* predicted from amino acid sequences. (**A**) Yellow Cheek Carp, (**B**) Human, (**C**) Grass Carp.

3.2. Tissue Expression of EbLep

The mRNA tissue expression levels of *EbLep* were analyzed by real-time PCR. The results show that *EbLep* can be expressed in liver, intestine, spleen, kidney, heart, gill, brain, head kidney, skin, and muscle. The mRNA expression level in the liver tissue was the highest, which was more than eight times that of other tissues. This was followed by the heart, intestine, skin, brain, muscle, kidney, gill, head, kidney, and spleen (Figure 5).

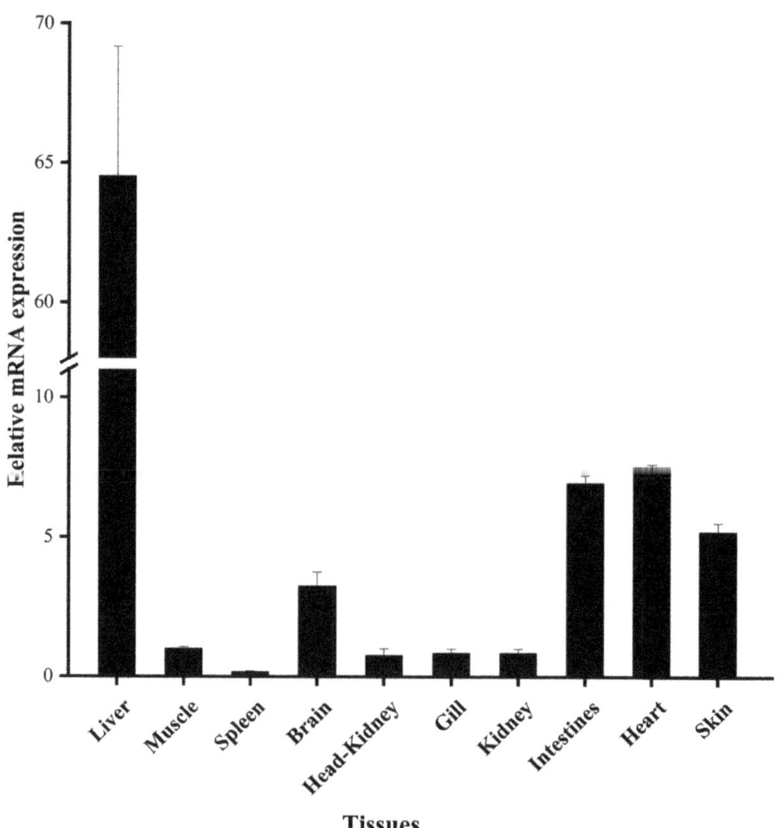

Figure 5. Relative mRNA expression of *EbLep* in various tissues. All values represent the mean ± S.D. (*n* = 6).

3.3. Expression of EbLep mRNA in the Liver and Brain under Fasting and Refeeding

The *EbLep* mRNA expression in the liver increased after fasting, as determined by real-time PCR (Figure 6). Compared with the control group, there was no significant difference after starvation for 3 days ($p > 0.05$), but there was a significant difference after starvation for 8 days ($p < 0.05$). After refeeding, the expression of *EbLep* mRNA in the liver showed a downward trend. One day after refeeding, it showed a significant decrease ($p < 0.05$), and 28 days after refeeding, it was significantly lower than the control group ($p < 0.05$) (Figure 6a).

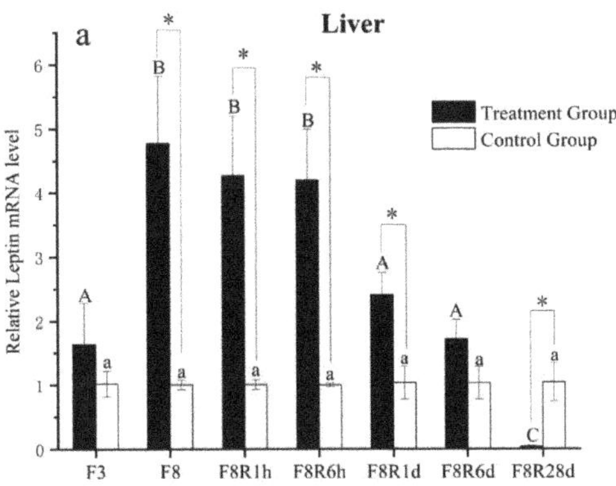

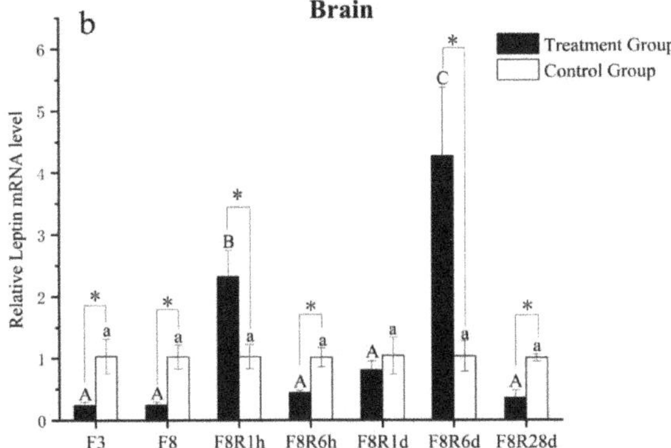

Figure 6. Relative Expression of *EbLep* mRNA in the liver (**a**) and brain (**b**) under fasting and refeeding. F3 and F8 indicate fasting for 3 and 8 days, respectively. F8R1h, F8R6h, F8R1d, F8R6d, and F8R28 indicate refeeding for 1 h, 6 h, 1 day, 6 days, and 28 days after 8 days of fasting, respectively. Significant differences among treatment groups (ANOVA, $p < 0.05$) are indicated by different capital letters (A, B, and C); no significant differences among control groups (ANOVA, $p > 0.05$) are indicated by different lowercase letters (a); and significant differences between control and treatment groups (ANOVA, $p < 0.05$) are indicated by *.

Fasting significantly decreased the expression of *EbLep* mRNA in the brain. After refeeding for 1 h, the expression of *EbLep* mRNA in the brain of the treatment group rapidly increased and became significantly higher than the control group ($p < 0.05$). After 6 h of refeeding, it gradually decreased and became significantly lower than that of the control group ($p < 0.05$). The expression of *EbLep* mRNA in the brain of the treatment group significantly increased and reached the highest level at 6 days of refeeding, followed by a significant decrease at 28 days ($p < 0.05$) (Figure 6b).

4. Discussion

Johnson et al., (2000) [31] first detected the presence of *leptin* in the blood, brain, heart, and liver of sunfish (*Lepomis macrochirus*), rainbow trout (*Oncorhynchus mykiss*), largemouth bass (*Pomonix annularis*), and microbleeker (*Ictalurus punctatus*). The *leptin* gene was later confirmed to exist in many other fish species [11,32,33]. *Leptin* in fish generally exists in two subtypes, named *lep-a* and *lep-b* [14,34]. This may be related to the fact that genome duplication occurred in the evolutionary process of fish [35]. The two isoforms encode two different products with low amino-acid identity (20–30%) [36]. In this study, the *Leptin* gene (*EbLep*) identified from the yellow cheek maybe was maybe lep-a by NCBI blast, but it cannot be determined whether other copies of this gene are present in the yellow cheek carp genome.

The yellow cheek carp shares only 23.20% sequence identity with human *leptin*. Similarly, goldfish (*Carassius auratus*) [37] and mandarin fish (*Siniperca chuatsi*) [15] have very low amino acid sequence homologies with mammalian *leptin*, both of which are less than 30%. This reflects the high evolutionary rate of the *leptin* gene. The amino acid sequences were quite different from those of other mammals [21]. The nucleic acid and amino acid sequences of *leptin* greatly differ from different fish species, but the tertiary structure of its protein is very conserved. They are highly similar in their predicted tertiary structure, when modelled based on the crystal structure of human *leptin* [11,38,39]. Our finding was in accordance with recent studies indicating that the tertiary structure of proteins is important for the main physiological function of the *leptin* system in fish. Kurokawa and Murashita (2009) [34] also had similar conclusions.

Leptin is expressed in various tissues; mammalian *leptin* is mainly expressed in adipose tissue [40]. Unlike mammals, the brain and heart are the main synthetic organs in *Xenopus* [7]. The expression level of *leptin* from Kermani sheep was highest in the adipose tissue and liver and lowest in the heart [20]. *Leptin* is slightly expressed in different tissues, such as the gut, adipose tissue, and brain in teleost fishes, but is mainly distributed in the liver [15,37]. However, some fishes have a high expression of *leptin* in other tissues. The expression of *leptin* in the kidneys, gills, intestines, and gonads of *Megalobrama amblycephala* was much higher than in the liver [41]. The *leptin* of Atlantic salmon (*Salmo salar*) had the highest expression in the brain and muscle [39]. In this paper, the tissue expression level of *EbLep* was highest in the liver, medium in the heart, intestine, skin, and brain, and lowest in the muscle, spleen, kidneys, gills, head, and kidney. The liver is the metabolic energy center of fish. Our findings match those observed in earlier studies [15,37], suggesting that *EbLep* may be involved in the regulation of energy balance. At the same time, *leptin* is commonly expressed in other tissues of yellow cheek carp. This suggests that the *EbLep* may be involved in many other physiological processes, and further studies are required.

To understand the physiological function of *leptin* in the yellow cheek carp, we compared the mRNA expression of *EbLep* in the liver and brain under different feeding states (short-term fasting and refeeding). We found that short-term fasting significantly increased the mRNA expression of *EbLep* in the liver, which returned to normal level after 6 days of refeeding and was significantly lower than the normal level after 28 days of refeeding. Similar findings were found in orange-spotted grouper [16], zebrafish (*leptin-A*) [14] and rainbow trout [42], and the opposite results were found in common carp [38], *Acrossocheilus fasciatus* [43], striped bass [25], and mandarin fish [15]. Through a comparison, it was found that the fasting time in the studies with similar results to this paper was more than 3 days, while the fasting time of the opposite results was less than 1 day. Therefore, we hypothesized that if the fasting time was less than 1 day, the mRNA expression of *leptin* in the liver would be reduced to promote appetite, and if the starvation time was more than 3 days, the mRNA expression of *leptin* in liver would be increased, in turn boosting glycolipid catabolism to provide energy for normal activities. The above conclusion needs to be verified in further research. It is worth noting that there were significant differences in the expression of *leptin* in the liver under different feeding states in the above studies. This suggests that *leptin* may play an important role in regulating energy status. The results of most fish studies show that specific *leptin* has an inhibitory effect on feeding [12,17,44]. It was speculated that *leptin* regulates the balance of feeding by regulating energy status in fish. The mechanism of the mRNA expression of *leptin* in the liver, which is adapted to fasting and refeeding, requires further research.

In fish, the endocrine signals in the hypothalamus region that influence the brain's regulation of food intake could be triggered by different nutritional and metabolic conditions [45,46]. Genes that control these endocrine signals (*leptin* is one of them) play an important role in appetite regulation [47]. Several studies have shown that *leptin* could inhibit food intake [12,17,44]. Murashita et al., (2008) [17] believed that *leptin* could regulate food intake by stimulating expression of the appetite inhibiting factor *proopiomelanocorein-A1/A2* (*POMC-A1/A2*) gene and reducing the expression level of the appetite stimulating factor *neuropeptide Y* (*NPY*) gene. At the same time, it has been reported that fasting could cause a downregulation of *leptin* gene expression in many aquatic animals and mammals [15,48,49]. In this study, the mRNA expression of *EbLep* in the brain was also significantly decreased during short-term fasting. Therefore, we thought that the downregulation of *leptin* caused by fasting may be related to the regulation of appetite in fish. Yuan et al., (2014) [50] had a similar view. In mammals, *leptin* could regulate feeding behavior by controlling the expression of some anorexia genes in the brain (e.g., *Cart, Crh, Mc4r, POMC*) [51–53]. In fish, regulation of feeding homeostasis is based on the hypothalamus' integration of metabolic and endocrine information [26]. This suggests that *EbLep* might regulate hypothalamic neuropeptides to regulate appetite during the fasting state. In this paper, the mRNA expression of *EbLep* in the brain significantly increased to higher than that in the control group after 1 h refeeding, and rapidly decreased to lower than that in the control group after 6 h of refeeding. My previous study found that the intestinal contents were less than 30% after 6 h of feeding [54]. This suggests that the *EbLep* expression in the brain may be related to the amount of food in the gut. In addition, the expression of *EbLep* in the brain returned to a normal level after 1 day of refeeding, and significantly decreased to a lower value than the control group after 28 days of refeeding. Several studies also thought that energy status could regulate food intake by regulating *leptin* expression in brain [55,56]. Therefore, the results of this study may be caused by the difference in energy status between 1 h, 6 h, 1 d, and 28 d after refeeding. In summary, *EbLep* expression in the brain may be related to the regulation of appetite. The change in the mRNA expression of *leptin* in the brain of yellow cheek carp may be an adaptive strategy for different energy levels.

5. Conclusions

In this study, we cloned the full-length cDNA sequence of *Eblep*, which was 1140 bp; the length of the open reading frame (ORF), which can encode a protein of 174 amino acids, was 525 bp. The signal peptide was predicted to contain 33 amino acids. The *EbLep* mRNA transcript was detected in all tested tissues with the highest expression in the liver and lowest expression in the spleen. By studying the expression of *EbLep* mRNA in the liver and brain under fasting and refeeding, we found that the change in the mRNA expression of EbLep may be an adaptive strategy for different energy levels.

Author Contributions: Conceptualization, M.X. and G.Z.; methodology, M.X., J.G., H.W. and X.C.; date collection, M.X., J.G., H.W., Z.Z. and R.S.; validation, J.Z. and C.L.; formal analysis, M.X., S.L. and G.Z.; investigation, M.X., J.G., H.W. and X.C.; resources, M.X., S.L. and G.Z.; writing-original draft preparation, M.X.; writing-review and editing, S.L. and G.Z.; visualization, M.X. and C.L. supervision, S.L. and G.Z. project administration, S.L. and G.Z.; funding acquisition, S.L. and G.Z. All authors have read and agreed to the published version of the manuscript.

Funding: This research was funded by the Changsha Natural Science Foundation for Changsha Science and Technology Bureau, grant number kq2202355, and the Earmarked Fund for China Agriculture Research System (CARS-45).

Institutional Review Board Statement: The animal research was approved by the Animal Care Committee of Hunan Fisheries Science Institute, Changsha, China (Approval Code: No. HFSI2021-06). All the experimental phases were strictly controlled.

Informed Consent Statement: Not applicable.

Data Availability Statement: The datasets generated and analyzed during the current study are available in the National Center for Biotechnology Information, USA National Library of Medicine repository, National Center for Biotechnology Information (nih.gov, accessed on 19 April 2023). The sequence of *Eblep* was submitted to the NCBI with GenBank Accession No. MW794324.

Conflicts of Interest: The authors declare no conflict of interest.

References

1. Zhang, Y.; Proenca, R.; Maffei, M.; Barone, M.; Leopold, L.; Friedman, J.M. Positional cloning of the mouse obese gene and its human homologue. *Nature* **1994**, *372*, 425–432. [CrossRef] [PubMed]
2. Ahima, R.S.; Osei, S.Y. Leptin signaling. *Physiol. Behav.* **2004**, *81*, 223–241. [CrossRef] [PubMed]
3. Schwartz, M.W.; Niswender, K.D. Adiposity signaling and biological defense against weight gain: Absence of protection or central hormone resistance. *J. Clin. Endocrinol. Metab.* **2004**, *89*, 5889–5897. [CrossRef] [PubMed]
4. Marty, N.; Dallaporta, M.; Thorens, B. Brain glucose sensing, counter-regulation, and energy homeostasis. *Physiology* **2007**, *22*, 241–251. [CrossRef]
5. Münzberg, H.; Heymsfield, S.B. New insights into the regulation of Leptin gene expression. *Cell Metab.* **2019**, *29*, 1013–1014. [CrossRef]
6. Procaccini, C.; La Rocca, C.; Carbone, F.; De Rosa, V.; Galgani, M.; Matarese, G. Leptin as immune mediator: Interaction between neuroendocrine and immune system. *Dev. Comp. Immunol.* **2017**, *66*, 120–129. [CrossRef]
7. Crespi, E.J.; Denver, R.J. *Leptin* (ob gene) of the South African clawed frog *Xenopus laevis*. *Proc. Natl. Acad. Sci. USA* **2006**, *103*, 10092–10097. [CrossRef]
8. Seroussi, E.; Knytl, M.; Pitel, F.; Elleder, D.; Krylov, V.; Leroux, S.; Morisson, M.; Yosefi, S.; Miyara, S.; Ganesan, S.; et al. Avian Expression Patterns and Genomic Mapping Implicate Leptin in Digestion and TNF in Immunity, Suggesting That Their Interacting Adipokine Role Has Been Acquired Only in Mammals. *Int. J. Mol. Sci.* **2019**, *20*, 4489. [CrossRef]
9. Wang, A.Z.; Husak, J.F.; Lovern, M. Leptin ameliorates the immunity, but not reproduction, trade-off with endurance in lizards. *J. Comp. Physiol. B* **2019**, *189*, 261–269. [CrossRef]
10. Sahu, A. Minireview: A hypothalamic role in energy balance with special emphasis on leptin. *Endocrinology* **2004**, *145*, 2613–2620. [CrossRef]
11. Kurokawa, T.; Uji, S.; Suzuki, T. Identification of cDNA coding for a homologue to mammalian leptin from pufferfish, Takifugu rubripes. *Peptides* **2005**, *26*, 745–750. [CrossRef]
12. Li, G.G.; Liang, X.F.; Xie, Q.; Li, G.; Yu, Y.; Lai, K. Gene structure, recombinant expression and functional characterization of grass carp leptin. *Gen. Comp. Endocrinol.* **2010**, *166*, 117–127. [CrossRef]
13. Tang, Y.; Yu, J.; Li, H.; Xu, P.; Li, J.; Ren, H. Molecular cloning, characterization and expression analysis of multiple leptin genes in Jian carp (*Cyprinus carpio* var. Jian). *Comp. Biochem. Physiol. Part B Biochem. Mol. Biol.* **2014**, *166*, 133–140. [CrossRef]
14. Gorissen, M.; Bernier, N.J.; Nabuurs, S.B.; Flik, G.; Huising, M.O. Two divergent leptin paralogues in zebrafish (*Danio rerio*) that originate early in teleostean evolution. *J. Endocrinol.* **2009**, *201*, 329–339. [CrossRef]
15. Yuan, X.; Li, A.; Liang, X.-F.; Huang, W.; Song, Y.; He, S.; Tao, Y. Leptin expression in mandarin fish *Siniperca chuatsi* (Basilewsky): Regulation by postprandial and short-term fasting treatment. *Comp. Biochem. Physiol. Part A Mol. Integr. Physiol.* **2016**, *194*, 8–18. [CrossRef]
16. Zhang, H.; Chen, H.; Zhang, Y.; Li, S.; Lu, D.; Zhang, H.; Meng, Z.; Liu, X.; Lin, H. Molecular cloning, characterization and expression profiles of multiple leptin genes and a leptin receptor gene in orange-spotted grouper (*Epinephelus coioides*). *Gen. Comp. Endocrinol.* **2013**, *181*, 295–305. [CrossRef]
17. Murashita, K.; Uji, S.; Yamamoto, T.; Rønnestad, I.; Kurokawa, T. Production of recombinant leptin and its effects on food intake in rainbow trout (*Oncorhynchus mykiss*). *Comp. Biochem. Physiol. Part B Biochem. Mol. Biol.* **2008**, *150*, 377–384. [CrossRef]
18. Yuan, X.-C.; Liang, X.-F.; Cai, W.-J.; Li, A.-X.; Huang, D.; He, S. Differential roles of two leptin gene paralogues on food intake and hepatic metabolism regulation in Mandarin fish. *Front. Endocrinol.* **2020**, *11*, 438. [CrossRef]
19. Tsakoumis, E.; Ahi, E.P.; Schmitz, M. Impaired leptin signaling causes subfertility in female zebrafish. *Mol. Cell. Endocrinol.* **2022**, *546*, 111595. [CrossRef]
20. Mohammadabadi, M.; Kord, M.; Nazari, M. Studying expression of leptin gene in different tissues of Kermani Sheep using Real Time PCR. *Agric. Biotechnol. J.* **2018**, *10*, 111–123.
21. Ayelén, M.B.; José, L.S. Leptin signalling in teleost fish with emphasis in food intake regulation. *Mol. Cell. Endocrinol.* **2021**, *526*, 111209. [CrossRef]
22. Martin, S.A.M.; Douglas, A.; Houlihan, D.F.; Secombes, C.J. Starvation alters the liver transcriptome of the innate immune response in Atlantic salmon (*Salmo salar*). *BMC Genom.* **2010**, *11*, 418. [CrossRef] [PubMed]
23. Power, D.M.; Melo, J.; Santos, C.R.A. The effect of food deprivation on the liver, tyroid hormones and transthyretin in sea bream. *J. Fish Biol.* **2000**, *56*, 374–387. [CrossRef]
24. Gambardella, C.; Gallus, L.; Amaroli, A.; Terova, G.; Masini, M.A.; Ferrando, S. Fasting and re-feeding impact on leptin and aquaglyceroporin 9 in the liver of European sea bass (*Dicentrarchus labrax*). *Aquaculture* **2012**, *354–355*, 1–6. [CrossRef]
25. Won, E.T.; Baltzegar, D.A.; Picha, M.E. Cloning and characterization of leptin in a Perciform fish, the striped bass (*Morone saxatilis*): Control of feeding and regulation by nutritional state. *Gen. Comp. Endocrinol.* **2012**, *178*, 98–107. [CrossRef]
26. Delgado, M.J.; Cerdá-Reverter, J.M.; Soengas, J.L. Hypothalamic integration of metabolic, endocrine, and circadian signals in fish: Involvement in the control of food intake. *Front. Neurosci.* **2017**, *11*, 354. [CrossRef]
27. Soengas, J.L.; Cerda-Reverter, J.M.; Delgado, M.J. Central regulation of food intake in fishes: An evolutionary perspective. *J. Mol. Endocrinol.* **2018**, *60*, R171–R199. [CrossRef]
28. Wan, S.L.; Wang, L.; Li, J.; Li, J.Z.; Liu, H.J.; Cai, C.M.; Lu, L.L.; Wen, Z.R. Flesh content and nutritive composition of yellow cheek carp (*Elopichthys bambusa*). *Freshw. Fish* **2008**, *273*, 27–29. [CrossRef]

29. Liang, Y.; Melack, J.M.; Wang, J. Primary production and fish yields in Chinese ponds and lakes. *Trans. Am. Fish. Soc.* **1981**, *110*, 346–350. [CrossRef]

30. Chen, Z.; Dong, S.; Dai, L.; Xie, M.; Fu, W.; Yuan, X.; Yuan, S.; Liu, J.; Peng, L.; Li, S.; et al. Effect of food domestication on the growth of *Elopichthys bambusa*. *Reprod. Breed.* **2021**, *1*, 157–166. [CrossRef]

31. Johnson, R.M.; Johnson, T.M.; Londraville, R.L. Evidence for leptin expression in fishes. *J. Exp. Zool.* **2000**, *286*, 718–724. [CrossRef]

32. Wen, Z.-Y.; Qin, C.-J.; Wang, J.; He, Y.; Li, H.-T.; Li, R.; Wang, X.-D. Molecular characterization of two leptin genes and their transcriptional changes in response to fasting and refeeding in Northern snakehead (*Channa argus*). *Gene* **2020**, *736*, 144420. [CrossRef]

33. Xu, Y.; Zhang, Y.; Wang, B.; Liu, X.; Liu, Q.; Song, X.; Shi, B.; Ren, K. Leptin and leptin receptor genes in tongue sole (*Cynoglossus semilaevis*): Molecular cloning, tissue distribution and differential regulation of these genes by sex steroids. *Comp. Biochem. Physiol. Part A Mol. Integr. Physiol.* **2018**, *224*, 11–22. [CrossRef]

34. Kurokawa, T.; Murashita, K. Genomic characterization of multiple leptin genes and a leptin receptor gene in the Japanese medaka, Oryzias latipes. *Gen. Comp. Endocrinol.* **2009**, *161*, 229–237. [CrossRef]

35. Glasauer, S.M.K.; Neuhauss, S.C.F. Whole-genome duplication in teleost fishes and its evolutionary consequences. *Mol. Genet. Genom.* **2014**, *289*, 1045–1060. [CrossRef]

36. Deck, C.A.; Honeycutt, J.L.; Cheung, E.; Reynolds, H.M.; Borski, R.J. Assessing the functional role of leptin in energy homeostasis and the stress response in vertebrates. *Front. Endocrinol.* **2017**, *8*, 63. [CrossRef]

37. Yan, A.; Li, J.; Liu, L.; Zhu, X.; Ren, C.; Hu, C.; Tang, D.; Chen, T. Tetraploid genes of leptin (leptin-AI,-AII,-BI and-BII) in goldfish: Molecular cloning, bioinformatics analysis, tissue distribution and differential regulation of transcript expression by glucocorticoids. *Aquac. Rep.* **2022**, *25*, 101191. [CrossRef]

38. Huising, M.O.; Geven EJ, W.; Kruiswijk, C.P.; Nabuurs, S.B.; Stolte, E.H.; Spanings FA, T.; Kemenade, L.V.; Flik, G. Increased leptin expression in common carp (*Cyprinus carpio*) after food intake but not after fasting or feeding to satiation. *Endocrinology* **2006**, *147*, 5786–5797. [CrossRef]

39. Rønnestad, I.; Nilsen, T.O.; Murashita, K.; Angotzi, A.R.; Gamst Moen, A.-G.; Stefansson, S.O.; Kurokawa, T. Leptin and leptin receptor genes in Atlantic salmon: Cloning, phylogeny, tissue distribution and expression correlated to long-term feeding status. *Gen. Comp. Endocrinol.* **2010**, *168*, 55–70. [CrossRef]

40. Bartha, T.; Sayed-Ahmed, A.; Rudas, P. Expression of leptin and its receptors in various tissues of ruminants. *Domest. Anim. Endocrinol.* **2005**, *29*, 193–202. [CrossRef]

41. Zhao, H.H.; Li, X.C.; Zeng CWang, W.M.; Dong, Z.J.; Gao, Z.X. Expression analysis of leptin genes in adults' tissue and during early development in Megalobrama amblycephala. *J. Huazhong Agric. Univ.* **2016**, *35*, 92–98. [CrossRef]

42. Kling, P.; Rønnestad, I.; Stefansson, S.O.; Murashita, K.; Kurokawa, T.; Björnsson, B.T.A. homologous salmonid leptin radioimmunoassay indicates elevated plasma leptin levels during fasting of rainbow trout. *Gen. Comp. Endocrinol.* **2009**, *62*, 307–312. [CrossRef] [PubMed]

43. Mu, F.S.; Miao, L.; Li, M.Y.; Hou, H.H.; Li, X.M.; Xu, Y.M. Cloning and expression of leptin gene in Acrossocheilus fasciatus during fasting and refeeding. *Oceanol. Limnol. Sin.* **2017**, *48*, 822–829.

44. Aguilar, A.J.; Conde-Sieira, M.; Polakof, S.; Míguez, J.M.; Soengas, J.L. Central leptin treatment modulates brain glucosensing function and peripheral energy metabolism of rainbow trout. *Peptides* **2010**, *31*, 1044–1054. [CrossRef]

45. Jeong, I.; Kim, E.; Kim, S.; Kim, H.; Lee, D.W.; Seong, J.Y.; Park, H.C. mRNA expression and metabolic regulation of npy and agrp1/2 in the zebrafish brain. *Neurosci. Lett.* **2018**, *668*, 73–79. [CrossRef]

46. Ahi, E.P.; Tsakoumis, E.; Brunel, M.; Schmitz, M. Transcriptional study reveals a potential leptin-dependent gene regulatory network in zebrafish brain. *Fish Physiol. Biochem.* **2021**, *47*, 1283–1298. [CrossRef]

47. Volkof, H. The neuroendocrine regulation of food intake in fish: A review of current knowledge. *Front. Neurosci.* **2016**, *10*, 540. [CrossRef]

48. Zhan, X.M.; Li, Y.L.; Wang, D.H. Effects of fasting and refeeding on body mass, thermogenesis, and serum leptin in Brandt's voles (*Lasiopodomys brandtii*). *J. Therm. Biol.* **2009**, *34*, 237–243. [CrossRef]

49. Chelikani, P.K.; Ambrose, J.D.; Keisler, D.H.; Kennelly, J.J. Effect of short-term fasting on plasma concentrations of leptin and other hormones and metabolites in dairy cattle. *Domest. Anim. Endocrinol.* **2004**, *26*, 33–48. [CrossRef]

50. Yuan, D.; Wang, T.; Zhou, C.; Lin, F.; Chen, H.; Wu, H.; Wei, R.; Xin, Z.; Li, Z. Leptin and cholecystokinin in *Schizothorax prenanti*: Molecular cloning, tissue expression, and mRNA expression responses to periprandial changes and fasting. *Gen. Comp. Endocrinol.* **2014**, *204*, 13–24. [CrossRef]

51. Thornton, J.E.; Cheung, C.C.; Clifton, D.K.; Steiner, R.A. Regulation of hypothalamic proopiomelanocortin mRNA by leptin in ob/ob mice. *Endocrinology* **1997**, *138*, 5063–5066. [CrossRef]

52. Ghamari-Langroudi, M.; Srisai, D.; Cone, R. Multinodal regulation of the arcuate/paraventricular nucleus circuit by leptin. *Proc. Natl. Acad. Sci. USA* **2011**, *108*, 355–360. [CrossRef]

53. Lee, S.J.; Verma, S.; Simonds, S.E.; Kirigiti, M.A.; Kievit, P.; Lindsley, S.R.; Loche, A.; Smith, M.S.; Cowley, M.A.; Grove, K.L. Leptin stimulates neuropeptide Y and cocaine amphetamine-regulated transcript co-expressing neuronal activity in the dorsomedial hypothalamus in diet-induced obese mice. *J. Neurosci.* **2013**, *33*, 15306–15317. [CrossRef]

54. Xie, M.; Xiang, J.G.; Guo, X.F. Measurement of empty-feeding rate of gut in yellow cheek carp (*Elopichthys bambusa*). *Sci. Fish Farming* **2015**, *10*, 50.

55. López, M.; Tovar, S.; Vazquez, M.J.; Nogueiras, R.; Senaris, R.; Dieguez, C. Sensing the fat: Fatty acid metabolism in the hypothalamus and the melanocortin system. *Peptides* **2005**, *26*, 1753–1758. [CrossRef]

56. Fuentes, E.N.; Kling, P.; Einarsdottir, I.E.; Alvarez, M.; Valdés, J.A.; Molina, A.; Björnsson, B.T. Plasma leptin and growth hormone levels in the fine flounder (*Paralichthys adspersus*) increase gradually during fasting and decline rapidly after refeeding. *Gen. Comp. Endocrinol.* **2012**, *177*, 120–127. [CrossRef]

biology

Article

Searching for a Home Port in a Polyvectic World: Molecular Analysis and Global Biogeography of the Marine Worm *Polydora hoplura* (Annelida: Spionidae)

Vasily I. Radashevsky [1,*], Vasily V. Malyar [1], Victoria V. Pankova [1], Jin-Woo Choi [2], Seungshic Yum [3] and James T. Carlton [4]

1. National Scientific Center of Marine Biology, Far Eastern Branch of the Russian Academy of Sciences, 17 Palchevsky Street, Vladivostok 690041, Russia
2. Blue Carbon Research Center, Seoul National University, Seoul 08826, Republic of Korea
3. Ecological Risk Research Division, Korea Institute of Ocean Science & Technology, Geoje 53201, Republic of Korea
4. Coastal and Ocean Studies Program, Williams College-Mystic Seaport, Mystic, CT 06355, USA
* Correspondence: radashevsky@mail.ru

Citation: Radashevsky, V.I.; Malyar, V.V.; Pankova, V.V.; Choi, J.-W.; Yum, S.; Carlton, J.T. Searching for a Home Port in a Polyvectic World: Molecular Analysis and Global Biogeography of the Marine Worm *Polydora hoplura* (Annelida: Spionidae). *Biology* **2023**, *12*, 780. https://doi.org/10.3390/biology12060780

Academic Editors: Cong Zeng and Deliang Li

Received: 6 March 2023
Revised: 22 May 2023
Accepted: 23 May 2023
Published: 27 May 2023

Simple Summary: Transoceanic shipping and global development of aquaculture are the main vectors for the introduction of marine organisms, as adults or their larvae, to new remote locations. Recent invasions by large species may be well known and documented, while older and smaller-bodied invasions are often hidden and more difficult to detect. In the present study, we investigated, using molecular methods, marine worms that bore into the shells of commercial molluscs on four continents. We have identified them as *Polydora hoplura*, a species originally described from Italy. The highest genetic diversity was detected in South African population. While high genetic diversity is often regarded as indicative of a species' natural range, our analysis of the worldwide discovery of *P. hoplura* calls into question its natural distribution in South Africa. The high genetic diversity of *P. hoplura* in this region may be the result of a complex dispersal history by ships and aquaculture. We tentatively propose the Northwest Pacific, or at the most the Indo–West Pacific, as its home region, not the Atlantic Ocean or the Eastern Pacific Ocean, and call for further exploration of this hypothesis.

Abstract: The spionid polychaete *Polydora hoplura* Claparède, 1868 is a shell borer widely occurring across the world and considered introduced in many areas. It was originally described in the Gulf of Naples, Italy. Adult diagnostic features are the palps with black bands, prostomium weakly incised anteriorly, caruncle extending to the end of chaetiger 3, short occipital antenna, and heavy sickle-shaped spines in the posterior notopodia. The Bayesian inference analysis of sequence data of four gene fragments (2369 bp in total) of the mitochondrial *16S* rDNA, nuclear *18S*, *28S* rDNA and *Histone 3* has shown that worms with these morphological features from the Mediterranean, northern Europe, Brazil, South Africa, Australia, Republic of Korea, Japan and California are genetically identical, form a well-supported clade, and can be considered conspecific. The genetic analysis of a *16S* dataset detected 15 haplotypes of this species, 10 of which occur only in South Africa. Despite the high genetic diversity of *P. hoplura* in South Africa, we tentatively propose the Northwest Pacific, or at the most the Indo–West Pacific, as its home region, not the Atlantic Ocean or the Eastern Pacific Ocean. The history of the discovery of *P. hoplura* around the world appears to be intimately linked to global shipping commencing in the mid-19th century, followed by the advent of the global movement of commercial shellfish (especially the Pacific oyster *Magallana gigas*) in the 20th century, interlaced with continued, complex dispersal by vessels and aquaculture. Given that *P. hoplura* has been detected in only a few of the 17 countries where Pacific oysters have been established, we predict that it may already be present in many more regions. As global connectivity through world trade continues to increase, it is likely that novel populations of *P. hoplura* will continue to emerge.

Keywords: polychaete; biological invasions; distribution; aquaculture; vessel biofouling; ballast; molecular systematics

1. Introduction

Polydoriases, "diseases" caused by shell-boring marine worms of the genus *Polydora* Bosc, 1802 and related genera (members of the tribe Polydorini Benham, 1896, collectively called polydorins), are often a plague in the cultivation of marine molluscs, especially oysters, abalone and clams [1–5]. Despite a long history of studies of these worms, the specific identification of the worms involved often remains problematic. Transported and introduced globally with aquaculture products [6–8] and other vectors, polydorins were often poorly described, misidentified or mistakenly redescribed as new species. Genetic characteristics are very helpful and sometimes crucial for the identification of shell-boring polydorins when morphological features are few and ambiguous [9]. In particular, specimens from the type locality of a species must be morphologically and genetically characterized in order to reliably anchor the name of a species to a precise lineage. While the type locality of a species is not necessarily within its native range [8,10], it often remains a fundamental aspect of understanding a species concept.

For many years most of the shell borers (also known as mud worms) around the globe were referred to as *Leucodore ciliatus* Johnston, 1838 (=*Polydora ciliata*), originally described from Berwick Bay (North Sea, Scotland). Radashevsky et al. [11], however, suggested that the systematic status and the mode of life of this species had been misinterpreted and that the global reports most likely include a complex of species.

As part of the global puzzle of mud worm identification, the spionid polychaete *Polydora hoplura* Claparède, 1868 is a shell borer widely occurring across the world and considered introduced in many areas. It was originally described from the Gulf of Naples (Tyrrhenian Sea, Italy) and later reported from the Mediterranean, Atlantic Europe, South Africa, New Zealand, Australia, Brazil and California (Figure 1; see reviews by Radashevsky et al. [12] and Radashevsky and Migotto [13]). Earlier records of *P. hoplura* from Kuwait [14] represent an undescribed species (Radashevsky unpublished). Identical-looking worms from Yamada Bay (Honshu, Japan) were described as a new species, *Polydora uncinata* Sato-Okoshi, 1998, and under this name reported from enclosed aquaculture operations in Chile [15], Republic of Korea [16] and Western Australia [17]. *Polydora hoplura* from Italy and *Polydora uncinata* from Japan were compared and found to be identical, and thus, *P. uncinata* was treated as a junior synonym of *P. hoplura* by Radashevsky et al. [12]. *Polydora hoplura/uncinata* from Japan, Australia, South Africa and Republic of Korea were sequenced by Sato-Okoshi and Abe [17], Teramoto et al. [18], David et al. [19], Williams et al. [9,20], Sato-Okoshi et al. [21], Lee et al. [22] and Abe and Sato-Okoshi [23]. Single *16S* sequence of an individual from Galicia (NW Spain) was provided by Almón et al. [24]. Genetic data for *P. hoplura* from the Mediterranean have remained unavailable.

We collected new and re-examined museum specimens of *Polydora* worms with morphological characteristics of *P. hoplura* worldwide, including its Gulf of Naples type locality. The purpose of the present study was to provide sequence data for worms from the type and other localities and verify by molecular analysis the conspecificity of disjunct populations. Based on the results of the genetic analysis, we further attempt to dissect and elucidate aspects of the global history of *P. hoplura* and ask whether the type locality is within the native range of this species.

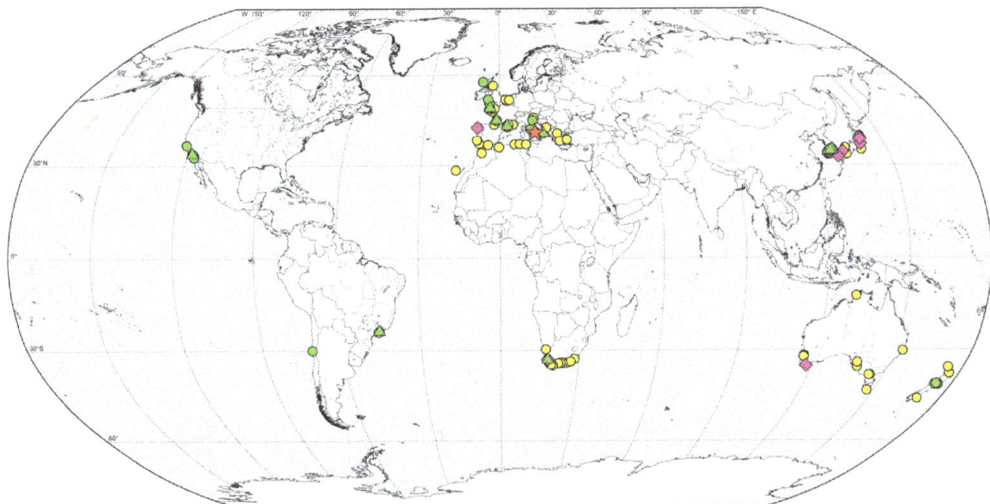

Figure 1. Map showing worldwide records of *Polydora hoplura* (see Supplementary Materials Tables S1 and S3). *Polydora hoplura*: red star—type locality: Gulf of Naples, Italy; green circles—worms identified based on the morphology in the present study; yellow circles—worms identified based on the morphology by other authors and not verified in the present study; green triangles—specimens sequenced in the present study; magenta rhombi—specimens sequenced by other authors.

2. Materials and Methods

2.1. Material

Collections were made in Italy (Tyrrhenian and Ionian Seas), France (both Atlantic and Mediterranean coasts), Republic of Korea (East Sea), USA (California), Brazil (São Paulo) and Chile (Figure 1). We collected bivalves, gastropods, barnacles, sponges and coralline algae from the intertidal zone manually and from shallow water using SCUBA equipment, grabs and trawls. Live *Polydora* worms were removed from infested shells or other substrata with a hammer and pliers, relaxed in isotonic magnesium chloride and then examined and photographed using light microscopes equipped with digital cameras. For molecular analysis, worm fragments were preserved in 95% ethanol.

After morphological examination in life, worms were fixed in 10% formalin solution, rinsed in fresh water, and transferred to 70% ethanol. Formalin-fixed specimens stored in ethanol were stained with Methylene Green (MG) to reveal specific staining pattern. We mainly followed the procedure described by Winsnes [25] (p. 19). The staining solution was made by adding MG to 70% ethanol to give a dark green color; thus, specimens in a dish could not be seen when covered by the solution. Specimens were soaked for 2–3 min and then moved to another dish containing clean 70% ethanol, where they were left to destain for 2–3 min until the excess color had disappeared. Stained specimens were examined and photographed using light microscopes equipped with digital cameras. Images of multiple focal layers were stacked using Zerene Stacker 1.04 software. Images of parts of worms were stitched into panoramas using PTGui 12.8 software. After complete examination, specimens were deposited in the polychaete collection of the Museum of the A.V. Zhirmunsky National Scientific Center of Marine Biology (MIMB), Vladivostok, Russia.

We also examined museum samples of various *Polydora* species worldwide, and collected for molecular comparison *Polydora brevipalpa* Zachs, 1933 from Peter the Great Bay, the Sea of Japan, Russia. Additional formalin- and/or ethanol-fixed *Polydora* specimens from Italy (Tyrrhenian Sea), France (Arcachon Bay and Gulf of Lion), Croatia (Adriatic Sea), Brazil (São Paulo), New Zealand, South Africa and California were provided by Maria

Cristina Gambi, Nicolas Lavesque, Céline Labrune, Barbara Mikac, João Nogueira, Sean Handley, Carol Simon and Sergey Nuzhdin.

To map the distribution of *P. hoplura*, we considered reliable records made by earlier authors based on morphological features and records by Sato-Okoshi and Abe [17], Teramoto et al. [18], David et al. [19], Williams et al. [9,20], Sato-Okoshi et al. [21], Lee et al. [22], Abe and Sato-Okoshi [23] and Almón et al. [24] based on genetic data. Complete information on newly collected material, museum samples examined during this study and by other authors and records by other authors for which no museum deposits were noted are provided in Supplementary Materials Tables S1 and S3. A list of the museums and other collections (and their acronyms) holding the examined or reported specimens of *P. hoplura* is provided in Supplementary Materials Table S2. The complete bibliography of the records provided by other authors and noted in Supplementary Materials Tables S1, S3 and S5 is provided in Supplementary Materials Table S6.

When no coordinates were provided for sampling sites from other studies, they were estimated using Google Earth Pro 7.3.6.9345 according to the original descriptions of the locations. Sampling locations of *P. hoplura* noted in Supplementary Materials Tables S1 and S3 are plotted on maps using QGIS 3.20.0 software and the geodata provided by the OpenStreetMap Project (https://osmdata.openstreetmap.de, accessed on 1 January 2022). Final maps and plates were prepared using CorelDRAW® 2019 software.

2.2. DNA Extraction, Amplification and Sequencing

We used the ReliaPrep gDNA Tissue Miniprep System (Promega Corporation, Madison, WI, USA) for DNA extraction and purification with standard protocol for animal tissue. Polymerase chain reaction (PCR) amplification of mitochondrial *16S* rDNA and nuclear *28S* rDNA, *18S* rDNA and *Histone 3* gene fragments were accomplished with the primers and the conditions described by Radashevsky et al. [26,27]. In addition, we used the D1R/D2C primer pair to amplify *28S* rDNA gene in some samples [28]. Purified PCR products were sequenced in both directions by the GnC Bio Company, Republic of Korea (www.gncbio.kr), and in the National Scientific Center of Marine Biology, Vladivostok, Russia, using the same primers as for PCR. Sequence editing and contig assembly were performed using SeqScape 2.5 (Applied Biosystems). GenBank accession numbers and brief information about sequences used in the present analysis are shown in Supplementary Materials Tables S3 and S5. To link sequences with complete corresponding data, unique numbers from the first author's database (VIR) are given to samples in Supplementary Materials Tables S1 and S3. These numbers precede the collecting location names on the phylogenetic trees shown in Figures 2 and 3A.

2.3. Data Analysis

In addition to new sequences obtained in the present study, we also included in the analysis sequences of *Polydora lingshuiensis* Ye et al., 2015 from China and *Polydora lingulicola* Abe and Sato-Okoshi, 2020 from Japan provided by Ye et al. [29] and Abe and Sato-Okoshi [23], respectively. These species were sequenced for the most complete set of gene fragments, including *16S*, *18S* and *28S*, and were used to evaluate genetic divergences between *Polydora* species and between distant populations of *P. hoplura*. The phylogenetic tree was rooted using the sequences of *P. brevipalpa*; this species was also sequenced for the most complete set of gene fragments (see Supplementary Materials Table S3).

We aligned DNA sequences using the MAFFT v7.2 software with the default settings (automatically chosen algorithms) [30,31]. Ambiguous positions and gaps for *16S* rDNA, *18S* rDNA and *28S* rDNA genes were excluded from subsequent analysis using trimAl 1.2 [32] with an automated heuristic approach. As the obtained sequences were similar, we chose to employ uncorrected values of sequence divergence (pairwise distances, p, see Nei and Kumar [33]) instead of complex distance measures (i.e., corrected by best-fit evolutionary model). Distances both within and between groups were calculated in MEGA 11.0 [34]. We concatenated DNA data partitions using SequenceMatrix 1.9 [35] and specified substitution models for each partition individually. The best-fitting nucleotide substitution models for Bayesian inference (BI) were selected in MrModeltest version 2.3 [36] using Akaike Information Criterion (AIC): SYM + G for each data partition.

We used MrBayes 3.2.7 [37,38] via the CIPRES web portal [39] for the Bayesian analysis of 10,000,000 generations, four parallel chains and sample frequencies set to 1000, in two separate runs. Based on the convergence of likelihood scores, 10% sampled trees were discarded as burn in. The same conditions were implemented for separate BI analysis of *16S* rDNA sequences.

The haplotype network was constructed based on *16S* rDNA dataset with a median-joining approach [40] using popART [41].

3. Results

3.1. Molecular Analysis

3.1.1. Phylogenetic Analysis of *Polydora Hoplura*

The combined aligned sequences of *Polydora* spp., with gaps excluded, comprised a total of 2364 bp, including 259 bp (98.5% of original sequences) for *16S* rDNA, 1578 bp (99.9% of original sequences) for *18S* rDNA, 228 bp (99.1% of original sequences) for *28S* rDNA and 299 bp (100% of original sequences) for *Histone 3*. The combined concatenated dataset contained 146 variable sites, 141 of which were parsimony informative; the frequencies of variable and informative sites were 6.17% and 5.96%, respectively. The frequency of variable sites in the aligned dataset of mitochondrial *16S* rDNA (19.7%) was greater than those in sequences of the nuclear genes: 2.72% for *18S*, 4.38% for *28S* and 13.37% for *Histone 3*.

The Bayesian analysis of the combined dataset of four gene fragments and the analysis of only *16S* rDNA sequences both showed that individuals from distant localities with diagnostic morphological features of *P. hoplura* form a well-supported clade (PP = 1 and PP = 0.97, respectively) (Figures 2 and 3A). The maximum uncorrected *p*-distance value for the most variable mitochondrial *16S* rDNA between individuals from different populations (p = 2.54) was comparable with the maximum genetic variability within these populations (p = 1.31) and essentially smaller than the minimum distance between *Polydora* species (p = 6.98) (see Supplementary Materials Table S4). Therefore, we consider the examined populations as conspecific and refer them to *P. hoplura*.

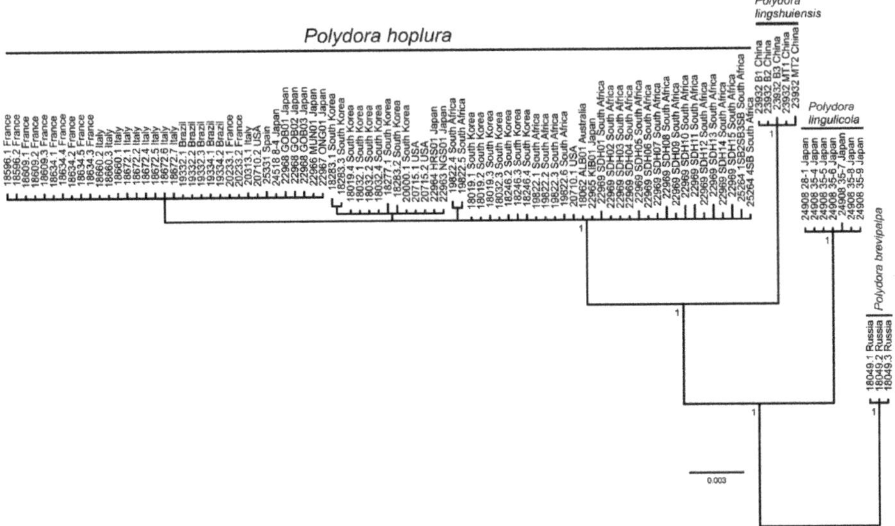

Figure 2. Majority rule consensus tree of the Bayesian inference analysis of the combined *16S* (263 bp), *18S* (1579 bp), *28S* (228 bp) and *Histone 3* (299 bp) sequences (2369 bp in total) of *Polydora* spp. rooted

with sequences of *Polydora brevipalpa*. Posterior probabilities of 0.95 and above are shown on the branches. The five-digit numbers preceding collecting locations are unique numbers from the VIR database linking the individuals on the tree with the sampling data in Supplementary Materials Tables S1 and S3; individual numbers are separated from sample numbers by dots.

3.1.2. Haplotype Network of *Polydora Hoplura*

The haplotype network analysis was based on 79 sequences of *16S* rDNA, of which 26 were previously obtained by Sato-Okoshi et al. [21], Abe and Sato-Okoshi [23] and Almón et al. [24]. The analysis recovered a single network comprising 15 haplotypes (Figure 3B, Supplementary Materials Table S5). Striking 10 unique haplotypes (H1-10; 21 individuals) were found only in South Africa's Western Cape. They formed a star-like structure with nine haplotypes (H2–H10) differing from the most common haplotype (H1, comprising 52% of South African individuals) by one mutational step. The majority of other individuals shared from one to four of the four most common haplotypes H11 (19%), H12 (25%), H14 (11%) and H15 (16%). Each of these haplotypes was detected in nine individuals from Japan; three of these haplotypes (H11, H14, H15) were detected in four individuals from California. All ten examined individuals from Italy (four from the Tyrrhenian Sea and six from the Ionian Sea) shared one haplotype (H12), which was also detected in the six individuals from Atlantic France (five from Arcachon Bay and one from Brittany, La Manche), three individuals from Japan and in one individual from Galicia, Spain. The haplotype H13 was detected only in one examined individual from Western Australia (LC101868, [21]). Two shared haplotypes (H14, H15) were detected in 16 individuals from Republic of Korea. Only one shared haplotype (H11) was detected in all five individuals from Brazil, all five individuals from Mediterranean France (Gulf of Lion) and in one individual from Atlantic France (Brittany, La Manche).

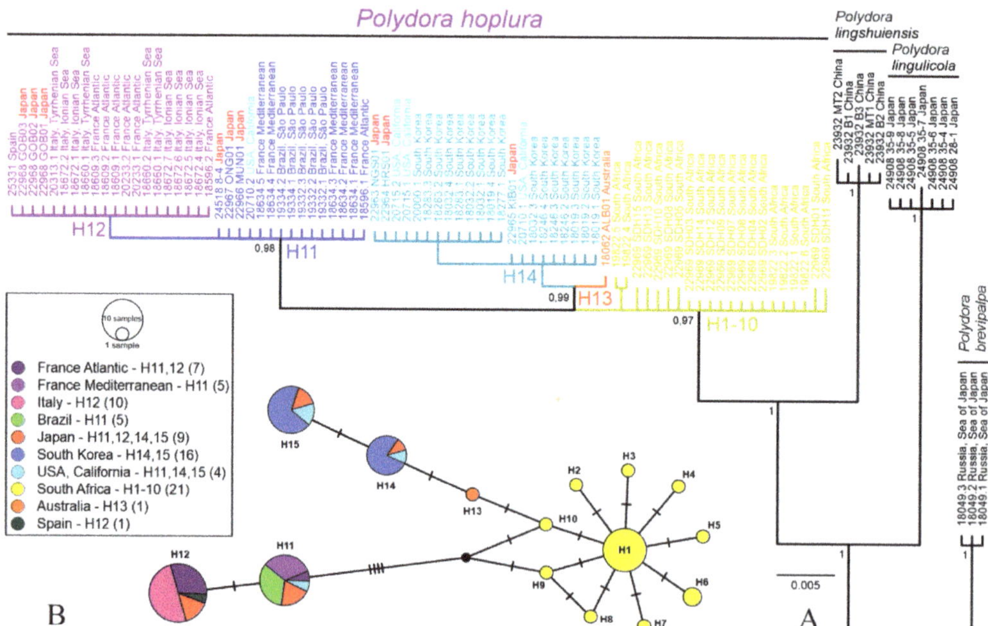

Figure 3. (**A**) Majority rule consensus tree of the Bayesian inference analysis of the *16S* rDNA (263 bp) sequences of *Polydora* spp. rooted with sequences of *Polydora brevipalpa*. Posterior probabilities of 0.95

and above are shown on the branches. The five-digit numbers preceding collecting locations are unique numbers from the VIR database linking the individuals on the tree with the sampling data in Supplementary Materials Tables S1 and S3; individual numbers are separated from sample numbers by dots. Colors indicate the unique South African haplotypes (H1–H10), most common shared haplotypes (H11, H12, H14, H15), and places with most diverse haplotypes (USA, California—3 haplotypes; Japan—4 haplotypes). (**B**) Haplotype network of the *16S* rDNA gene in *Polydora hoplura*. Haplotypes are represented by colored circles whose size is proportional to the number of corresponding individuals. Multiple colors indicate haplotypes shared by individuals from more than one sampling locality, with sections scaled according to their frequency. Each hatch-mark on the lines connecting haplotypes symbolizes one mutational step. A small black circle represents a missing haplotype. Total number of examined individuals shown in parentheses after the names of localities.

3.2. Morphology and biology

Polydora hoplura Claparède, 1868

Figure 4

Polydora hoplura Claparède [42] pp. 318–319, pl. XXII, figure 2; [43] pp. 58–59, pl. XXII, figure 2; [44] pp. 58–59, pl. XXII, figure 2. Andreu [45] pp. 87–91. Leloup and Polk [46] pp. 71–72. Gravina et al. [47] p. 165. David et al. [48] pp. 890–894, figures 4–6 (larval morphology). Radashevsky et al. [12] pp. 545–551, figures 2–4 (references). Radashevsky and Migotto [13] pp. 860–865, figures 2–5 (adult and larval morphology). Sato-Okoshi et al. [21] pp. 1677–1680, figures 6 and 7. Lee et al. [49] pp. 461–463, figure 2. Abe and Sato-Okoshi [23] p. 52, figure 8F. Almón et al. [24] p. 3, figures 1 and 2.

Polydora (Polydora) hoplura: Rioja [50] (*Part.*) p. 70, pl. 19, figures 8–13. Hartmann-Schröder [51] p. 305; [52] p. 318.

Polydora hoplura hoplura: Day [53] p. 468, figure 18.2k–m.

Leucodora sanguinea Giard [54] pp. 71–73. *Fide* Dollfus [55] p. 17; [56] p. 275.

Polydora uncinata Sato-Okoshi [57] pp. 278–280, figure 1; [58] p. 835. Radashevsky and Olivares [15] pp. 491–494, figures 2–4. Sato-Okoshi et al. [59] pp. 493–495, figures 2 and 3; [16] p. 87, figure 4A,B,D. Sato-Okoshi and Abe [17] pp. 43–44, figure 3. *Fide* Radashevsky et al. [12] p. 545.

3.2.1. Diagnostic Features

The adult morphology of *P. hoplura*, including the neotype and other individuals from the type locality in the Gulf of Naples, Italy, was recently redescribed by Radashevsky et al. [12]. Worms from Chile, Republic of Korea and Brazil were described in detail by Radashevsky and Olivares [15] (as *P. uncinata*), Radashevsky et al. [12], and Radashevsky and Migotto [13], respectively. Worms from Japan and South Africa were described by Sato-Okoshi et al. [21]. The diagnostic features of the adults include palps with black bands, prostomium weakly incised anteriorly, caruncle extending to the end of chaetiger 3, short occipital antenna, and two kinds of spines (heavy sickle-shaped and slender awl-like) in addition to slender capillary chaetae in the posterior notopodia, and cup-shaped to disc-like pygidium with middorsal gap (Figure 4A–E). Moreover, all the populations examined by us were parthenogenic (Radashevsky unpublished). Some characters are variable, however. Pigmentation on palps and anterior chaetigers can be well developed or totally lacking. Prostomial incision in some individuals is weakly developed and can be seen in the ventral view only or not developed at all. The caruncle is short in small worms and the occipital antenna is occasionally small and hardly discernible. Slender awl-like spines in the posterior notopodia are arranged differently in individuals of different sizes, and in large individuals are present in a few posterior branchiate chaetigers only [12]. None of these characters is unique; all of them are shared by some other spionids. Nevertheless, the combination of these characters defines the species unambiguously, and the heavy sickle-shaped spines in the postbranchiate chaetigers are the most diagnostic. Such types of spines are also present in *Polydora colonia* Moore,

1907 [60], some *Boccardiella* and some other spionids [61], which, however, can easily be distinguished by other characters.

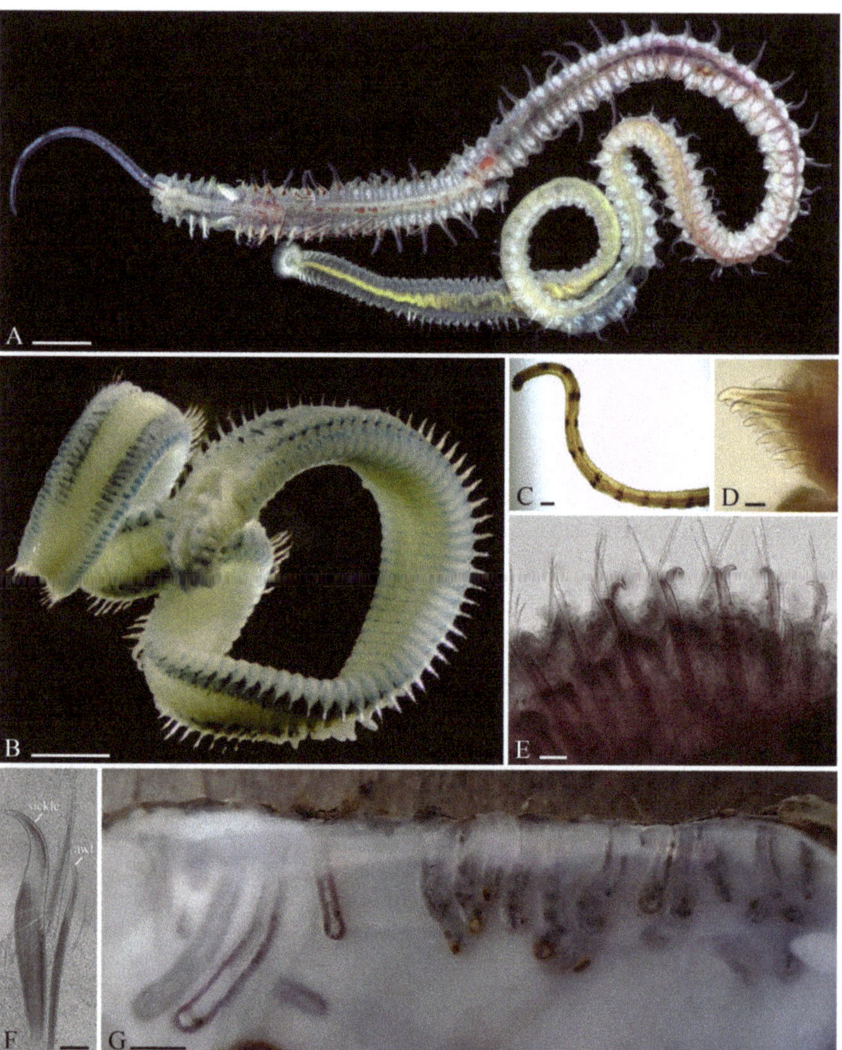

Figure 4. Adult morphology and habitat of *Polydora hoplura*. (**A**) Complete live individual, dorsal view; left palp missing. (**B**) Complete individual fixed in formalin and stained with methylene green; palps missing. (**C**) Palp fragment, showing paired black bands on sides of frontal groove. (**D**) Heavy falcate spines of chaetiger 5, left side, ventral view. (**E**) Posterior notopodia with heavy recurved spines in addition to slender capillary chaetae, left lateral view. (**F**) Two kinds of spines (heavy sickle-shaped and slender awl-like) and slender capillary chaetae from a posterior notopodium. (**G**) Edge of a shell of the Pacific oyster *Magallana gigas* showing U-shaped burrows of worms. Scale bars: (**A**), (**B**)—1 mm, (**C**)—200 μm, (**D**)—30 μm, (**E**), (**F**)—50 μm, (**G**)—5 mm. (**A**)—California, USA (LACM-AHF Poly 12859); (**B**)—Mar Grande of Taranto, Italy (MIMB 33029); (**C**), (**F**)—Wando, Republic of Korea (MIMB 33038, 33035); (**D**)—Gulf of Naples, Italy (MIMB 28148); (**E**), (**G**)—California, USA (LACM-AHF Poly 12862, MIMB 39108). (**A**)—photo by Leslie H. Harris.

3.2.2. Habitat

Adults of *P. hoplura* make U-shaped burrows in shells of barnacles, bivalves and gastropods, including clams, oysters and cultured abalone (Figure 4F). Occasionally, worms bore into sponges [13,20,62], coralline algae [20,62–64] and the stony coral *Mussismilia hispida* (Verrill) (present study) (Supplementary Materials Table S1). Branches of each U are situated close to each other and interconnected by a narrow space all along their length, thus the burrow in cross-section appears as a characteristic 8-shaped hollow in a shell. The walls of the burrows are lined with detritus; the median space between branches is also filled with detritus forming a medial detrital wall. Each burrow opens to the outside via two joined apertures and is extended by two smooth silty tubes, each up to 5 mm long.

3.2.3. Remarks

The larval morphology of the species was described from the Plymouth area (English Channel, UK) [65], Chile [15] (as *P. uncinata*) and Brazil [13]. Poecilogony with females producing two kinds of larvae, planktotrophic and adelphophagic, was demonstrated in South African populations [48].

Infestation of cultured oysters and especially abalone by *P. hoplura* has been a serious problem for aquaculture for a long time. Commercial interest stimulated studies on the morphology, genetic structure, reproductive biology and dispersal ability of this species [9,19,20,48,49,66–74]. Some control of polychaete pests on cultured molluscs has been suggested [67,75–80], but the problem of infestation continues.

3.2.4. Distribution

Complete information about worldwide records of *P. hoplura* is provided in Supplementary Materials Table S1 (mapped on Figure 1).

4. Discussion

4.1. Population Genetics of Polydora hoplura

We used the set of genes originally chosen for phylogenetic analysis of Spionidae [27,81]. The *16S* rDNA is the most variable among examined markers and has been proven to be highly suitable for molecular identification at the species level for spionids [26,27,81–86]. Our genetic analysis of *16S* dataset, however, does not reveal strongly robust biogeographic patterns of *P. hoplura* populations worldwide, since haplotype analysis was not the primary objective of our study, nor were our analyzed sample sizes large enough to reveal the full extent of haplotype diversity in any one region.

4.2. Historical and Modern Biogeography and Vectors of Dispersal

Potential explanations of the observed haplotype network are confounded in part by the majority of our analyzed specimens representing worms recovered from cultured oysters and abalones (in sea farms, in sea cages, or land tanks) (Supplementary Materials Table S1). These cultured molluscs likely represent, in most cases (even in Japan), populations that have been mixed and remixed for many decades through aquaculture movements and (for oysters) historical transport in vessel-fouling communities (as we discuss below). Our sequenced material from Italy and Brazil was from natural substrates and our material from Atlantic France was from both natural substrates and commercial oysters, while the rest of our sequences are derived from commercial molluscan populations.

That said, we suggest that our preliminary data, when combined with historical knowledge of *P. hoplura*'s absence in certain regions, may hint at its home port—its potential endemic region—even as our data also reveal an intriguing concentration of unique haplotypes in one region with a relatively new, but intensive, history of imported mollusc culture. We briefly review the global history of *P. hoplura* before suggesting a pathway to its native province (see Table 1).

Table 1. Possible introduction mechanisms and subsequent anthropogenic dispersal vectors by region, arranged chronologically by year of discovery.

Region	Year First Collected	Possible Method of Initial Introduction	Probable Secondary Anthropogenic Dispersal Mechanisms throughout Continent, Country, or State	Established in Wild?	Comments
Mediterranean	Italy: 1866–1867; France: 1873–1874	Vessel fouling	Fouling, aquaculture, ballast water	Yes	Arrival in Mediterranean preceded importations of Pacific oysters (*Magallana gigas*) and the common use of ballast water internationally
Atlantic Europe: France	circa 1880	Vessel fouling	Fouling, aquaculture, ballast water	Yes	Arrival in Atlantic Europe preceded importations of Pacific oysters and common use of ballast water
South Africa: Langebaan lagoon	1946	Vessel fouling or ballast water	Aquaculture, fouling, ballast water	Yes	Probable arrival in South Africa preceded commercial mollusc importations
Australia: Victoria and South Australia	1971 *	Aquaculture	Vessel fouling; ballast water	Yes	Pacific oyster importations commenced in 1940s
New Zealand: Wellington Harbour	1972	Vessel fouling or ballast water?	Aquaculture; ballast water	Yes	See discussion relative to timing of Pacific oyster introductions
Brazil: São Paulo	1995	Aquaculture	Vessel fouling	Yes	Pacific oyster importations commenced in 1970s
Chile: Coquimbo	2002	Aquaculture		No	Japanese abalone (*Haliotis discus hannai*) importations commenced in 1987
Republic of Korea: Wando	2004	Aquaculture	Aquaculture	Uncertain	Pacific oyster and Japanese abalone importations
USA: California: San Diego Bay	2004	Aquaculture	Vessel fouling (north from southern CA to Monterey Bay)	Yes	Pacific oyster importations prior to 2000

* Carazzi [87] suggested that Whitelegge [88] has misidentified *P. hoplura* as *P. ciliata* in Australia. Our study of Whitelegge's description and figures suggests that he had neither species.

Polydora hoplura was first collected in 1866–1867 in Italy [42] in barnacle shells. It was found in many different habitats in the decades thereafter, or had already spread before being collected in Italy, in the northern and western Mediterranean and in Western Europe. It was first found in 1946 in South Africa in Langebaan Lagoon [89] and in 1947 in Cape Town fouling communities [90]. After a long but not unusual hiatus typical of the global spread of marine invasions [91], it was collected in 1971 in Australia (Supplementary Materials Table S1) and in 1972 in New Zealand [92], followed by another multidecade gap before being recovered in Brazil in 1995 (Supplementary Materials Table S1) and again in 2015 [13], in Japan in 1997 [57], in Chile in 2002 [15] and in Republic of Korea and California in 2004 (Supplementary Materials Table S1). In all areas of the world, *P. hoplura* is now established in the wild, except for Chile and perhaps Republic of Korea, as discussed below.

Polydora hoplura has long been recorded from molluscan hosts and other calcareous substrates (such as barnacles and serpulid tubeworms) as well as from fouling communities on docks and pilings (Supplementary Materials Table S1). This broad habitat diversity, combined with its known larval biology, suggests three probable human-mediated global transport vectors: (1) the movement of commercial shellfish, (2) vessel fouling and (3) vessel ballast water.

Relative to (1), the Pacific oyster *Magallana gigas* (Thunberg) (commonly known as *Crassostrea gigas*) is one of the most "globalised" species dominating bivalve production in many regions [6,93]. Transoceanic movement of *M. gigas* commenced over 100 years ago [93]. Native to the cool-to-warm temperate shallow waters of the Asian Pacific (Japan, Republic of Korea and Russia), *M. gigas* "have been introduced to 66 countries outside their native range, mainly for aquaculture, and there are now established self-sustaining populations in at least 17 countries" [94] (p. 2836). Biofouled adult oysters were typically transported long distances (to establish new populations or for commercial outplanting) both historically [95] and into modern times [96], providing ample opportunity for numerous accompanying nontarget epibiotic and endobiotic species to be introduced.

Relative to (2), classic hull-fouling communities on both coastal and ocean-going vessels include abundant calcareous substrates such as barnacles, oysters, serpulid polychaetes and also noncalcareous habitats such as sponges [97]. Spionid polychaetes, including *Polydora* spp., have been found in the fouling communities of ocean-going and coastal vessels [98–100], (JTC personal observations 2000). *Polydora hoplura* is commonly found in fouling communities on harbor pilings and docks (as reviewed below, and in Supplementary Materials Table S1), which we interpret here to represent individuals likely dislodged from their calcareous substrates or from sponges.

Relative to (3), ballast water may have provided an additional means of dispersal for *P. hoplura* (as we detail further below). *Polydora* larvae have been regularly found in ballast water [101–104], (JTC and VIR personal observations), with Smith et al. [105] reporting that larval spionids dominated the polychaete fraction in the ballast water of vessels arriving from around the world in Chesapeake Bay. In turn, David et al. [48] found that the average larval life of planktotrophic larvae of *P. hoplura* is 40 days, sufficient time (for those populations with planktotrophic larvae) for both transoceanic and interoceanic transport of larvae in ballast water.

Given the historical records of *P. hoplura*'s reports around the world, combined with the known range of potential anthropogenic vectors, we review here the probable interplay of the chronological history of reports with the probable transport vectors at the times and places of discovery (see Table 1).

Polydora hoplura was found in Italy and France prior to the introduction of commercial molluscs from elsewhere in the world [93,106]. We suggest that it arrived in the Mediterranean in vessel fouling, perhaps in barnacles, serpulid tubeworms or oysters attached to hulls. Importantly, ballast water was not a common vector for species in international vessel traffic until the 1880s and later [107]. Its spread around the European theater may then have been facilitated by coastal vessel traffic (both fouling and, later, ballast water) and by the intracountry movement of commercial oysters. Its occasional appearances in the Netherlands [108] and Belgium [109] have been suggested to be directly linked to oyster importations.

The appearance of *P. hoplura* in the late 1940s in South Africa similarly predated the introduction of commercial molluscs, with the importation of the Pacific oyster *M. gigas* not commencing until 1950 [93]. During and after World War II, there was increased commercial and military vessel traffic connecting Cape Town, Western Europe and Asia, suggesting that shipping provided a steady source of *P. hoplura* (albeit sources not yet supported by genetic connectivity, due, as we suggest, below, to population genetic undersampling in Europe and the Western Pacific).

The first collection of *P. hoplura* in 1971 in Australia followed a long history of the importation of the Pacific oyster *M. gigas* from Japan commencing in the 1940s to Victoria, Tasmania, Western Australia, and New South Wales [93]. The first collections of *P. hoplura* at about the same time, in 1972, in southern ports of New Zealand's North Island, link to a more complicated vector history. Dinamani [110,111] suggested that Japanese vessels with oyster fouling accidentally introduced *M. gigas* to the northern part of the North Island in the early 1960s, followed by its southward spread. Earlier, however, Dromgoole and Foster [112] suggested that *M. gigas* may have been intentionally introduced. The location (Wellington Harbor), habitat (native wild pen shells, oysters, scallops and abalones), and timing (1972–1974) of the first collections of *P. hoplura* in New Zealand are not, however, clearly related to the first appearance of alien oyster populations in the same area, but suggestive of the introduction of *P. hoplura* by vessels.

Polydora hoplura was next collected in Brazil in the native stony coral *Mussismilia hispida* in 1995 at Ilha de Alcatrazes and in shells of the native oyster *Crassostrea rhizophorae* (Guilding) (Supplementary Materials Table S1) and in fouling on pier pilings in 2015 [13]. *Magallana gigas* introductions began in Brazil in 1974, resumed in 1986, and importations continue [96]. While the collection sites of *P. hoplura* in Brazil (São Paulo) are not oyster culture sites, *M. gigas* has been imported and outplanted both to the north and south of Sao Sebastião and Alcatrazes [96].

Polydora hoplura was transported from Japan to Coquimbo, Chile, perhaps commencing as early as 1987 [15], in the shells of the abalone *Haliotis discus hannai* Ino imported for land-based culture. *Polydora hoplura* were found in these abalones in 2005. Radashevsky and Migotto [13] noted that the "progeny of the abalone were ... going to be introduced into coastal waters for further commercial cultivation", the possibility of which was also discussed by Stotz et al. [113], but there is no evidence that any abalones were ever released into the sea (JTC based on a visit to and interviews in Coquimbo in 2019).

In Republic of Korea, *P. hoplura* was first found in 2004 in cultured oysters *M. gigas* and tank-cultured abalone *Haliotis discus discus* Reeve presumably imported from Japan in 2002; Sato-Okoshi et al. [16] noted that it had been absent prior to 2004. It is also now found in Republic of Korea in abalones cultured in cages in the sea [12] (Supplementary Materials Table S1).

Polydora hoplura was first found in California in 2004 (Supplementary Materials Table S1), and collected again in 2011 [13], all in biofouling communities on docks and pilings. It was found again in 2017 in the shells of *M. gigas* in the wild on a pier and in a commercial oyster farm (Supplementary Materials Table S1). Crooks et al. [114] reviewed the history of the discovery of wild populations of *M. gigas* in southern California commencing in 2000, likely linked to unreported and unrecorded aquaculture introductions. Its presence to the north in Monterey Bay in 2011 is likely due to coastal vessel traffic. Among the various global regions in which *P. hoplura* is now established, southern California offers perhaps one of the most robust historical baselines, in terms of polychaete collections. Southern California coastal polychaetes were extensively sampled between the 1930s and 1980s by the well-known polychaetologists Olga Hartman [115,116], Keith Woodwick [117] and Donald Reish [118,119], and it is highly unlikely that these careful workers would have overlooked *P. hoplura* in biofouling communities, which they regularly sampled.

In summary (Table 1), we suggest a chronological progression of the driving global dispersal vectors of *P. hoplura*. The earliest record of *P. hoplura*, in what appears to be its first long-distance colonization episode, we relate to vessel fouling, before the advent of the international movement of ballast water and of commercial oyster movements. Following widespread dispersal through Western Europe, it was not until the mid-20th century that *P. hoplura* was found in South Africa, by which time both vessel fouling and ballast water were in play. While we do not set aside the continued role of vessel traffic, we relate the appearance of *P. hoplura* in Australia, Brazil and California to the vast uptick of the movement of Pacific oysters since the 1970s around the world, with vessel fouling and ballast water likely playing roles in the secondary movement of populations (Table 1). The evidence in hand does not permit us to separate out the relative roles of vessel traffic *versus* oyster culture in the arrival of *P. hoplura* in New Zealand.

4.3. Maritime History Considerations and Invasion Timing

In the 1730s and 1740s, the East Indiaman cargo vessel *Schellak* made regular voyages from the Netherlands around South Africa to Jakarta in Indonesia, with occasional trips as far north as Japan [120]. By the late 1700s, regular passages by ships had commenced between Britain and Australia [121]. Could hundreds—and eventually thousands—of such early maritime voyages, fluidly connecting Europe, South Africa, the Indian Ocean, Australasia, and the North Pacific, have transported and introduced marine worms many centuries ago? Could *P. hoplura* specifically have been introduced by ships to South Africa and Australia as early as, for example, the 1700s, and overlooked in these countries until the 20th century?

There is no doubt that ocean-going vessels moved and introduced marine animals and plants interoceanically, often undetected, for hundreds of years, resulting in the obfuscation of the historical biogeography of a great many species [10,122–125]. That said, a great many variables contribute to an often strong disconnect between the existence of a vector and the probability of successful introductions [126], as witnessed by continued new invasions of many biofouling species around the world, despite the fact that the same species have been carried around the world over many centuries.

Polydora hoplura is a comparatively large, distinctive worm, not easily overlooked; it reaches 55 mm in length and bears large conspicuous recurved spines in the posterior notopodia, easily recognized by polychaete workers [12]. *Polydora hoplura* also has broad habitat diversity, being found in a wide range of calcareous substrates (molluscs, tubeworms, and barnacles) and in sponges. Collecting *P. hoplura* is not limited to examining calcareous substrates; biofouling collections from harbor docks and pilings produce individuals dislodged from their substrates (Radashevsky unpublished).

In both South Africa [127] and Australia [128], collections of shallow-water marine worms began in the 1850s and were well underway by the turn of the 20th century. In Australia, the zoologist William Haswell worked out of a laboratory on the shores of Sydney Harbour commencing in the 1890s [128]. Haswell [129], Whitelegge [88] and Roughley [130] investigated oyster-boring spionids, reported as *Polydora ciliata*, in Australia. Blake and Kudenov [131] suggested that Haswell's and Whitelegge's reports may refer to *Polydora websteri* Hartman in Loosanoff and Engle, 1943. As we note in Table 1, Carazzi [87] suggested that Whitelegge [88] had misidentified *P. hoplura* as *P. ciliata*, but Whitelegge's description and figures suggests that he had neither species. Roughley [130] reported boring worms in oysters, also identified as *P. ciliata*, but no longer than 2.5 cm, indicating that these were not *P. hoplura*. Extensive polychaete work in Australia resumed in the 1950s and 1960s, without *P. hoplura* being reported. Thus, for Australia, we have reasonably thorough indications that *P. hoplura* was not likely present prior to the mid-20th century. In South Africa, worm collections in the Cape Town area were underway by the early 1900s, without the conspicuous *P. hoplura* being reported. Day [132] specifically reported on spionids in South Africa, and, again, this conspicuous worm was not recorded.

In clear contrast, as we have detailed above, large numbers of Pacific oysters, a well-known host of *P. hoplura*, began to be imported in the 20th century into both Australia and South Africa. We suggest it is not a coincidence that these commercial aquaculture operations were a major vector in introducing *P. hoplura* to these countries.

We recognize a distinction between when a non-native species is first collected and when it may have been introduced. In many cases there are well-known lag times between initial arrival and colonization of a species and its population growth to the point where it can be detected (collected) [133]. Nevertheless, for species that are relatively large and conspicuous in regions with a relatively long history of investigation, the lag time between a species becoming established and encountering it in the field is rarely on the order of centuries, except for those invasions going unrecognized under other names. We have no evidence that this is the case for *P. hoplura* in the regions where it has now been found.

This said, we do not know when *P. hoplura* first arrived in the Mediterranean: it may have been introduced to the Gulf of Naples well before the 1860s. Similarly, South African populations may of course have been present some years before 1946, and Australian and New Zealand populations may have become established by the 1960s. None of these adjustments would alter the larger picture that *P. hoplura* is a species whose global voyages commenced and continued through the 19th and 20th centuries.

Finally, the uptick in the globalization of *P. hoplura* commencing in the last half of the 20th century and continuing into the early 21st century falls into a now well-recognized pattern of a world sustaining more and more invasions paralleling a significant parallel uptick in maritime trade. Examples abound: the well-known European shore crab *Carcinus maenas* (Linnaeus), although introduced by ships to Atlantic North America and Australia in the 19th century, began resuming its global voyages in the late 20th century [91]. The Asian mitten crab *Eriocheir sinensis* H. Milne Edwards, introduced to Europe in the early 1900s, also did not resume its voyages until the late 20th century [134]. The abundant intertidal barnacle of the Northeast Pacific, *Balanus glandula* Darwin, despite hundreds of years of opportunity to spread around the world, only began colonizing foreign shores (Argentina, Japan, South Africa) in the late 20th century [135]. The Japanese caprellid amphipod *Caprella mutica* Schurin similarly began spreading globally only in the 1970s [136]. There are scores of additional cases. We suggest that the increasing number of recognized populations of *P. hoplura* nests well within this modern pattern of increased global invasions.

4.4. Possible Native Range of Polydora Hoplura

We suggest that the global history of detection of *P. hoplura*, combined with the albeit limited haplotype data in hand, permit preliminary identification of its potential native region. Haplotypes H11, 12, 14 and 15 are found in Japan. We know that *P. hoplura* appeared in approximately 2004 in Republic of Korea, and the presence of H14 and 15 there, combined with the history of importations of *M. gigas* from Japan, suggest that South Korean populations are derived from there. California populations possess H11, 14 and 15, as does Japan. Italy and Atlantic France possess H12, as does Japan. Brazil populations to date present the H11 haplotype, of yet uncertain origin. In short, the shared haplotypes in the North Pacific, versus only two shared haplotypes (H11, H12) detected in the Northeast Atlantic and Mediterranean (both of which also occur in the North Pacific), suggest that *P. hoplura* originated in the Pacific Ocean, not in the Mediterranean Sea. The demonstrable historical absence of *P. hoplura* from California (above), versus our inability to demonstrate the same for Japan, narrows the possible origin to the Western Pacific, or at the most the Indo–West Pacific.

The late detection of *P. hoplura* in Japan may be explained by the lack of earlier extensive studies within the potential range of this species. The shell-boring spionids in Japan were first studied in the scallop *Mizuhopecten yessoensis* (Jay) cultured in northern Hokkaido and northeastern Honshu [137–140], areas likely too cold for the warmer water *P. hoplura*. Prior to the 1990s, there is only one report of shell-boring spionids in southern Japan (Tokushima Prefecture, Shikoku Island), that of Kojima and Imajima [141], who identified *Polydora ciliata* and *Polydora websteri* in the wild abalone *Haliotis diversicolor* Reeve. Sato-Okoshi [57] described *P. uncinata* (=*P. hoplura*) burrowing in shells of the wild gastropod *Omphalius rusticus* (Gmelin) from Kochi Prefecture (Shikoku Is.) and in cultured *M. gigas* from Iwate Prefecture (Honshu Island). Soon after that, Sato-Okoshi [58] reported 13 shell-boring polydorins from Japan, including seven *Polydora* species.

We are left with the apparent enigma of the intriguing genetic diversity of *P. hoplura* in South Africa. All 10 *16S* haplotypes detected in the Western Cape Province were unique and not yet recovered in other global populations. David et al. [19] recovered 42 haplotypes (seven shared, 35 unique) for the mitochondrial *cytochrome b* (*Cyt b*) gene, and 44 haplotypes (14 shared, 30 unique) for the ATP-synthetase nuclear-encoded protein complex (ATPSα) gene from seven localities encompassing the reported distributional range of *P. hoplura* in South Africa. No geographic pattern of haplotypes was detected in that study; frequent movement of oysters along with coastal shipping (such as worms burrowing into sponges and barnacles on ship hulls) were suggested as factors homogenizing South African populations [19] (p. 608) [142,143]. In addition, we note that documenting unique haplotypes in introduced populations—that is, genes not yet detected in the known home range of a species—is a recognized phenomenon [144–146].

South African populations of *P. hoplura* could thus reflect 75 or more years of introductions from around the world via vessel hull fouling and oyster importations. In particular, many decades of oyster imports from a host of different countries would serve to both "collect" and concentrate many haplotypes not yet detected elsewhere. Williams et al. [20] (text and especially their Figure 5) have underscored the scale of anthropogenic introductions of *P. hoplura* in South Africa from potentially many genetically distinct source populations. Our current sample sizes from Europe, Australia, Japan, and elsewhere are too small to capture the global *16S* haplotype diversity of *P. hoplura*. We thus make two predictions: one, that more thorough modern and historical (museum) sampling (below) will reveal the presence in the Pacific Ocean of the 10 haplotypes currently registered only for South Africa (as well as H13 in regions outside of Australia) and, two, that sampling wild native host populations will reveal allochthonous haplotypes in South Africa already known from elsewhere. Our predictions aside, the concentration of haplotypes in South Africa and in the Pacific would suggest at the very most an Indo–West Pacific, if not specifically a Northwest Pacific, origin for *P. hoplura*, not the Atlantic Ocean or the Eastern Pacific.

5. Conclusions

In moving around the world, *P. hoplura* has acquired a rich diversity of hosts, including a wide number of bivalve and gastropod mollusc species, serpulid tubeworms, bryozoans, barnacles, sponges and others. The world's museums possess equally rich historical resources of all of these substrates from all of the countries discussed here. Along with harvesting and sequencing larger numbers of individuals globally, we predict that thorough mining of these museum resources will produce historical material of *P. hoplura* and further elucidate the global biogeographic history of this globe-trotting worm.

Carlton and Ruiz [147] described many of the complexities involved in analyzing the human-mediated dispersal history of marine species. They noted that species that have likely endured multiple mechanisms of global transport are *polyvectic*. The history of the discovery of *P. hoplura* around the world appears to be intimately linked to global shipping that commenced in the mid-19th century, followed by the advent of the global movement of commercial shellfish in the 20th century, interlaced with continued dispersal by vessels and aquaculture. Many populations over the decades may thus have originated from sustained infusions from widespread populations by both vectors. Given that *M. gigas* has been successfully established in no fewer than 17 countries, of which *P. hoplura* has been detected in only a few of those to date, we predict that it may already be established in many more regions. Further, as global connectivity continues to increase in world trade of both products and seafood, it is likely that novel populations of *P. hoplura* will continue to emerge.

Supplementary Materials: The following supporting information can be downloaded at: https://www.mdpi.com/article/10.3390/biology12060780/s1 which includes the following Tables S1–S6: Table S1. Sampling location data and museum registration numbers of *Polydora hoplura*. Table S2. List of the museums and collections (and their acronyms) holding the examined or reported samples of *Polydora* spp. Table S3. Taxa, sampling location data, museum registration numbers and GenBank accession numbers of sequences obtained in the present study and also provided by earlier authors. Complete information about samples is given in Table S1. Table S4. Uncorrected pairwise average distances (*p*, in %) between populations for *16S* (259 bp), *18S* (1578 bp), *28S* (228 bp), and *Histone 3* (299 bp) sequences. Table S5. *16S* haplotypes recognized in the present study. Table S6. References cited in Tables ESM1, ESM3 and ESM5 [148–179].

Author Contributions: V.I.R. formulated the idea, collected and examined most of the material, wrote the first draft of the manuscript. V.V.M. and V.V.P. sequenced material and analyzed molecular data. J.-W.C. organized sampling in Republic of Korea and financial support for this study, collected essential material. S.Y. provided laboratory facilities and financial support for sequencing in Republic of Korea. J.T.C. provided additional records, new insights on the biogeography, invasion history, discussion and final editing of the manuscript. All authors have read and agreed to the published version of the manuscript.

Funding: This study was funded by the Korean Ministry of Oceans and Fisheries (MOF) for the Korea Institute of Marine Science & Technology Promotion (Project KIMST 20220526 "Development of living shoreline technology based on blue carbon science toward climate change adaptation").

Institutional Review Board Statement: Not applicable.

Informed Consent Statement: Not applicable.

Data Availability Statement: All the research data are presented in the text of the paper and in the Supplementary Materials.

Acknowledgments: Our sincere thanks to Maria Cristina Gambi (Italy), Nicolas Lavesque (France), Céline Labrune (France), Barbara Mikac (Croatia), João Nogueira (Brazil), Sean Handley (New Zealand), Carol Simon (South Africa), and Sergey Nuzhdin (USA) for providing material for this study. Vyatcheslav V. Potin (Russia), Leslie H. Harris (USA), and Stephen Keable (Australia) helped with museum collections. Leslie H. Harris also provided an excellent photo of *P. hoplura* from California. To all these colleagues, we express our sincere gratitude.

Conflicts of Interest: The authors declare no conflict of interest.

References

1. Lauckner, G. Diseases of Mollusca: Gastropoda. In *Diseases of Marine Animals*; Kinne, O., Ed.; John Wiley & Sons: Chichester, UK, 1980; Volume I, pp. 311–424.
2. Lauckner, G. Diseases of Mollusca: Bivalvia. In *Diseases of Marine Animals*; Kinne, O., Ed.; Biologische Anstalt Helgoland: Hamburg, Germany, 1983; Volume II, pp. 477–961.
3. Handley, S.J. Mudworm infestations and oyster production. *Aquacult. Update MAF Fish.* **1993**, *8*, 6–7.
4. Diggles, B.K.; Hine, P.M.; Handley, S.J.; Boustead, N.C. A handbook of diseases of importance to aquaculture in New Zealand. *NIWA Sci. Technol. Ser.* **2002**, *49*, 1–200.
5. Sato-Okoshi, W.; Okoshi, K.; Abe, H.; Dauvin, J.-C. Polydorid species (Annelida: Spionidae) associated with commercially important oyster shells and their shell infestation along the coast of Normandy, in the English Channel, France. *Aquac. Int.* **2023**, *31*, 195–230. [CrossRef]
6. Simon, C.A.; Sato-Okoshi, W. Polydorid polychaetes on farmed molluscs: Distribution, spread and factors contributing to their success. *Aquac. Environ. Interact.* **2015**, *7*, 147–166. [CrossRef]
7. Alvarez-Aguilar, A.; Van Rensburg, H.; Simon, C.A. Impacts of alien polychaete species in marine ecosystems: A systematic review. *J. Mar. Biol. Assoc. United Kingd.* **2022**, *102*, 3–26. [CrossRef]
8. Radashevsky, V.I.; Pankova, V.V.; Malyar, V.V.; Carlton, J.T. Boring can get you far: Shell-boring *Dipolydora* from Temperate Northern Pacific, with emphasis on the global history of *Dipolydora giardi* (Mesnil, 1893) (Annelida: Spionidae). *Biol. Invasions* **2023**, *25*, 741–772. [CrossRef]
9. Williams, L.-G.; Karl, S.A.; Rice, S.; Simon, C. Molecular identification of polydorid polychaetes (Annelida: Spionidae): Is there a quick way to identify pest and alien species? *Afr. Zool.* **2017**, *52*, 105–117. [CrossRef]
10. Carlton, J.T. Deep invasion ecology and the assembly of communities in historical time. In *Biological Invasions in Marine Ecosystems. Ecological Studies 204*; Rilov, G., Crooks, J.A., Eds.; Springer Verlag: Berlin, Germany, 2009; pp. 13–56.
11. Radashevsky, V.I.; Pankova, V.V. The morphology of two sibling sympatric *Polydora* species (Polychaeta: Spionidae) from the Sea of Japan. *J. Mar. Biol. Assoc. United Kingd.* **2006**, *86*, 245–252. [CrossRef]
12. Radashevsky, V.I.; Choi, J.-W.; Gambi, M.C. Morphology and biology of *Polydora hoplura* Claparède, 1868 (Annelida: Spionidae). *Zootaxa* **2017**, *4282*, 543–555. [CrossRef]
13. Radashevsky, V.I.; Migotto, A.E. First report of the polychaete *Polydora hoplura* (Annelida: Spionidae) from North and South America and Asian Pacific. *Mar. Biodivers.* **2017**, *47*, 859–868. [CrossRef]
14. Mohammad, M.-B.M. Intertidal polychaetes from Kuwait, Arabian Gulf, with descriptions of three new species. *J. Zool.* **1971**, *163*, 285–303. [CrossRef]
15. Radashevsky, V.I.; Olivares, C. *Polydora uncinata* (Polychaeta: Spionidae) in Chile: An accidental transportation across the Pacific. *Biol. Invasions* **2005**, *7*, 489–496. [CrossRef]
16. Sato-Okoshi, W.; Okoshi, K.; Koh, B.S.; Kim, Y.H.; Hong, J.-S. Polydorid species (Polychaeta, Spionidae) associated with commercially important mollusk shells in Korean waters. *Aquaculture* **2012**, *350–353*, 82–90. [CrossRef]
17. Sato-Okoshi, W.; Abe, H. Morphological and molecular sequence analysis of the harmful shell boring species of *Polydora* (Polychaeta: Spionidae) from Japan and Australia. *Aquaculture* **2012**, *368–369*, 40–47. [CrossRef]
18. Teramoto, W.; Sato-Okoshi, W.; Abe, H.; Nishitani, G.; Endo, Y. Morphology, 18S rRNA gene sequence and life history of a new *Polydora* species (Polychaeta: Spionidae) from northeastern Japan. *Aquat. Biol.* **2013**, *18*, 31–45. [CrossRef]
19. David, A.A.; Matthee, C.A.; Loveday, B.R.; Simon, C.A. Predicting the dispersal potential of an invasive polychaete pest along a complex coastal biome. *Integr. Comp. Biol.* **2016**, *56*, 600–610. [CrossRef] [PubMed]
20. Williams, L.; Matthee, C.A.; Simon, C.A. Dispersal and genetic structure of *Boccardia polybranchia* and *Polydora hoplura* (Annelida: Spionidae) in South Africa and their implications for aquaculture. *Aquaculture* **2016**, *465*, 235–244. [CrossRef]
21. Sato-Okoshi, W.; Abe, H.; Nishitani, G.; Simon, C.A. And then there was one: *Polydora uncinata* and *Polydora hoplura* (Annelida: Spionidae), the problematic polydorid pest species represent a single species. *J. Mar. Biol. Assoc. United Kingd.* **2017**, *97*, 1675–1684. [CrossRef]
22. Lee, D.-C.; Kim, K.-Y.; Han, J.; Kim, D.; Kim, B.-H. Genetic classification of polychaetes isolated from cultured abalone, *Haliotis discus hannai* shells. *Korean J. Malacol.* **2020**, *36*, 157–165.
23. Abe, H.; Sato-Okoshi, W. Molecular identification and larval morphology of spionid polychaetes (Annelida: Spionidae) from northeastern Japan. *Zookeys* **2021**, *1015*, 1–86. [CrossRef]

24. Almón, B.; Pérez-Dieste, J.; de Carlos, A.; Bañón, R. Identification of the shell-boring parasite *Polydora hoplura* (Annelida: Spionidae) on wild stocks of *Pecten maximus* in Galician waters, NW Spain. *J. Invertebr. Pathol.* **2022**, *190*, 107750. [CrossRef]
25. Winsnes, I.M. The use of methyl green as an aid in species discrimination in Onuphidae (Annelida, Polychaeta). *Zool. Scr.* **1985**, *14*, 19–23. [CrossRef]
26. Radashevsky, V.I.; Neretina, T.V.; Pankova, V.V.; Tzetlin, A.B.; Choi, J.-W. Molecular identity, morphology and taxonomy of the *Rhynchospio glutaea* complex with a key to *Rhynchospio* species (Annelida, Spionidae). *Syst. Biodivers.* **2014**, *12*, 424–433. [CrossRef]
27. Radashevsky, V.I.; Pankova, V.V.; Neretina, T.V.; Stupnikova, A.N.; Tzetlin, A.B. Molecular analysis of the *Pygospio elegans* group of species (Annelida: Spionidae). *Zootaxa* **2016**, *4083*, 239–250. [CrossRef]
28. Scholin, C.A.; Herzog, M.; Sogin, M.; Anderson, D.M. Identification of group- and strain-specific genetic markers for globally distributed *Alexandrium* (Dinophyceae). II. Sequence analysis of a fragment of the LSU rRNA gene. *J. Phycol.* **1994**, *30*, 999–1011. [CrossRef]
29. Ye, L.; Tang, B.; Wu, K.; Su, Y.; Wang, R.; Yu, Z.; Wang, J. Mudworm *Polydora lingshuiensis* sp. n is a new species that inhabits both shell burrows and mudtubes. *Zootaxa* **2015**, *3986*, 88–100. [CrossRef] [PubMed]
30. Katoh, K.; Misawa, K.; Kuma, K.i.; Miyata, T. MAFFT: A novel method for rapid multiple sequence alignment based on fast Fourier transform. *Nucleic Acids Res.* **2002**, *30*, 3059–3066. [CrossRef] [PubMed]
31. Katoh, K.; Standley, D.M. MAFFT Multiple Sequence Alignment Software Version 7: Improvements in Performance and Usability. *Mol. Biol. Evol.* **2013**, *30*, 772–780. [CrossRef]
32. Capella-Gutiérrez, S.; Silla-Martínez, J.M.; Gabaldón, T. trimAl: A tool for automated alignment trimming in large-scale phylogenetic analyses. *Bioinformatics* **2009**, *25*, 1972–1973. [CrossRef]
33. Nei, M.; Kumar, S. *Molecular Evolution and Phylogenetics*; Oxford University Press: Oxford, UK, 2000; p. 333.
34. Tamura, K.; Peterson, D.; Peterson, N.; Stecher, G.; Nei, M.; Kumar, S. MEGA5: Molecular evolutionary genetics analysis using maximum likelihood, evolutionary distance, and maximum parsimony methods. *Mol. Biol. Evol.* **2011**, *28*, 2731–2739. [CrossRef]
35. Vaidya, G.; Lohman, D.J.; Meier, R. SequenceMatrix: Concatenation software for the fast assembly of multi-gene datasets with character set and codon information. *Cladistics* **2011**, *27*, 171–180.
36. Nylander, J.A.A. *MrModeltest v2. Program Distributed by the Author*; Evolutionary Biology Centre, Uppsala University: Uppsala, Sweden, 2004.
37. Huelsenbeck, J.P.; Ronquist, F. MRBAYES: Bayesian inference of phylogenetic trees. *Bioinformatics* **2001**, *17*, 754–755. [CrossRef]
38. Ronquist, F.; Huelsenbeck, J.P. MRBAYES 3: Bayesian phylogenetic inference under mixed models. *Bioinformatics* **2003**, *19*, 1572–1574. [CrossRef]
39. Miller, M.A.; Pfeiffer, W.; Schwartz, T. Creating the CIPRES Science Gateway for inference of large phylogenetic trees. In Proceedings of the Gateway Computing Environments Workshop (GCE), New Orleans, LA, USA, 14 November 2010; IEEE: New Orleans, LA, USA, 2010; pp. 1–8.
40. Bandelt, H.-J.; Forster, P.; Röhl, A. Median-joining networks for inferring intraspecific phylogenies. *Mol. Biol. Evol.* **1999**, *16*, 37–48. [CrossRef] [PubMed]
41. Leigh, J.W.; Bryant, D. PopART: Full-feature software for haplotype network construction. *Methods Ecol. Evol.* **2015**, *6*, 1110–1116. [CrossRef]
42. Claparède, E. *Les Annélides Chétopodes du Golfe de Naples*; Ramboz et Schuchardt: Genève, Switzerland, 1868; p. 500.
43. Claparède, E. Les Annélides Chétopodes du Golfe de Naples. Seconde partie. *Mém. Soc. Phys. Hist. Nat. Genève* **1869**, *20*, 1–225.
44. Claparède, E. *Les Annélides Chétopodes du Golfe de Naples. Annélides Sédentaires*; Ramboz et Schuchardt: Genève, Switzerland, 1870; p. 225, 15 plates (17–31).
45. Andreu, B. Abundancia estacional de *Polydora* en las rías gallegas y problemas que plantea en el cultivo de la ostra. *Reun. Product. Pesq.* **1957**, *3*, 87–91.
46. Leloup, E.; Polk, P. La flore et la faune du bassin de chasse d'Ostende (1960–1961): III. Étude zoologique. *Mém. Inst. Roy. Sci. Nat. Belg.* **1967**, *157*, 1–114.
47. Gravina, M.F.; Ardizzone, G.D.; Belluscio, A. Polychaetes of an artificial reef in the central Mediterranean Sea. *Estuar. Coast. Shelf Sci.* **1989**, *28*, 161–172. [CrossRef]
48. David, A.A.; Matthee, C.A.; Simon, C.A. Poecilogony in *Polydora hoplura* (Polychaeta: Spionidae) from commercially important molluscs in South Africa. *Mar. Biol.* **2014**, *161*, 887–898. [CrossRef]
49. Lee, S.J.; Kwon, M.-G.; Lee, S.-R. Molecular detection for two abalone shell-boring species *Polydora haswelli* and *P. hoplura* (Polychaeta, Spionidae) from Korea using 18S rDNA and cox1 Markers. *Ocean Sci. J.* **2020**, *55*, 459–464. [CrossRef]
50. Rioja, E. Estudio de los poliquetos de la Península Ibérica. *Mem. Real Acad. Cienc. Exact. Fís. Nat. Madrid Ser. Cienc. Nat.* **1931**, *2*, 1–471.
51. Hartmann-Schröder, G. Annelida, Borstenwürmer, Polychaeta. *Tierwelt Dtschl.* **1971**, *58*, 1–594.
52. Hartmann-Schröder, G. Annelida, Borstenwürmer, Polychaeta. *Tierwelt Dtschl.* **1996**, *58*, 1–645, 2., neubearbeitete Auflage.
53. Day, J.H. *A Monograph on the Polychaeta of Southern Africa. Part 2. Sedentaria*; The British Museum (Natural History): London, UK, 1967; pp. i–xvii, 459–878.
54. Giard, A. Fragmentes biologiques. II. Deux ennemis de l'ostréiculture. *Bull. Sci. Dépt. Nord. Pays Voisins* **1881**, *13*, 70–73.
55. Dollfus, R.-P. Résumé de nos principales connaissances pratiques sur les maladies et les enemies de l'huître. *Off. Scient. Tech. Pêches Marit. Paris Notes Et Mém.* **1921**, *7*, 1–46.

56. Dollfus, R.-P. Sur l'attaque de la coquille des bigoneaux *Littorina littorea* (L.) de Hollande par *Polydora*. *Rev. Trav. Off. Pêches Marit. Paris* **1932**, *5*, 273–277.

57. Sato-Okoshi, W. Three new species of polydorids (Polychaeta, Spionidae) from Japan. *Species Divers.* **1998**, *3*, 277–288. [CrossRef]

58. Sato-Okoshi, W. Polydorid species (Polychaeta: Spionidae) in Japan, with descriptions of morphology, ecology and burrow structure. 1. Boring species. *J. Mar. Biol. Assoc. United Kingd.* **1999**, *79*, 831–848. [CrossRef]

59. Sato-Okoshi, W.; Okoshi, K.; Shaw, J. Polydorid species (Polychaeta: Spionidae) in south-western Australian waters with special reference to *Polydora uncinata* and *Boccardia knoxi*. *J. Mar. Biol. Assoc. United Kingd.* **2008**, *88*, 491–501. [CrossRef]

60. David, A.A.; Williams, J.D. Morphology and natural history of the cryptogenic sponge associate *Polydora colonia* Moore, 1907 (Polychaeta: Spionidae). *J. Nat. Hist. London* **2012**, *46*, 1509–1528. [CrossRef]

61. Blake, J.A.; Maciolek, N.J.; Meißner, K. Spionidae Grube, 1850. In *Handbook of Zoology: Annelida. Volume 2: Pleistoannelida, Sedentaria II*; Purschke, G., Böggemann, M., Westheide, W., Eds.; De Gruyter: Berlin, Germany, 2020; pp. 1–103.

62. David, A.A.; Simon, C.A. The effect of temperature on larval development of two non-indigenous poecilogonous polychaetes (Annelida: Spionidae) with implications for life history theory, establishment and range expansion. *J. Exp. Mar. Biol. Ecol.* **2014**, *461*, 20–30. [CrossRef]

63. Aguirrezabalaga, F. Contribución al estudio de los Anélidos Poliquetos de la costa de Guipúzcoa. *Munibe* **1984**, *36*, 119–130.

64. San Martín, G.; Aguirre, O. Ciclo anual de los poliquetos asociados al alga *Mesophyllum lichenoides* (Ellis) en una playa mediterránea. *Bol. Inst. Esp. Oceanogr.* **1991**, *7*, 157–170.

65. Wilson, D.P. The larvae of *Polydora ciliata* Johnston and *Polydora hoplura* Claparède. *J. Mar. Biol. Assoc. United Kingd.* **1928**, *15*, 567–603. [CrossRef]

66. Labura, Z.; Hrs-Brenko, M. Infestation of European flat oyster *(Ostrea edulis)* by polychaete *(Polydora hoplura)* in the northern Adriatic Sea. *Acta Adriat.* **1990**, *31*, 173–181.

67. Nel, R.; Coetzee, P.S.; Van Niekerk, G. The evaluation of two treatments to reduce mud worm *(Polydora hoplura* Claparède) infestation in commercially reared oysters *(Crassostrea gigas* Thunberg). *Aquaculture* **1996**, *141*, 31–39. [CrossRef]

68. Handley, S.J.; Bergquist, P.R. Spionid polychaete infestations of intertidal Pacific oysters *Crassostrea gigas* (Thunberg), Mahurangi harbour, northern New Zealand. *Aquaculture* **1997**, *153*, 191–205. [CrossRef]

69. Lleonart, M. *Australian Abalone Mudworms: Avoidance & Identification. A Farm Manual*; Fisheries Research & Development Corporation: Tasmania, Australia, 2001; p. 00.

70. Lleonart, M.; Handlinger, J.; Powell, M. Spionid mud worm infestation of farmed abalone *(Haliotis* spp.). *Aquaculture* **2003**, *221*, 85–96. [CrossRef]

71. Simon, C.A.; Ludford, A.; Wynne, S. Spionid polychaetes infesting cultured abalone, *Haliotis midae*, in South Africa. *Afr. J. Mar. Sci.* **2006**, *28*, 167–171. [CrossRef]

72. Boonzaaier, M.K.; Neethling, S.; Mouton, A.; Simon, C.A. Polydorid polychaetes (Spionidae) on farmed and wild abalone *(Haliotis midae)* in South Africa: An epidemiological survey. *Afr. J. Mar. Sci.* **2014**, *36*, 369–376. [CrossRef]

73. Simon, C.A. Observations on the composition and larval developmental modes of polydorid pests of farmed oysters *(Crassostrea gigas)* and abalone *(Haliotis midae)* in South Africa. *Invertebr. Reprod. Dev.* **2015**, *59*, 124–130. [CrossRef]

74. David, A.A. Climate change and shell-boring polychaetes (Annelida: Spionidae): Current state of knowledge and the need for more experimental research. *Biol. Bull.* **2021**, *241*, 4–15. [CrossRef] [PubMed]

75. Mackenzie, C.L.; Shearer, L.W. Chemical control of *Polydora websteri* and other annelids inhabiting oyster shells. *Proc. Natl. Shellfish. Ass.* **1959**, *50*, 105–111.

76. Ghode, G.; Kripa, V. *Polydora* infestation on *Crassostrea madrasensis*: A study on the infestation rate and eradication methods. *J. Mar. Biol. Assoc. India* **2001**, *43*, 1110–1119.

77. Handley, S.J. Optimizing intertidal Pacific oyster (Thunberg) culture, Houhora Harbour, northern New Zealand. *Aquac. Res.* **2002**, *33*, 1019–1030. [CrossRef]

78. Handlinger, J.; Lleonart, M.; Powell, M. Development of an Integrated Management Program for the Control of Spionid Mudworms in Cultured Abalone. FRDC Project No. 98/307. Report to the Fisheries Research & Development Corporation (FRDC). 2004, pp. 1–132. Available online: https://www.frdc.com.au/Archived-Reports/FRDC%20Projects/1998-307-DLD.PDF (accessed on 17 June 2019).

79. Simon, C.A.; Bentley, M.G.; Caldwell, G.S. 2,4-Decadienal: Exploring a novel approach for the control of polychaete pests on cultured abalone. *Aquaculture* **2010**, *310*, 52–60. [CrossRef]

80. Bilbao, A.; Núñez, J.; Viera, M.D.P.; Sosa, B.; Fernández-Palacios, H.; Hernández-Cruz, C.M. Control of shell-boring polychaetes in *Haliotis tuberculata coccinea* (Reeve 1846) aquaculture: Species identification and effectiveness of mebendazole. *J. Shellfish Res.* **2011**, *30*, 331–336. [CrossRef]

81. Radashevsky, V.I.; Malyar, V.V.; Pankova, V.V.; Nuzhdin, S.V. Molecular analysis of six *Rhynchospio* Hartman, 1936 species (Annelida: Spionidae) with comments on the evolution of brooding within the group. *Zootaxa* **2016**, *4127*, 579–590. [CrossRef]

82. Radashevsky, V.I.; Malyar, V.V.; Pankova, V.V.; Gambi, M.C.; Giangrande, A.; Keppel, E.; Nygren, A.; Al-Kandari, M.; Carlton, J.T. Disentangling invasions in the sea: Molecular analysis of a global polychaete species complex (Annelida: Spionidae: *Pseudopolydora paucibranchiata*). *Biol. Invasions* **2020**, *22*, 3621–3644. [CrossRef]

83. Radashevsky, V.I.; Pankova, V.V.; Malyar, V.V.; Neretina, T.V.; Choi, J.-W.; Yum, S.; Houbin, C. Molecular analysis of *Spiophanes bombyx* complex (Annelida: Spionidae) with description of a new species. *PLoS ONE* **2020**, *15*, e0234238. [CrossRef]

84. Radashevsky, V.I.; Al-Kandari, M.; Malyar, V.V.; Pankova, V.V. *Pseudopolydora* (Annelida: Spionidae) from the Arabian Gulf, Kuwait. *Europ. J. Taxon.* **2021**, *773*, 120–168. [CrossRef]
85. Radashevsky, V.I.; Pankova, V.V.; Neretina, T.V.; Tzetlin, A.B. Canals and invasions: A review of the distribution of *Marenzelleria* (Annelida: Spionidae) in Eurasia, with a key to *Marenzelleria* species and insights on their relationships. *Aquat. Invasions* **2022**, *17*, 186–206. [CrossRef]
86. Simon, C.A.; Sato-Okoshi, W.; Abe, H. Hidden diversity within the cosmopolitan species *Pseudopolydora antennata* (Claparède, 1869) (Spionidae: Annelida). *Mar. Biodivers.* **2019**, *49*, 25–42. [CrossRef]
87. Carazzi, D. Revisione del genere *Polydora* Bosc e cenni su due specie che vivono sulle ostriche. *Mitt. Zool. Stn Neapel* **1893**, *11*, 4–45.
88. Whitelegge, T. Report on the worm disease affecting the oysters on the coast of New South Wales. *Rec. Aust. Mus.* **1890**, *1*, 41–54. [CrossRef]
89. Day, J.H. The Polychaeta of South Africa. Part 3. Sedentary species from Cape shores and estuaries. *J. Linn. Soc. London Zool.* **1955**, *42*, 407–452. [CrossRef]
90. Mead, A.; Carlton, J.T.; Griffiths, C.L.; Rius, M. Introduced and cryptogenic marine and estuarine species of South Africa. *J. Nat. Hist. London* **2011**, *45*, 2463–2524. [CrossRef]
91. Carlton, J.T.; Cohen, A.N. Episodic global dispersal in shallow water marine organisms: The case history of the European shore crabs *Carcinus maenas* and *C. aestuarii*. *J. Biogeogr.* **2003**, *30*, 1809–1820. [CrossRef]
92. Read, G.B. Systematics and biology of polydorid species (Polychaeta: Spionidae) from Wellington Harbour. *J. Roy. Soc. N. Z.* **1975**, *5*, 395–419. [CrossRef]
93. Ruesink, J.L.; Lenihan, H.S.; Trimble, A.C.; Heiman, K.W.; Micheli, F.; Byers, J.E.; Kay, M.C. Introduction of non-native oysters: Ecosystem effects and restoration implications. *Annu. Rev. Ecol. Evol. Syst.* **2005**, *36*, 643–689. [CrossRef]
94. Herbert, R.J.H.; Humphreys, J.; Davies, C.J.; Roberts, C.; Fletcher, S.; Crowe, T.P. Ecological impacts of non-native Pacific oysters (*Crassostrea gigas*) and management measures for protected areas in Europe. *Biodivers. Conserv.* **2016**, *25*, 2835–2865. [CrossRef]
95. Naylor, R.L.; Williams, S.L.; Strong, D.R. Aquaculture—A gateway for exotic species. *Science* **2001**, *294*, 1655–1656. [CrossRef]
96. Tavares, M. On *Halicarcinus planatus* (Fabricius) (Brachyura, Hymenosomatidae) transported from Chile to Brazil along with the exotic oyster *Crassostrea gigas* (Thunberg). *Nauplius* **2003**, *11*, 45–50.
97. Hutchins, L.W. Species recorded from fouling. Chapter 10. In *Marine Fouling and Its Prevention*; United States Naval Institute: Annapolis, MD, USA, 1952; pp. 165–297.
98. Zevina, G.B. Fouling of vessels docking in Kol'skii Bay (Barents Sea). *Okeanologiya* **1962**, *2*, 126–133.
99. Davidson, I.; Ashton, G.; Zabin, C.; Ruiz, G. *Aquatic Invasive Species Vector Risk Assessments: The Role of Fishing Vessels as Vectors for Marine and Estuarine Species in California*; Final Report; Aquatic Bioinvasion Research & Policy Institute; Portland State University and the Smithsonian Environmental Research Center: Edgewater, NJ, USA, 2012; p. 70.
100. Meloni, M.; Correa, N.; Pitombo, F.B.; Chiesa, I.L.; Doti, B.; Elías, R.; Genzano, G.; Giachetti, C.B.; Giménez, D.; López-Gappa, J. In-water and dry-dock hull fouling assessments reveal high risk for regional translocation of nonindigenous species in the southwestern Atlantic. *Hydrobiologia* **2021**, *848*, 1981–1996. [CrossRef]
101. Locke, A.; Reid, D.M.; Sprules, W.G.; Carlton, J.T.; van Leeuwen, H.C. Effectiveness of mid-ocean exchange in controlling freshwater and coastal zooplankton in ballast water. *Can. Tech. Rep. Fish. Aquat. Sci.* **1991**, *1822*, 1–93.
102. Carlton, J.T.; Geller, J.B. Ecological roulette: The global transport and invasion of nonindigenous marine organisms. *Science* **1993**, *261*, 78–82. [CrossRef]
103. Zvyagintsev, A.Y.; Ivin, V.V.; Kashin, I.A.; Orlova, T.Y.; Selina, M.S.; Kasyan, V.V.; Korn, O.M.; Kornienko, E.S.; Kulikova, V.A.; Bezverbnaya, I.P.; et al. Ships' ballast water organisms in the port of Vladivostok. *Biol. Morya Vladivostok* **2009**, *35*, 29–40.
104. Chan, F.T.; MacIsaac, H.J.; Bailey, S.A. Relative importance of vessel hull fouling and ballast water as transport vectors of nonindigenous species to the Canadian Arctic. *Can. J. Fish Aquat. Sci.* **2015**, *72*, 1230–1242. [CrossRef]
105. Smith, L.D.; Wonham, M.J.; McCann, L.D.; Ruiz, G.M.; Hines, A.H.; Carlton, J.T. Invasion pressure to a ballast-flooded estuary and an assessment of inoculant survival. *Biol. Invasions* **1999**, *1*, 67–87. [CrossRef]
106. Zibrowius, H. Ongoing modification of the Mediterranean marine fauna and flora by the establishment of exotic species. *Mesogee* **1991**, *51*, 83–107.
107. Carlton, J.T. Transoceanic and interoceanic dispersal of coastal marine organisms: The biology of ballast water. *Oceanogr. Mar. Biol. Annu. Rev.* **1985**, *23*, 313–371.
108. Wolff, W.J. Non-indigenous marine and estuarine species in the Netherlands. *Zool. Med. Leiden* **2005**, *79*, 1–116.
109. Kerckhof, F.; Haelters, J.; Gollasch, S. Alien species in the marine and brackish ecosystem: The situation in Belgian waters. *Aquat. Invasions* **2007**, *2*, 243–257. [CrossRef]
110. Dinamani, P. The Pacific oyster, *Crassostrea gigas* (Thunberg, 1793), in New Zealand. In *Estuarine and Marine Bivalve Mollusk Culture*; Menzel, W., Ed.; CRC Press: Boca Raton, FL, USA, 1991; pp. 343–352.
111. Dinamani, P. Introduced Pacific oysters in New Zealand. In *The Ecology of Crassostrea gigas in Australia, New Zealand, France and Washington State*; Leffler, M., Greer, J., Eds.; Maryland Sea Grant College Program: College Park, MD, USA, 1992; pp. 9–11.
112. Dromgoole, F.I.; Foster, B.A. Changes to the marine biota of the Auckland Harbour. *Tane* **1983**, *29*, 79–96.
113. Stotz, W.B.; Caillaux, L.; Aburto, J. Interactions between the Japanese abalone *Haliotis discus hannai* (Ino 1953) and Chilean species: Consumption, competition, and predation. *Aquaculture* **2006**, *255*, 447–455. [CrossRef]

114. Crooks, J.A.; Crooks, K.R.; Crooks, A.J. Observations of the non-native Pacific oyster (*Crassostrea gigas*) in San Diego County, California. *Calif. Fish Game* **2015**, *101*, 101–107.
115. Hartman, O. Polychaetous annelids from California. *Allan Hancock Pac. Exped.* **1961**, *25*, 1–226.
116. Hartman, O. *Atlas of the Sedentariate Polychaetous Annelids from California*; Allan Hancock Foundation, University of Southern California: Los Angeles, CA, USA, 1969; p. 812.
117. Blake, J.A.; Woodwick, K.H. New species of *Polydora* (Polychaeta: Spionidae) from the coast of California. *Bull. South. Calif. Acad. Sci.* **1972**, *70*, 72–79.
118. Reish, D.J. A biological survey of Bahia de Los Angeles, Gulf of California, Mexico. II. Benthic polychaetous annelids. *Trans. San Diego Soc. Nat. Hist.* **1968**, *15*, 67–106. [CrossRef]
119. Reish, D.J. Invertebrates, especially benthic annelids in outer Anaheim Bay. *Fish Bull. Calif. Dept. Fish Game* **1975**, *165*, 73–78.
120. De Balbian Verster, J.F.L. A Japanese print of a Dutch East Indiaman. *Mar. Mirror* **1914**, *4*, 257–260. [CrossRef]
121. Hughes, R. The fatal shore. In *The Epic of Australia's Founding*; Vintage: New York, NY, USA, 1988; p. 628.
122. Carlton, J.T. Patterns of transoceanic marine biological invasions in the Pacific Ocean. *Bull. Mar. Sci.* **1987**, *41*, 452–465.
123. Carlton, J.T. Community assembly and historical biogeography in the North Atlantic Ocean: The potential role of human-mediated dispersal vectors. *Hydrobiologia* **2003**, *503*, 1–8. [CrossRef]
124. Darling, J.A.; Carlton, J.T. A framework for understanding marine cosmopolitanism in the Anthropocene. *Front. Mar. Sci.* **2018**, *5*, 1–25. [CrossRef]
125. Ojaveer, H.; Galil, B.S.; Carlton, J.T.; Alleway, H.; Goulletquer, P.; Lehtiniemi, M.; Marchini, A.; Miller, W.; Occhipinti-Ambrogi, A.; Peharda, M.; et al. Historical baselines in marine bioinvasions: Implications for policy and management. *PLoS ONE* **2018**, *13*, e0202383. [CrossRef]
126. Carlton, J.T. Pattern, process, and prediction in marine invasion ecology. *Biol. Conserv.* **1996**, *78*, 97–106. [CrossRef]
127. Day, J.H. *A Monograph on the Polychaeta of Southern Africa. Part 1. Errantia*; The British Museum (Natural History): London, UK, 1967; p. 458.
128. Hutchings, P.A.; Glasby, C.J. History of discovery. In *Polychaetes & Allies: The Southern Synthesis; Fauna of Australia; Vol. 4A Polychaeta, Myzostomida, Pogonophora, Echiurida, Sipunculida*; Beesley, P.L., Ross, G.J.B., Glasby, C.J., Eds.; CSIRO Publishing: Melbourne, Australia, 2000; pp. 3–9.
129. Haswell, W.A. On a destructive parasite of the rock oyster. *Proc. Linn. Soc. N. S. W.* **1885**, *10*, 273–275.
130. Roughley, T.C. The perils of an oyster. *Aust. Mus. Mag.* **1925**, *2*, 277–284.
131. Blake, J.A.; Kudenov, J.D. The Spionidae (Polychaeta) from southeastern Australia and adjacent areas with a revision of the genera. *Mem. Natl. Mus. Vic.* **1978**, *39*, 171–280. [CrossRef]
132. Day, J.H. On a collection of South African Polychaeta, with a catalogue of the species recorded from South Africa, Angola, Mosambique and Madagascar. *J. Linn. Soc. London Zool.* **1934**, *39*, 15–82. [CrossRef]
133. Crooks, J.A. Lag times and exotic species: The ecology and management of biological invasions in slow-motion. *Écoscience* **2005**, *12*, 316–329. [CrossRef]
134. Ruiz, G.M.; Fegley, L.; Fofonoff, P.; Cheng, Y.; Lemaitre, R. First records of *Eriocheir sinensis* H. Milne Edwards, 1853 (Crustacea: Brachyura: Varunidae) for Chesapeake Bay and the mid-Atlantic coast of North America. *Aquat. Invasions* **2006**, *1*, 137–142. [CrossRef]
135. Kerckhof, F.; De Mesel, I.; Degraer, S. First European record of the invasive barnacle *Balanus glandula* Darwin, 1854. *Bioinvasions Rec.* **2018**, *7*, 21–31. [CrossRef]
136. Ashton, G.V.; Willis, K.J.; Cook, E.J.; Burrows, M. Distribution of the introduced amphipod, *Caprella mutica* Schurin, 1935 (Amphipoda: Caprellida: Caprellidae) on the west coast of Scotland and a review of its global distribution. *Hydrobiologia* **2007**, *590*, 31–41. [CrossRef]
137. Okuda, S. Spioniform polychaetes from Japan. *J. Fac. Sci. Hokkaido Imp. Univ. Ser. VI Zool.* **1937**, *5*, 217–254.
138. Imajima, M.; Sato, W. A new species of *Polydora* (Polychaeta, Spionidae) collected from Abashiri Bay, Hokkaido. *Bull. Natl. Sci. Mus. Tokyo Ser. A* **1984**, *10*, 57–62.
139. Mori, K.; Sato, W.; Nomura, T.; Imajima, M. Infestation of the Japanese scallop *Patinopecten yessoensis* by the boring polychaetes, *Polydora*, on the Okhotsk Sea coast of Hokkaido, especially in Abashiri waters. *Bull. Jpn. Soc. Sci. Fish.* **1985**, *51*, 371–380. [CrossRef]
140. Sato-Okoshi, W.; Nomura, T. Infestation of the Japanese scallop *Patinopecten yessoensis* by the boring polychaetes *Polydora* on the coast of Hokkaido and Tohoku district. *Nippon. Suisan Gakkaishi* **1990**, *56*, 1593–1598. [CrossRef]
141. Kojima, H.; Imajima, M. Burrowing polychaetes in the shells of the abalone *Haliotis diversicolor aquatilis* chiefly on the species of *Polydora*. *Bull. Jpn. Soc. Sci. Fish.* **1982**, *48*, 31–35. [CrossRef]
142. Haupt, T.M.; Griffiths, C.L.; Robinson, T.B. Intra-regional translocations of epifaunal and infaunal species associated with cultured Pacific oysters *Crassostrea gigas*. *Afr. J. Mar. Sci.* **2012**, *34*, 187–194. [CrossRef]
143. David, A.A.; Loveday, B.R. The role of cryptic dispersal in shaping connectivity patterns of marine populations in a changing world. *J. Mar. Biol. Assoc. United Kingd.* **2018**, *98*, 647–655. [CrossRef]
144. Blakeslee, A.M.H.; Byers, J.E.; Lesser, M.P. Solving cryptogenic histories using host and parasite molecular genetics: The resolution of *Littorina littorea*'s North American origin. *Mol. Ecol.* **2008**, *17*, 3684–3696. [CrossRef]

145. Collin, R.; Ramos-Espla, A.; Izquierdo, A. Identification of the South Atlantic spiny slipper limpet *Bostrycapulus odites* Collin, 2005 (Caenogastropoda: Calyptraeidae) on the Spanish Mediterranean coast. *Aquat. Invasions* **2010**, *5*, 197–200. [CrossRef]

146. Pineda, M.C.; Lopez-Legentil, S.; Turon, X. The whereabouts of an ancient wanderer: Global phylogeography of the solitary ascidian *Styela plicata*. *PLoS ONE* **2011**, *6*, e25495. [CrossRef] [PubMed]

147. Carlton, J.T.; Ruiz, G.M. Vector science and integrated vector management in bioinvasion ecology: Conceptual frameworks. In *Invasive Alien Species: A New Synthesis*; Mooney, H.A., Mack, R.N., McNeely, J.A., Neville, L.E., Schei, P.J., Waage, J.K., Eds.; Island Press: Covelo, CA, USA, 2005; pp. 36–58.

148. Ayari, R.; Muir, A.; Paterson, G.; Afli, A.; Aissa, P. An updated list of polychaetous annelids from Tunisian coasts (Western Mediterranean Sea). *Cah. Biol. Mar.* **2009**, *50*, 33–45.

149. Bakalem, A.; Gillet, P.; Pezy, J.-P.; Dauvin, J.-C. Inventory and the biogeographical affinities of Annelida Polychaeta in the Algerian coastline (Western Mediterranean). *Mediterr Mar Sci* **2020**, *21*, 157–182. [CrossRef]

150. Bénard, F. La faunule associée au Lithophyllum incrustans Phyl des cuvettes de la région de Roscoff. *Cah. Biol. Mar.* **1960**, *1*, 89–102. [CrossRef]

151. Correia, M.J.; Costa, J.L.; Chainho, P.; Félix, P.M.; Chaves, M.L.; Medeiros, J.P.; Silva, G.; Azeda, C.; Tavares, P.; Costa, A.; et al. Inter-annual variations of macrobenthic communities over three decades in a land-locked coastal lagoon (Santo André, SW Portugal). *Estuar. Coast. Shelf Sci.* **2012**, *110*, 168–175. [CrossRef]

152. Çinar, M.E.; Dagli, E. Bioeroding (boring) polychaete species (Annelida: Polychaeta) from the Aegean Sea (eastern Mediterranean). *J. Mar. Biol. Assoc. U.K.* **2021**, *101*, 309–318. [CrossRef]

153. Douvillé, H. Perforations d'Annélides. *Bull. Soc. Géol. France sér. 4* **1907**, *7*, 361–370.

154. Fischer, P.-H. Association occasionnelle du *Purpura lapillus* L. avec un Annélide Polychète (Polydora hoplura Claparède). *J. Conch.* **1930**, *74*, 35–38.

155. Giard, A. Le laboratoire de Wimereux en 1889, (Recherches fauniques). *Bull. Sci. Fr. Belg.* **1890**, *22*, 60–87.

156. Gillet, P. Structure des peuplements intertidaux d'Annélides Polychètes de l'estuaire du Bou Regreg (Maroc). *Bull. Ecol.* **1988**, *19*, 33–42.

157. Graeffe, E. Übersicht der Fauna des Golfes von Triest nebst Notizen über Vorkommen, Lebensweise, Erscheinungs- und Laichzeit der einzelnen Arten. X. Vermes. I. Teil. *Arb. zool. Inst. Univ. Wien Zool. Stat. Trieste* **1905**, *15*, 317–332.

158. Handley, S.J. Spionid polychaetes in Pacific oysters, Crassostrea gigas (Thunberg) from Admiralty Bay, Marlborough Sounds, New Zealand. *N. Z. J. Mar. Fresh. Res.* **1995**, *29*, 305–309. [CrossRef]

159. Hutchings, P.A.; Turvey, S.P. The Spionidae of South Australia (Annelida: Polychaeta). *Trans. R. Soc. S. Aust.* **1984**, *108*, 1–20.

160. Inglis, G.; Gust, N.; Fitridge, I.; Floerl, O.; Woods, C.; Hayden, B.; Fenwick, G. Port of Nelson. Baseline survey for non-indigenous species. *Biosec. N. Z. Tech. Pap.* **2006**, *2005/02*, 1–79.

161. Jimi, N.; Yasuoka, N.; Kajihara, H. Polychaetes collected from floats of oyster-farming rafts in Kure, the Seto Inland Sea, Japan, with notes on the pest species Polydora hoplura (Annelida: Spionidae). *Bull. Kitakyushu Mus. Nat. Hist. Hum. Hist., Ser. A (Nat. Hist.)* **2018**, *16*, 1–4.

162. Karalis, P.; Antoniadou, C.; Chintiroglou, C. Structure of the artificial hard substrate assemblages in ports in Thermaikos Gulf (North Aegean Sea). *Oceanol. Acta* **2003**, *26*, 215–224. [CrossRef]

163. Korringa, P. The shell of Ostrea edulis as a habitat. Observations on the epifauna of oysters living in the Oosterschelde, Holland, with some notes on polychaete worms occurring there in other habitats. *Arch. Neerland. Zool.* **1951**, *10*, 32–152. [CrossRef]

164. Marion, A.F.; Bobretzky, N. Étude des Annélides du golfe de Marseille. *Ann. Sci. Nat. Zool. Biol. Anim.* **1875**, *2*, 1–106.

165. McIntosh, W.C. Notes from the Gatty Marine Laboratory, St. Andrews. No. XXXI. 2. On the British Spionidae. 3. On the Spionidae dredged by H.M.S. 'Porcupine' in 1869 and 1870. *Ann. Mag. Nat. Hist. London, Ser. 8* **1909**, *3*, 156–180.

166. McIntosh, W.C. *A Monograph of the British Marine Annelids. Vol. 3. Part 1.—Text. Polychæta. Opheliidæ to Ammocharidæ*; Ray Society: London, UK, 1915; p. 368.

167. Pruvot, G. Essai sur les fonds et la faune de la Manche occidentale (côtes de Bretagne) comparés à ceux du Golfe du Lion. *Arch. Zool. Expér. Génér. sér. 3* **1897**, *5*, 511–660.

168. Rioja, E. Nota sobre algunos anélidos recogidos en Málaga. *Bol. R. Soc. Esp. Hist. Nat.* **1917**, *1917*, 176–185.

169. Russell, B.; Hewitt, C. Baseline Survey of the Port of Darwin for Introduced Marine Species. 2000. Available online: https://researchportal.murdoch.edu.au/esploro/outputs/report/Baseline-survey-of-the-Port-of/991005544972007891 (accessed on 17 June 2019).

170. Saint-Joseph, A.d. Les Annélides Polychètes des côtes de Dinard, pt. 3. *Ann. Sci. Nat. VII Sér. Zool. Paris* **1894**, *17*, 1–395.

171. Simon, C.A. Polydora and Dipolydora (Polychaeta: Spionidae) associated with molluscs on the south coast of South Africa, with descriptions of two new species. *Afr. Invertebr.* **2011**, *52*, 39–50. [CrossRef]

172. Simon, C.A.; Booth, A.J. Population structure and growth of polydorid polychaetes that infest the cultured abalone, Haliotis midae. *Afr. J. Mar. Sci.* **2007**, *29*, 499–509. [CrossRef]

173. Soulier, A. Revision des Annélides de la région de Cette. pt. 2. *Mém. Acad. Sci. Lett. Montpellier. Sec. Sci. 2e Sér.* **1903**, *3*, 193–278.

174. Stjepčević, J. Ekologija dagnje (Mytilus galloprovincialis Lamk.) i kamenice (*Ostrea edulis* L.) u gajilištima Bokokotorskog zaliva. *Stud. Mar.* **1974**, *7*, 5–164.

175. Vasconcelos, P.; Cúrdia, J.; Castro, M.; Gaspar, M.B. The shell of Hexaplex (Truncanlariopsis) trunculus (Gastropoda: Muricidae) as a mobile hard substratum for epibiotic polychaetes (Annelida: Polychaeta) in the Ria Formosa (Algarve coast-southern Portugal). *Hydrobiologia* **2007**, *575*, 161–172. [CrossRef]
176. Villalba, A.; Viéitez, J.M. Estudio de la fauna de anélidos poliquetos de sustrato rocoso intermareal de una zona contaminada de la ría de Pontevedra (Galicia). I Resultados biocenóticos. *Cah. Biol. Mar.* **1985**, *26*, 359–377.
177. Wilson, D.P.; Tebble, N. Annelida. In *Plymouth Marine Fauna*; Bruce, J.R., Colman, J.S., Jones, N.S., Eds.; Latimer, Trend & Co.: Plymouth, UK, 1957; pp. 109–151.
178. Zarkanellas, A.J.; Bogdanos, C.D. Benthic studies of a polluted area in the Upper-Saronikos Gulf. *Thalassographica* **1977**, *2*, 155–177.
179. Zintzen, V.; Massin, C. Artificial hard substrata from the Belgian part of the North Sea and their influence on the distributional range of species. *Belg. J. Zool.* **2010**, *140*, 20–29.

Article

Composition and Long-Term Variation Characteristics of Coral Reef Fish Species in Yongle Atoll, Xisha Islands, China

Jinfa Zhao [1,2,3,4], Chunhou Li [1,2,3,4], Teng Wang [1,2,3,4,*], Juan Shi [1,2,3,4], Xiaoyu Song [1,2,3,4] and Yong Liu [1,2,3,4,*]

[1] Key Laboratory of South China Sea Fishery Resources Exploitation and Utilization, Ministry of Agriculture and Rural Affairs, South China Sea Fisheries Research Institute, Chinese Academy of Fishery Sciences, Guangzhou 510300, China; zhaojf2019@163.com (J.Z.); scslch@vip.163.com (C.L.); sjuan0917@163.com (J.S.); sxy1289667672@163.com (X.S.)

[2] Scientific Observation and Research Station of Xisha Island Reef Fishery Ecosystem of Hainan Province, Key Laboratory of Efficient Utilization and Processing of Marine Fishery Resources of Hainan Province, Sanya Tropical Fisheries Research Institute, Sanya 572018, China

[3] Guangdong Provincial Key Laboratory of Fishery Ecology Environment, Guangzhou 510300, China

[4] Observation and Research Station of Pearl River Estuary Ecosystem, Guangzhou 510300, China

[*] Correspondence: wangteng@scsfri.ac.cn (T.W.); liuyong@scsfri.ac.cn (Y.L.); Tel.: +86-18-929-597-042 (T.W.); +86-13-632-252-885 (Y.L.)

Simple Summary: The coral reef ecosystem not only brings enormous economic value to humans but also provides livelihoods and a major source of protein for millions of people. Moreover, coral reefs provide refuge and food sources for many fish species and are also breeding grounds and spawning grounds for various fish species. However, due to climate and human factors, the coral reef ecosystem has been destroyed, and its ecological function has been damaged. Fishery resources have also been affected, with changes in fish species composition and community structure. This study analyzed the fish community structure of the largest atoll in the Xisha Islands in the South China Sea, summarized a list of coral reef fish in Yongle Atoll, and analyzed the reasons for the change in the fish community and the future variation trend. The completion of this study will contribute to the better protection and recovery of coral reef fish and provide an important reference for the enhancement and restoration of coral reef habitats in the Xisha Islands.

Citation: Zhao, J.; Li, C.; Wang, T.; Shi, J.; Song, X.; Liu, Y. Composition and Long-Term Variation Characteristics of Coral Reef Fish Species in Yongle Atoll, Xisha Islands, China. *Biology* **2023**, *12*, 1062. https://doi.org/10.3390/biology12081062

Academic Editor: M. Gonzalo Claros

Received: 11 June 2023
Revised: 18 July 2023
Accepted: 27 July 2023
Published: 28 July 2023

Abstract: Yongle Atoll was the largest atoll in the Xisha Islands of the South China Sea, and it was a coral reef ecosystem with important ecological and economic values. In order to better protect and manage the coral reef fish resources in Yongle Atoll, we analyzed field survey data from artisanal fishery, catches, and underwater video from 2020 to 2022 and combined historical research to explore the changes in fish species composition and community structure in Yongle Atoll over the past 50 years. The results showed that a total of 336 species of fish were found on Yongle Atoll, belonging to 17 orders and 60 families. Among them, Perciformes had the most fish species with 259 species accounting for 77.08% of the total number of species. The number of fish species in the coral reef of Yongle Atoll was exponentially correlated with its corresponding maximum length and significantly decreases with its increase. The fish community structure of Yongle Atoll changed, and the proportion of large carnivorous fish decreased significantly, while the proportion of small-sized and medium-sized fish increased. At the same time, Yongle Atoll has 18 species of fish listed on the IUCN Red List, 15 of which are large fish. The average taxonomic distinctness (Delta+, $\Delta+$) and the variation taxonomic distinctness (Lambda+, $\Lambda+$) in 2020–2022 were lower than the historical data, and the number of fish orders, families, and genera in Yongle Atoll has decreased significantly, which indicates that the current coral reef fish species in Yongle Atoll have closer relatives and higher fish species uniformity. In addition, the similarity of fish species in Yongle Atoll was relatively low at various time periods, further proving that the fish community structure has undergone significant variation. In general, due to multiple impacts, such as overfishing, fishing methods, environmental changes, and habitat degradation, the fish species composition of Yongle Atoll may have basically evolved from carnivorous to herbivorous, from large fish to small fish, and from complexity to simplicity, leaving Yongle Atoll in an unstable state. Therefore, we need to strengthen the continuous monitoring of the coral reef ecosystem in Yongle Atoll to achieve the protection and restoration of its ecological environment and fishery resources, as well as sustainable utilization and management.

Keywords: overfishing; habitat decline; individual size; food habits; conservation status

1. Introduction

Coral reefs are one of the most diverse marine ecosystems, known as the "undersea tropical rainforest", which nourishes one-third of the world's marine fish and generates enormous economic value for humanity [1]. Coral reef fish provide jobs for 6 million fishermen and direct food sources for hundreds of millions of people [2,3]. However, in recent decades, the health and stability of coral reef ecosystems have been under great pressure, and climate change has gradually become the main threat to coral reef ecosystems. Researchers have studied the changes and causes of coral bleaching events that occur repeatedly on the Great Barrier Reef in Australia and found that coral reefs have little resistance to extremely high temperatures, with global warming being the main factor contributing to their bleaching [4]. Moreover, ocean warming also indirectly threatens coral reefs by increasing the intensity of global cyclones, which can lead to coral reef destruction and associated loss of biological abundance and diversity [5]. Marine pollution caused by human activities poses a great threat to coral reef ecosystems. Plastic waste can be decomposed into microplastics in the ocean, which can enter the body of coral polyps and cause coral reef diseases [6]. In addition, unsustainable human fishing patterns have also caused great damage to the structure of coral reef fish populations. Studies have evaluated the distribution of coral reef fish communities and their interactions with the environment off the coast of East Africa and found that overfishing led to changes in the structure of coral reef fish populations with large fish populations declining rapidly and gradually being replaced by small fish with low economic value [7]. The decline of fish species will also lead to the proliferation of algae and corallivores (such as sea urchins and starfish). The balance between corals and algae will be broken, and a large number of algae will occupy the living space of corals, reducing coral coverage and further degrading coral reef habitats [8].

Yongle Atoll was located in the core area of the western islands of China's Xisha Islands. It was a typical coral atoll and the largest atoll in the Xisha Islands. The length of the atoll was 24 km from northeast to southwest, the width was 17 km from north to south, and the water depth was about 40 m. The lagoon area was 187.22 km², and the total area was 274 km² (excluding the Money Island reef). The atoll consists of 13 islands, of which, 9 are inhabited islands, including Drummond Island, Yagong Island, Robert Island, Observation Bank, etc. [9,10]. Its abundant fishery resources and high diversity of fish provide refuge and a food source for many fish in the surrounding sea, and it was also the spawning ground and nursery ground for a variety of fish. Since the 1970s, researchers have conducted research on Yongle Atoll, including fish species composition [11], early fish resources [12], fish biology [9], and coral reef community ecology [10]. However, the long-term variation of fish community structure under anthropogenic pressures and climate change was still poorly documented.

Therefore, in this study, we conducted a retrospective analysis of the survey fish data of Yongle Atoll from 1970 to 2022, compiled a list of coral reef fish in Yongle Atoll, analyzed the changes in fish species composition, community structure, feeding habits, and individual size, and analyzed the reasons for this change. This study will provide data for the study of coral reef fisheries in Yongle Atoll, theoretical reference for the protection and management of fisheries and the restoration and protection of coral reefs in the Xisha Islands, and basic data for the study of the geographical distribution of coral reef fishes.

2. Materials and Methods

2.1. Data Acquisition

The data in this study were collected from the historical research of coral reef fishes in Yongle Atoll (16°25′ N–16°36′ N, 111°34′ E–111°48′ E). The historical data of "Ichthyology of South China Sea Islands" [11] in 1970s and the historical archive data of 1998–1999, 2003, and 2005 of the South China Sea Fisheries Research Institute of Chinese Academy of Fishery Sciences were compiled, and the coral reef fish list of Yongle Atoll was formed by combining the field investigation of this study from 2020 to 2022. The survey method used in 1998–1999 was a combination of hand fishing and gill nets. Gill nets were used in the 2003 and 2005 surveys, and nets were generally lowered in the evening and collected in the morning. The survey methods from 2020 to 2021 were diving fishing and underwater video. The diving fishing time was generally from 20:00 to 24:00 in the evening, and the underwater video was conducted during the day. The obtained samples were frozen and kept back in the laboratory for anatomical experiments to preliminarily understand their food classification. The underwater video data were analyzed, identified, and sorted out according to the video image information (Figure 1).

Figure 1. Photos of coral reef fishes extracted from underwater video records of Yongle Atoll.

2.2. Feeding Habits

Fish feeding habits were divided into three groups, which were carnivorous, herbivorous, and omnivorous. Fish feeding habits comprehensively reference the Fishbase Database (https://www.fishbase.se/search.php, (accessed on 11 May 2023)) [13], the Ecology of Fishes on Coral Reefs [14], and the information on the Ichthyology of the South China Sea Islands [11]. It was found that the clastic feeding recorded by the Fishbase Database had a high overlap with the herbivorous studied by the Ecology of Fishes on Coral Reefs, and it was found that the detritivorous feeding classified by the Fishbase Database was basically herbivorous in anatomy. Therefore, the clastic feeding in the Fishbase Database was classified as herbivorous, and the distinction was determined by referring to the description in the Ichthyology of the South China Sea Islands [11].

2.3. Individual Size

The individual size of the fish could be divided into three types, which were large-sized fish: the maximum length was $\geq$65 cm; medium-sized fish: 65 cm > maximum length $\geq$ 35 cm; and small-sized fish: maximum length < 35 cm [15]. The maximum total length of fish was obtained from the Fishbase Database, which is the maximum total length of an adult individual of a fish species. And for very few fish that have not obtained the maximum full length, refer to the values for fish of the same genus.

2.4. Conservation Status

The conservation status of fish species was based on the International Union for the Conservation of Nature and Natural Resources Red List (IUCN Red List: https://www.iucnredlist.org/, accessed on 11 May 2023) [16]. The protection level of fish species was classified from low to high as Not Evaluated (NE), Data Deficient (DD), Least Concerned (LC), Near Threatened (Near Threatened). NT), Vulnerable (VU), Endangered (EN), Critically Endangered (CR), Extinct in the Wild (EW), and Extinction (EX).

2.5. Data Analyses

SPSS 24.0 software was used to analyze the correlation between the maximum length of coral reef fish and the number of species, and linear function, exponential function, power function, and logistic function equation were used to fit, with the largest correlation coefficient R^2 as the fitting equation to analyze the correlation characteristics. At the same time, an independent sample T-test was used to analyze the significance of differences in the maximum length of coral reef fish of different feeding habits, and the statistical significance criterion was set at 0.05. In addition, Excel 2019 and Origin 2021 were used for data statistical analysis, while Origin 2021 and Photoshop 2019 were used for image processing. The average taxonomic distinctness (Delta+, Δ+) and the variation taxonomic distinctness (Lambda+, Λ+) were used to analyze the diversity of coral reef fish communities [17]. Lambda (+) refers to the theoretical average of the average classification distance path between any pair of species in the species list, which does not change with the number of species. Delta (+) represents the average deviation degree of Lambda (+), represents the degree of difference in path length between species, and reflects the degree of uniform distribution of species composition and kinship. Compared to traditional diversity calculation methods, Lambda (+) and Delta (+) based on taxonomic status can explain the interspecific relationships of communities without being affected by sampling methods and natural changes of habitat ecological types [18]. The taxonomic diversity index Lambda (+) and Delta (+) were calculated using the Taxdtest software package of Primer5.2.

The average taxonomic distinctness (Delta+, Δ+) and the variation taxonomic distinctness (Lambda+, Λ+) [18]:

$$\Delta^+ = \frac{\sum \sum_{i<j} \omega_{ij}}{S(S-1)/2} \tag{1}$$

$$\Lambda^+ = \frac{\sum\sum_{ij}\left(\omega_{i<j} - \Delta^+\right)^2}{S(S-1)/2} \tag{2}$$

where ω_{ij} is the path length of the i and j species in the classification system tree; S is the number of species. The weights of the weighted path length among the 6 classification levels of phylum, class, order, family, genus, and species were set as 100.000, 83.333, 66.667, 50.000, 33.333, 16.667.

Jaccard (J_s) [19]:

$$J_s = \frac{c}{a+b-c} \tag{3}$$

where a is the number of fish species or the number of orders, families, and genera in year a; b is the number of fish species or the number of orders, families, and genera in year b; c is the number of fish species or the number of orders, families, and genera in common in the two years investigated. The similarity level was as follows: $0 < J_s < 0.25$ was extremely dissimilar; $0.25 \le J_s < 0.50$ was not similar; $0.50 \le J_s < 0.75$ was moderate similarity; $0.75 \le J_s < 1.00$ was extremely similar.

Coral Fish Diversity Index (CFDI) [20]:

$$N = 3.39(\text{CFDI}) - 20.595 \tag{4}$$

where N is the predicted total number of species, and CFDI is the total number of species in the six families of Acanthuridae, Chaetodontidae, Labridae, Scaridae, Pomacanthidae, and Pomacentridae.

3. Results

3.1. Fish Species Composition

Based on historical data and archival data of the South China Sea Fisheries Research Institute, this study shows that a total of 336 fish species have been found in Yongle Atoll (Figure 2, Table S1). The composition of fish species found in the surveys has increased significantly from the 1970s to 2022, with 123 more species in 2005 than in the 1970s and 93 more in 2022 than in 2005. Overall, the increase in fish species composition in each survey showed a downward trend. In particular, only nine new species were discovered in 2022 compared with 2021, which is a sharp decline from the previous data.

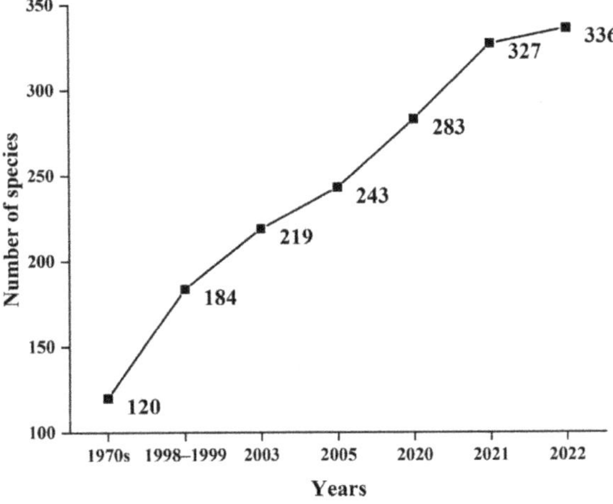

Figure 2. Temporal variation of fish species on the coral reef fishes of Yongle Atoll.

The coral reef fish in Yongle Atoll belong to the order Actinopterygii (322 species) and the order Chondrichthyes (14 species), consisting of 17 orders, namely, Perciformes, Anguilliformes, Tetraodontiformes, Beryciformes, Carcharhiniformes, Beloniformes, Myliobatiformes, Gasterosteiformes, Scorpaeniformes Squaliformes, Aulopiformes, Mugiliformes, Ophidiiformes, Pleuronectiformes, Polymixiiformes, Hexanchiformes, and Rajiformes. Among them, the Perciformes has the highest number of fish species with 259 species accounting for 77.08% of the total, followed by 17 species in the order Anguilliformes, 17 species in the order Tetraodontiformes, 14 species in the order Beryciformes, and less than 10 species in other orders (Figure 3). The coral reef fish in Yongle Atoll were composed of 60 families, among which, Labridae has the highest number of species, accounting for 33 species. The following were Pomacentridae (26 species), Scaridae (25 species), Chaetodontidae (24 species), Acanthuridae (21 species), Serranidae (21 species), Lutjanidae (18 species), Muraenidae (14 species), Holocentridae (14 species), Carangidae (12 species), and Lethrinidae (11 species), while the number of species in other families was lower than 10 (Figure 4).

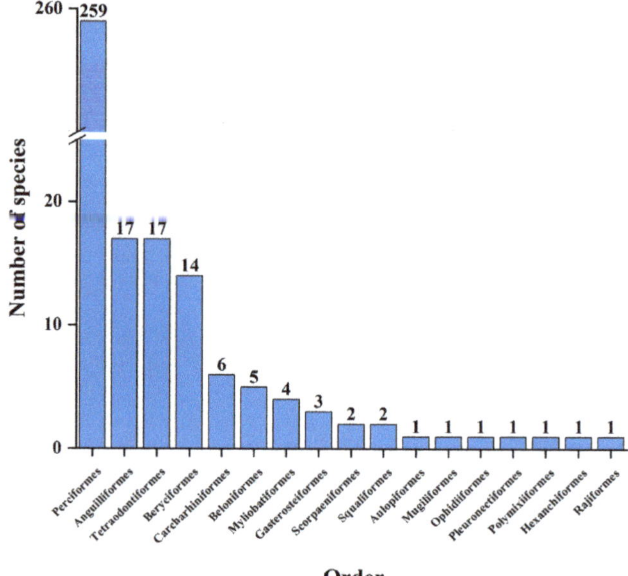

Figure 3. The distribution characteristics of species number at the order level of coral reef fishes in Yongle Atoll.

In the past 50 years, although Perciformes were the first dominant order in both periods, the composition of orders also showed significant changes, among which, the proportion of Perciformes increased to 79.10% (193 species) during 1970–2005 and 79.90% (155 species) during 2020–2022 (Figure 5). During 2020–2022, compared to 1970–2005, Carcharhiniformes, Myliobatiformes, Pleuronectiformes, Hexanchiformes, Aulopiformes, Polymixiiformes, Rajiformes, and Ophidiiformes did not reappear, while Mugiliformes emerged. During the two periods, the second dominant order of fishes also showed changes, from Anguilliformes (4.51%) during 1970–2005 to Tetraodontiformes (7.22%) during 2020–2022 and from carnivorous fishes to omnivorous fishes (Figure 5). In both periods, the dominant family of fish in Yongle Atoll was Labridae, and its proportion increased from 9.84% (24 species) during 1970–2005 to 10.82% (21 species) during 2020–2022. Compared to 1970–2005, there were 22 families that did not reappear in 2020–2022, while 9 new families emerged. Among the top 20 most species-rich families in both periods, Carcharhinidae,

Blenniidae, and Caesionidae lost their positions in the top 20 during 1970–2005 and were replaced by Diodontidae, Tetraodontidae, and Monacanthidae (Figure 6). In the past 50 years, the changes in fish composition in Yongle Atoll have basically presented the variation trend of miniaturization and "vegetarianization".

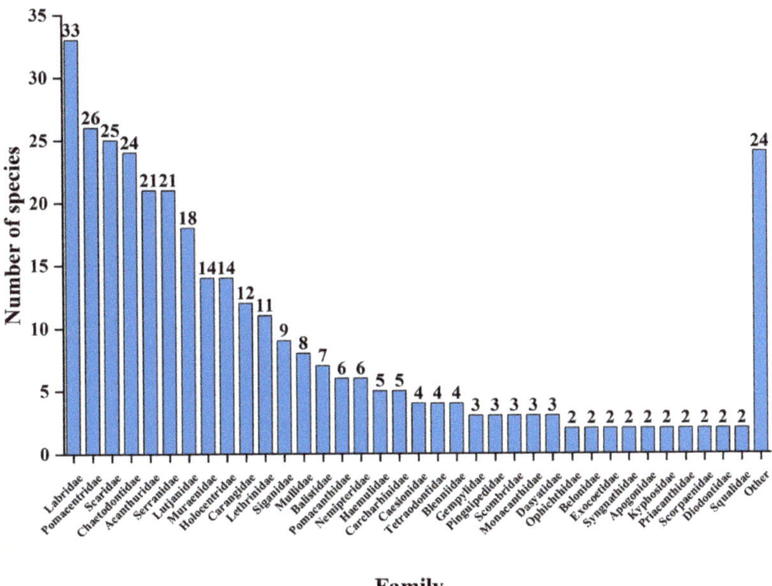

Figure 4. The distribution characteristics of species number at the family level of coral reef fishes in Yongle Atoll.

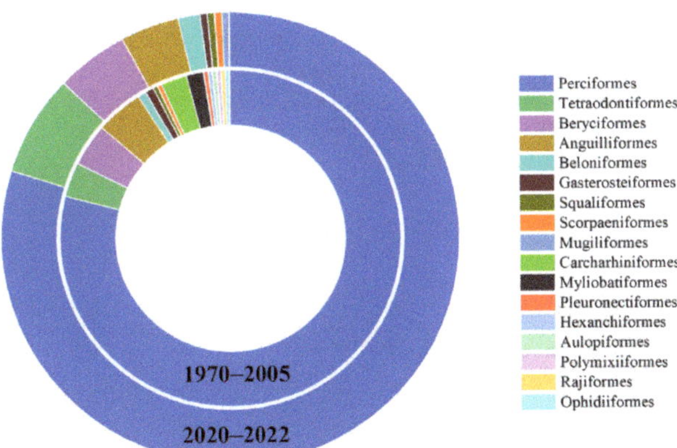

Figure 5. The distribution characteristics of species number at the order level of coral reef fishes in Yongle Atoll in different years.

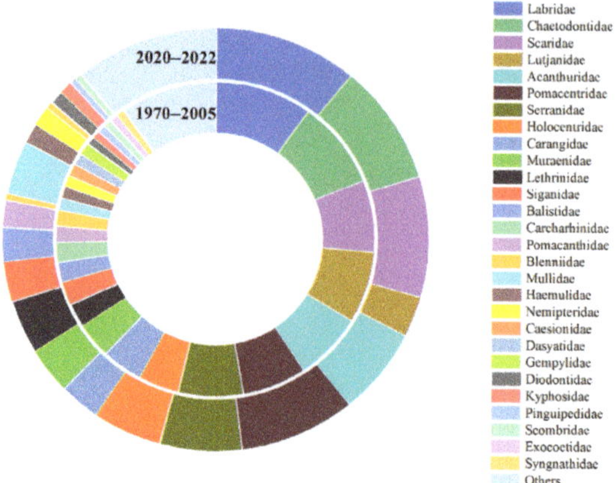

Figure 6. The distribution characteristics of species number at the family level of coral reef fishes in Yongle Atoll in different years.

3.2. Fish Community Structure and Changes

The maximum total length of coral reef fish in Yongle Atoll was distributed from 3.3 cm to 750 cm. The maximum total length of coral reef fish in Yongle Atoll was divided from 0 cm into an interval of 10 cm to calculate the number of species in each interval, and the median represented the maximum total length of this interval. There were no more than 1 species in the interval above 400 cm, and only 2 groups of intervals had species distribution. Therefore, these outliers were discarded in the analysis of species numbers with maximum total length distribution. The overall results showed that there was an exponential correlation between the number of species and their maximum total length. When the maximum total length was less than 120 cm, the number of species decreased significantly with the increase of the maximum total length, while when the maximum total length was greater than 120 cm, the number of species decreased slowly with the increase of the maximum total length. The results showed that the fish species in Yongle Atoll were mainly concentrated in the interval with a maximum total length of 0–120 cm, and the number of fish species in this interval was up to 299, accounting for 88.99% of the fish species in Yongle Atoll (Figure 7).

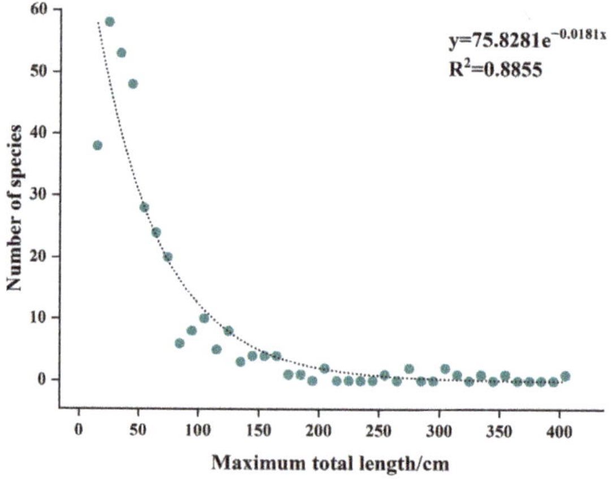

Figure 7. The distribution characteristics of species number at the maximum total length of coral reef fishes in Yongle Atoll.

In terms of body size, Yongle Atoll has 138 species of small-sized fish, accounting for 41.07% of the total number of fish species. Next came 107 species of medium-sized fishes, accounting for 31.85% of the total. There were 91 species of large-sized fish, accounting for 27.08% of the total. There was little difference in the number of fish species of different body types, and the whole fish group was mainly small-sized fish and medium-sized fish. In terms of feeding types, the number of carnivorous fish species was 203, accounting for 60.42% of the total fish species. Next were 74 species of omnivorous fish, accounting for 22.02% of the total, and 59 species of herbivorous fish, accounting for 17.56% of the total. Carnivorous fishes dominated the large-sized and medium-sized fish groups, accounting for 84.62% and 62.62%, respectively, followed by herbivorous fishes, accounting for 12.09% and 31.78%, respectively, and omnivorous fishes were the least, accounting for 3.30% and 5.61%, respectively. Among the small-sized fishes, omnivorous and carnivorous fish occupy an absolute advantage, accounting for 47.10% and 42.75% of their respective groups, respectively, and herbivorous fish were the least, accounting for only 10.14% (Figure 8).

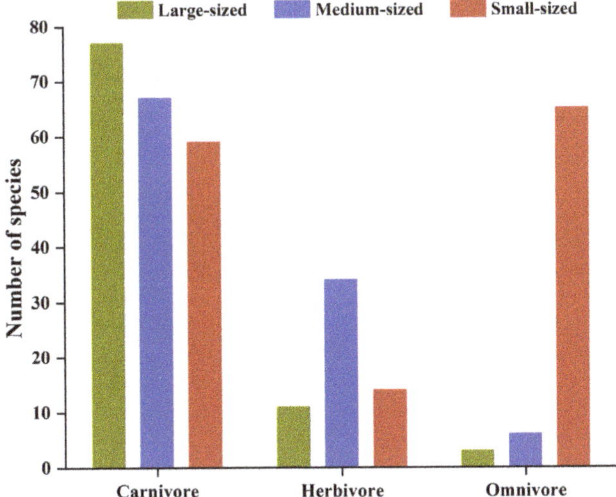

Figure 8. Distribution characteristics of coral reef fishes with different sizes and feeding habits in Yongle Atoll.

In carnivorous fish, the distribution of the three different sizes is more uniform, while medium-sized fish dominate in the herbivorous fish. Omnivorous fish were dominated by small-sized fish; the proportion was as high as 87.84% (Figure 8). The maximum total length distribution range of the three feeding habits of fish was significantly different. The average total length of the carnivorous fish was 76.8 ± 86.10 cm, and the distribution range was the largest, ranging from 3.32–750.00 cm. The individual omnivorous fish was the smallest with an average maximum total length of 23.6 ± 19.81 cm and a distribution range of 8.00–122.70 cm. The average maximum total length of phytophagous fish was 48.4 ± 21.47 cm, and the distribution range was minimal, ranging from 20.00–130.00 cm (Figure 9).

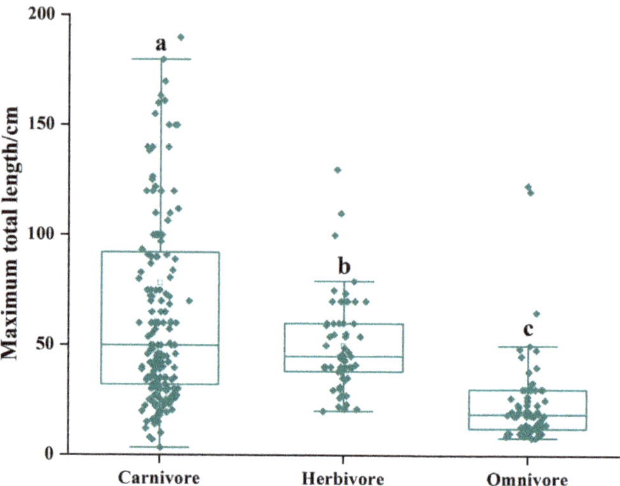

Figure 9. Body size characteristics of coral reef fishes with different feeding habits in Yongle Atoll. (a, b, c: different letters indicate significant differences.)

In the current survey (2020–2022), the highest proportion of small-sized fish in Yongle Atoll was 42.29% with a small increase compared to the historical data (1970s–2005). The proportion of large-sized fish was 21.65%, which changed obviously compared to the history, and the number of species decreased by 56, and the proportion decreased by 6.22%. The proportion of medium-sized fish was 36.08%, an increase of 6.16% compared to the historical level (Figure 10A). At the same time, the number of carnivorous fish species decreased significantly, decreasing by 58 species (9.00%) compared to the historical number, while the proportion of herbivorous fish and omnivorous fish increased by 3.62% and 5.39% compared to the historical data (Figure 10B). Moreover, compared to the historical data, the number of orders, families, and genera of coral reef fish in Yongle Atoll has significantly decreased, with the number of objects reduced by 7, the number of families reduced by 13, and the number of genera decreased by 43. In addition, the Lambda+(Λ+) and Delta+(Δ+) of coral reef fish in Yongle Atoll during 1970–2005 were 165.7 and 56.83, respectively. The Lambda+(Λ+) and Delta+(Δ+) values for coral reef fish in 2020–2022 were 114.7 and 54.85.

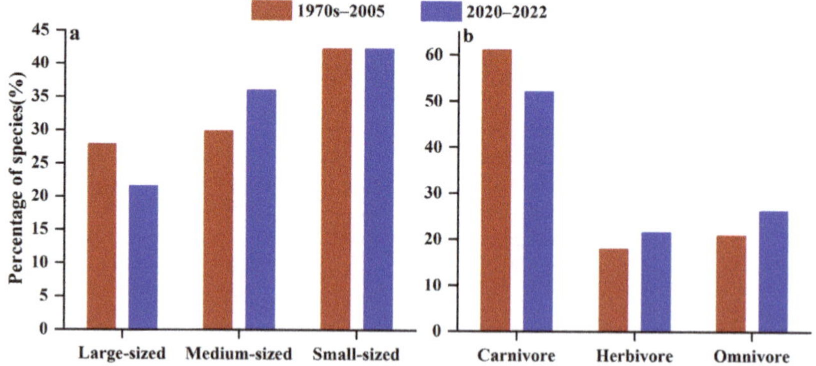

Figure 10. Historical comparative distribution characteristics of coral reef fishes of different sizes and feeding habits in Yongle Atoll. ((**a**). Different individual sizes, (**b**). Different feeding habits.)

From the perspective of similarity, the similarity index between the different time periods was relatively low. The similarity of order, family, genus, and species between the 1970s and 1998–2005 was 0.17, 0.22, 0.12, and 0.10, respectively. The similarity of order, family, genus, and species between 1970s and 2020–2022 was 0.25, 0.29, 0.28, and 0.22, respectively. The similarity of order, family, genus, and species between 1998–2005 and 2020–2022 were 0.27, 0.31, 0.24, and 0.21, respectively (Table 1).

Table 1. Fish species similarity at Yongle Atoll in different years.

Years	1998–2005				2020–2022			
	Order	Family	Genus	Species	Order	Family	Genus	Species
1970s	0.17	0.22	0.12	0.10	0.25	0.29	0.28	0.22
1998–2005	/	/	/	/	0.27	0.31	0.24	0.21

3.3. Disappearing Fish Community Structure

Compared to the historical survey, 142 species of fish were not found in this study, especially carnivorous fish (102 species), accounting for 71.83% of the total number of undiscovered fish followed by herbivorous and omnivorous fish 11.97% (17 species) and 16.20% (23 species), respectively. Moreover, all three different sizes of fish are predominantly carnivorous; 91.55% of these undiscovered fish belong to the order Perciformes, Anguilliformes, Beryciformes, Tetraodontiformes, Carcharhiniformes, and Myliobatiformes.

There was a total of 142 fish species that have not been rediscovered on the coral reef of Yongle Atoll, accounting for 42.26% of the total fish species. In terms of different body types, the disappearance rate of large-sized fish was the highest (53.85%) followed by small-sized fish (40.58%), and the disappearance rate of medium-sized fish was the lowest (34.58%). In terms of different feeding habits, the disappearance rate of carnivorous fish was the highest, accounting for 50.25% of the total number of carnivorous fish, followed by omnivorous fish, accounting for 31.08% of the total number of omnivorous fish, and the disappearance rate of herbivorous fish was the lowest, accounting for 28.81% of the total number of herbivorous fish (Figure 11). In addition, 18 species of fish were listed on the IUCN Red List in the coral reef of Yongle Atoll, of which, 15 species have not been rediscovered. There were three critically endangered species, *Carcharhinus longimanus*, *Rhynchobatus djiddensis*, and *Glyphis gangeticus*, which have not been found again. There were five endangered species, namely, *Cheilinus undulatus*, *Squalus japonicus*, *Mustellus griseus*, *Dasyatis sinensis*, and *Squalus brevirostris*. Among them, four species have not been rediscovered. Vulnerable species include six, including *Carcharhinus falciformis*, *Taeniura meyeni*, *Urugymnus asperrmius*, *Epinephelus fuscoguttatus*, and *Bolbometoton muriatum*, among which, five have not been found again. There were four species near threatened, namely, *Galeocerdo cuvier*, *Hexanchus griseus*, *Carcharhinus amblyrhynchoides*, and *Chaetodon trifascialis*. Three of them have not been found again (Figure 12). The above fish species, except for *Bolbometoton muriatum*, *Chaetodon trifascialis*, and *Squalus japonicus*, have not been found, and *Squalus japonicus* was a newly discovered fish in Yongle Atoll.

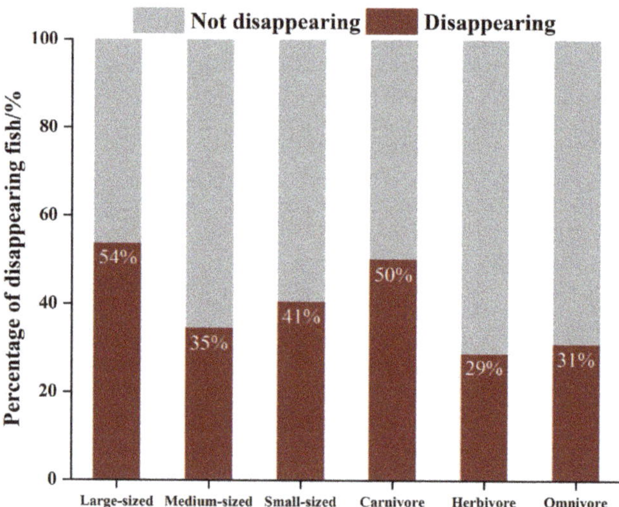

Figure 11. The proportion of the disappearing coral reef fishes in each functional group in Yongle Atoll.

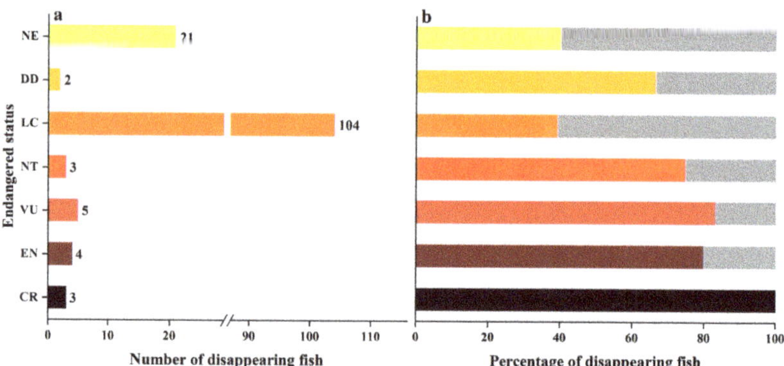

Figure 12. The Conservation state of the disappearing coral reef fishes in Yongle Atoll. ((**a**). Number of disappearing fish, (**b**). Percentage of disappearing fish.)

4. Discussion

4.1. Causes of Changes in Fish Species Composition

This study combined historical data to summarize the coral reef fish species in Yongle Atoll and found that there were a relatively rich number of 336 species of coral reef fish in Yongle Atoll. Compared with other islands of the Xisha Islands in the South China Sea, it was relatively rich in fish species (such as Dongdao Island (235 species) [21], Qilianyu Islands (315 species) [15], Yongxing Island (341 species) [22]), probably because Yongle Atoll has open lagoons, complex coral reef structures, more feeding microhabitats, greater off-reef transport rates, and more reliable food access. Moreover, it was more suitable as a spawning ground and nursery ground for fish [12,23]. At the same time, according to Allen's Coral Fish Diversity Index (CFDI) [20], which predicts the total species of coral reef fish in Yongle Atoll, including 6 families, including Acanthuridae, Chaetodontidae, Labridae, Scaridae, Pomacanthidae, and Pomacentridae, the total species number was 3.39 (CFDI)-20.595, resulting in a possible 437 species of coral reef fish in Yongle Atoll. Nearly 101 species have not been discovered, possibly due to their extinction or extremely low resource levels. In addition, the composition of the fish community in Yongle Atoll has undergone significant changes in the past 50 years, with the Anguilliformes gradually losing their second dominant position and being replaced by the Tetraodontiformes. Anguilliformes were typical cave-dwelling fish that have certain requirements for the three-dimensional structure of coral reefs. Their extinction may be caused by the destruction of the three-dimensional structure of coral reefs, habitat decline, or overfishing. Moreover, the species similarity in several historical periods of this study was particularly low, indicating that the fish succession and changes in Yongle Atoll were very significant, which also preliminarily supports the above results.

The coral reef waters of the Indian Pacific Ocean were extremely rich in coral reef fish, inhabiting over 100 families of fish. The research shows that 29 families of fish cover most of these coral reef fish, which were Serranidae, Chaetodontidae, Labridae, Scaridae, Gobiidae, Acanthuridae, Pomacentridae, Mullidae, Siganidae, Lutjanidae, Haemulidae, Lethrinidae, Apogonidae, Pseudochoromidae, Cirrhitidae, Pomacanthidae, Carangidae, Blenniidae, Holocentridae, Tetraodontidae, Balistidae, Nemipteridae, Monacanthidae, Scorpaenidae, Muraenidae, Syngnathidae, Pinguipedidae, Caesinidae, and Microdesmidae [24]. In this study, it was found in all the other families except Gobiidae and Microdesmidae, which accounted for 85.1 of the total species. The results of this study were similar to other islands of the Xisha Islands in the South China Sea. For example, 4 families were not found on Dongdao Island, and the remaining families accounted for 82.13% of the total species [21]. A total of 1 family was not found in the Qilianyu islands, and the rest accounted for 89.52% of the total species [15]. However, the proportion of these 29 families of coral reef fish to the total species was higher than that of Redang Islands in Malaysia (63.50%) [25], Mayotte Island in the Southwest Indian Ocean (73.80%) [26], and Ouvea Atoll in New Caledonia (79.60%) [27]. The average taxonomic distinctness (Delta+, Δ+) reflects the proximity of relatives between fish communities [28]. Compared to the historical study (1970–2005), the average taxonomic distinctness (Delta+, Δ+) of coral reef fish in Yongle Atoll in 2020–2022 was smaller, indicating that the coral reef fish in 2020–2022 were more closely related. The variation in taxonomic distinctness (Lambda+, Λ+) reflects the evenness of the taxonomic relationship, which was similar to the variation trend of the average taxonomic distinctness. Compared to historical studies (1970–2005), the coral coverage and habitat diversity of Yongle Atoll were significantly reduced, and the lack of environmental heterogeneity and stability reduced the opportunities for the coexistence of different species [29]. Therefore, the variation taxonomic distinctness (Lambda+, Λ+) decreased; that is, the uniformity of species taxonomic relationship among fish communities increased. Similar evolution trends also appeared in Terminos Lagoon [30] in the Gulf of Mexico and the Mauritanian coast [31] in the northwest of the Atlantic Ocean, and the taxonomic diversity index of fish communities showed a monotonous downward trend. In addition, the number of fish orders, families, and genera in Yongle Atoll has decreased significantly, which further verifies the above view. These all indicate that the current fish species have closer relatives and the evenness of fish species was higher, which will lead to the increase of instability of the ecosystem and reduce the resilience of the ecosystem [32,33] and further prove the necessity of monitoring and protection of their habitat changes.

4.2. Causes of Changes in Fish Community Structure

After years of long-term investigation, the species composition of coral reef fish in Yongle Atoll can basically represent the original state of this area. The coral reef fishes in Yongle Atoll were overwhelmingly carnivorous, which was consistent with the earlier research that most coral reef fishes were carnivorous [34]. However, the current survey found that compared to the historical data, carnivore fish had the most obvious changes. Among the undiscovered fish, carnivore fish accounted for the highest proportion, reaching 71.83%. The nutrition pyramid changed to a low trophic level but did not reverse directly, which was consistent with the rule that fish communities were initially disturbed [35]. In addition, 15 of the 18 kinds of fish listed in the IUCN Red List were the top carnivore fish in the ocean, which further indicates that carnivore fish were more likely to die out, which was in line with the law that the current fishery activities lead to the decline of the trophic level of the ecosystem. At the same time, both herbivorous and omnivorous fishes have increased to a certain extent compared to previous studies, especially the number of omnivorous fishes, which indicates that the coral reef ecosystem was in the process of succession from carnivorous fishes to herbivorous fishes. The results of this study were similar to those of other islands and reefs in the Xisha Islands (Qilianyu Islands [15], Yongxing Island [22], and Dongdao Island [21]) in the South China Sea, which was consistent with the succession process of coral reef fish in the Xisha Islands. A similar evolution has also occurred in coral reefs in Tanzania [36], Australia's Great Barrier Reef [37], and the Caribbean [38]. The change in the coral reef ecosystem will lead to an increase in herbivorous fish. In the short term, the change of fishery fishing to herbivorous fish has significant value for fishery discovery, but in the long term, overfishing will destroy the balance and function of the ecosystem [39].

Under the influence of climate and human activities, large-sized coral reef fish will decline and die out first. Because large fish have high economic value and are the main target species for fishing, the characteristics of large fish, such as slow growth and late sexual maturity, make it difficult to quickly replenish when the population is impacted [40,41]. This study just verified this theory; the highest proportion of large-sized fish disappearing among Yongle Atoll reef fish was 53.85%, while small-sized and medium-sized fish have increased. There were three possible explanations for the fact that smaller fish were better adapted to the current reef environment than larger fish. First, small fish have a higher population turnover rate and speciation rate, shorter life cycles, high reproduction rate, and abundant complementary groups, which can better adapt to the fragmentation of coral reef habitats [42]. Second, small fish live in a small range, their long-distance swimming ability is weak, the demand for food resources is low, and the same area can accommodate more small species [43]. Third, environmental factors can also affect the size of fish. Studies have analyzed the reproductive biological characteristics of female *Larimichthys polyactis* in the Yellow Sea and Bohai Sea from 1960 to 2020 and found that in addition to fishing pressure, the rise in water temperature was also the reason for the early sexual maturity and size reduction of *Larimichthys polyactis* [44]. It was worth noting that small-sized fish also account for a large proportion of undiscovered fish, especially reef fish. Most of these fish are highly dependent on the coral reef ecosystem, and some can only survive in the living coral habitat. The complex reef structure not only provides refuge for coral reef fish but also provides rich food for them [45,46]. Therefore, the decline of habitat may directly lead to the loss of habitat for these small reef-dwelling fish and lead to extinction. Coral cover in the Xisha Islands has also declined significantly in recent years [47]. In addition, we also found traces of *Acanthaster planci* in the underwater video, which also confirmed that the Yongle Atoll habitat was declining. To prevent further deterioration of the fish composition in the coral reef of Yongle Atoll, it was necessary to control the local fishing intensity, monitor and understand changes in its habitat, and take necessary measures to prevent further decline in this habitat.

Yongle Atoll was the largest atoll in the Xisha Islands. Affected by a large number of human activities in recent years, the species diversity of fish was generally declining, and the succession of fish composition was obvious. The main reasons for this phenomenon were as follows: Firstly, overfishing was a direct factor leading to a decrease in fish abundance and changes in community structure [48]. With the rapid development of marine fish, fishing techniques have become more progressive, and fishing intensity has increased year by year and far exceeds the replenishment capacity of marine fish resources, causing changes in fish community structure [49]. The research results of some studies and the Ecological Environment Bulletin of Hainan Province show that the density of fish in the coral reef of the Xisha Islands will decline from 310 fish/100 m^2 in 2005 to 146.7 fish/100 m^2 in 2021, further supporting the impact of human activities on fish in the islands [47]. Secondly, the change in fishing methods, especially selective fishing, was a key factor leading to changes in fish community structure. Large carnivorous fish were the key fishing objects of the fishery, while a large number of non-economic or low-economic value fish would not be selected [40,50]. As a result, large coral reef fish were more likely to be caught than smaller fish [51]. The apparent decline of large carnivorous fish in the Qilianyu Islands [15], Dongdao Island [21], and Yongxing Island [22] in the Xisha Islands, as well as reefs in Australia's Great Barrier Reef [37] and the Caribbean [38], further supports this view. Thirdly, there were changes in marine environmental factors, and ocean warming was a factor that affects fish diversity and resource levels. Rising water temperature can alter the rhythm of feeding, digestion, and movement activities of marine fish and reduce their reproductive capacity, as fish need to consume additional energy to cope with extreme temperature events [52,53]. Studies have used the Ecoath with Ecosim model to explore the impact of ocean warming on the coral reef ecosystem of the Xisha Islands in the South China Sea. The results showed that by the middle of this century, ocean warming will lead to a decline of 3.79% in total catch compared to 2009 [54]. Moreover, the sea surface temperature of the Xisha Islands rose faster than that of the Zhongsha and Nansha Islands [55]. Fourthly, the habitat of fish has been destroyed. Coral reefs provide habitats for a large number of marine fish, but with the increasing severity of human activities and climate change, a large number of coral reefs have undergone bleaching and death and have lost their three-dimensional structure, thereby losing their function as fish habitats, leading to the disappearance of a large number of marine fish [56–58]. Changes in the fish community in Australia's Great Barrier Reef over a period of 15 years following habitat degradation and restoration have been studied and found that the richness of fish communities was consistent with the trend of coral species restoration [46]. These results indicate that the restoration of coral communities and the complex interaction of coral reef frameworks determine the functional structure of related fish communities and further demonstrate the importance of coral reef habitats for fish communities.

In order to better understand and grasp the evolution dynamics of fish communities in the Yongle Atoll area in the future, it is necessary to increase the frequency of surveys as much as possible while protecting the ecological environment and use a unified survey method to obtain more high-quality continuous observation data. At the same time, combined with long-term historical data, the evolution trend of the fish community structure was analyzed from a long-term scale analysis to provide a scientific basis for sustainable development of fishery resources and policy management.

5. Conclusions

This study analyzed the evolution of the coral reef fish community structure in Yongle Atoll over the past 50 years and compiled a relatively complete list of coral reef fish in Yongle Atoll. At present, the coral reef fish in Yongle Atoll were mainly affected by factors such as overfishing, habitat degradation, and rising water temperature, resulting in a decrease in the richness and diversity of coral reef fish, changes in the community structure, and obvious succession of fish composition. This study was of great significance for the protection and restoration of coral reef fish in Yongle Atoll and provides an important reference for the enhancement and restoration of the coral reef habitat in the Xisha Islands.

Supplementary Materials: The following supporting information can be downloaded at: https://www.mdpi.com/article/10.3390/biology12081062/s1, Table S1: The checklist of coral reef fishes in Yongle Atoll.

Author Contributions: Conceptualization, T.W. and Y.L.; data curation, J.Z. and T.W.; formal analysis, T.W. and X.S.; investigation, J.Z., T.W., and Y.L.; methodology, J.Z. and T.W.; resources, C.L., T.W., and Y.L.; software, T.W. and J.S.; supervision, C.L. and T.W.; validation, C.L., T.W. and Y.L.; visualization, J.Z. and T.W.; writing—original draft, J.Z. and T.W.; writing—review and editing, C.L., T.W., and Y.L. All authors have read and agreed to the published version of the manuscript.

Funding: This study was funded by the Fundamental and Applied Fundamental Research Major Program of Guangdong Province (2019B030302004-05); the Hainan Provincial Natural Science Foundation (323MS124); the Hainan Provincial Natural Science Foundation (322CXTD530); the Central Public-interest Scientific Institution Basal Research Fund, CAFS (2020TD16); and the Financial Fund of the Ministry of Agriculture and Rural Affairs, China (NFZX2021).

Institutional Review Board Statement: Not applicable.

Informed Consent Statement: Not applicable.

Data Availability Statement: Part of the data presented in this study are available in the Supplementary Material. The remaining data presented in this study are available upon reasonable request from the corresponding author.

Conflicts of Interest: The authors declare that they have no known competing financial interests or personal relationships that could appear to have influenced the work reported here.

References

1. Fisher, R.; O'Leary, R.A.; Low-Choy, S.; Mengersen, K.; Knowlton, N.; Brainard, R.E.; Caley, M.J. Species Richness on Coral Reefs and the Pursuit of Convergent Global Estimates. *Curr. Biol.* **2015**, *25*, 500–505. [CrossRef] [PubMed]
2. Robinson, J.P.W.; Wilson, S.K.; Robinson, J.; Gerry, C.; Lucas, J.; Assan, C.; Govinden, R.; Jennings, S.; Graham, N.A.J. Productive instability of coral reef fisheries after climate-driven regime shifts. *Nat. Ecol. Evol.* **2019**, *3*, 183–190. [CrossRef] [PubMed]
3. Shi, J.; Li, C.H.; Wang, T.; Zhao, J.F.; Liu, Y.; Xiao, Y.Y. Distribution Pattern of Coral Reef Fishes in China. *Sustainability* **2022**, *14*, 15107. [CrossRef]
4. Hughes, T.P.; Kerry, J.T.; Alvarez-Noriega, M.; Alvarez-Romero, J.G.; Anderson, K.D.; Baird, A.H.; Babcock, R.C.; Beger, M.; Bellwood, D.R.; Berkelmans, R.; et al. Global warming and recurrent mass bleaching of corals. *Nature* **2017**, *543*, 373–377. [CrossRef]
5. Cheal, A.J.; Macneil, M.A.; Emslie, M.J.; Sweatman, H. The threat to coral reefs from more intense cyclones under climate change. *Glob. Change Biol.* **2017**, *23*, 1511–1524. [CrossRef]
6. Lamb, J.B.; Willis, B.L.; Fiorenza, E.A.; Couch, C.S.; Howard, R.; Rader, D.N.; True, J.D.; Kelly, L.A.; Ahmad, A.; Jompa, J.; et al. Plastic waste associated with disease on coral reefs. *Science* **2018**, *359*, 460–462. [CrossRef]
7. McCallen, E.; Knott, J.; Nunez-Mir, G.; Taylor, B.; Jo, I.; Fei, S.L. Trends in ecology: Shifts in ecological research themes over the past four decades. *Front. Ecol. Environ.* **2019**, *17*, 109–116. [CrossRef]
8. Pratchett, M.S.; Cumming, G.S. Managing cross-scale dynamics in marine conservation: Pest irruptions and lessons from culling of crown-of-thorns starfish (*Acanthaster* spp.). *Biol. Conserv.* **2019**, *238*, 108211. [CrossRef]
9. Wang, T.; Liu, Y.T.; Liu, Y.; Li, C.H.; Lin, L.; Xiao, Y.Y.; Wu, P.; Li, C.R. Reproductive biological characteristics of *Chlorurus sordidus* from the Yongle Atoll and Meiji Reef. *J. Fish. Sci. China* **2022**, *29*, 1366–1374.
10. Zhao, M.X.; Yu, K.F.; Shi, Q.; Yang, H.Q.; Riegl, B.; Zhang, Q.M.; Yan, H.Q.; Chen, T.R.; Liu, G.H.; Lin, Z.Y. The coral communities of Yongle atoll: Status, threats and conservation significance for coral reefs in South China Sea. *Mar. Freshw. Res.* **2016**, *67*, 1888–1896. [CrossRef]
11. South China Sea Fisheries Research Institute of the State Administration of Fisheries; Xiamen Fisheries College; South China Sea Institute of Oceanography, Chinese Academy of Sciences; South China Sea Fisheries Research Institute, Chinese Academy of Fishery Sciences. *Ichthyology of the South China Sea Islands*; Science Press: Beijing, China, 1979.
12. Zhao, J.F.; Liu, Y.; Li, C.H.; Wang, T.; Shi, J.; Xiao, Y.Y.; Wu, P.; Song, X.Y. Study on species composition and distribution of fish eggs in Yongle Atoll and Dongdao Island by high-throughput sequencing technology. *J. Trop. Oceanogr.* **2023**, 1–10. [CrossRef]
13. Fishbase Database. Available online: www.fishbase.se/search.php (accessed on 26 October 2022).
14. Peter, F.S. *The Ecology of Fishes on Coral Reefs*; Academic Press: San Diego, CA, USA, 1991.
15. Wang, T.; Liu, Y.; Quan, Q.M.; Xiao, Y.Y.; Wu, P.; Li, C.H. Species composition characteristics analysis of Qilianyu reef fishes of Xisha Islands. *J. Fish. Sci. China* **2022**, *29*, 102–117.
16. IUCN. Available online: www.iucnredlist.org (accessed on 29 October 2022).
17. Mouillot, D.; Laune, J.; Tomasini, J.A.; Aliaume, C.; Brehmer, P.; Dutrieux, E.; Chi, T.D. Assessment of coastal lagoon quality with taxonomic diversity indices of fish, zoobenthos and macrophyte communities. *Hydrobiologia* **2005**, *550*, 121–130. [CrossRef]
18. Clarke, K.R.; Warwick, R.M. A further biodiversity index applicable to species lists: Variation in taxonomic distinctness. *Mar. Ecol. Prog. Ser.* **2001**, *216*, 265–278. [CrossRef]
19. Zintzen, V.; Anderson, M.J.; Roberts, C.D.; Diebel, C.E. Increasing variation in taxonomic distinctness reveals clusters of specialists in the deep sea. *Ecography* **2011**, *34*, 306–317. [CrossRef]
20. Allen, G.R. Reef and shore fishes of Milne Bay Province, Papua New Guinea. In *A Rapid Biodiversity Assessment of the Coral Reefs of Milne Bay Province, Papua New Guinea*; Werner, T.B., Allen, G.R., Eds.; RAP working papers; Conservation International: Arlington County, VA, USA, 1988.
21. Wang, T.; Li, C.H.; Zhao, J.F.; Shi, J.; Yu, Y.F.; Xiao, Y.Y.; Liu, Y. The characteristica of species composition and succession of coral reef fishes in Xisha East Island. *Acta Hydrobiol. Sin.* **2023**, *47*, 1456–1463.
22. Wang, T.; Liu, Y.; Li, C.H.; Xiao, Y.Y.; Lin, L.; Li, C.R.; Xie, Y.F.; Wu, P. Characteristics of fish community structure in coral reefs adjacent to yongxing island of xisha islands. *Acta Hydrobiol. Sin.* **2023**, *47*, 674–683.

23. Gladstone, W. Selection of a spawning aggregation site by Chromis hypsilepis (Pisces: Pomacentridae): Habitat structure, transport potential, and food availability. *Mar. Ecol. Prog. Ser.* **2007**, *351*, 235–247. [CrossRef]

24. Allen, G.R. Indo-Pacific Coral-Reef Fishes as Indicators of Conservation Hotspots. In Proceedings of the Ninth International Coral Reef Symposium, Bali, Indonesia, 23–27 October 2000; Ministry of Environment of the Republic of Indonesia: Jakarta, Indonesia, 2002; Volume 2, pp. 921–926. [CrossRef]

25. Du, J.G.; Loh, K.H.; Hu, W.J.; Zheng, X.Q.; Affendi, Y.A.; Ooi, J.L.S.; Ma, Z.Y.; Rizman-Idid, M.; Chan, A.A. An updated checklist of the marine fish fauna of Redang Islands, Malaysia. *Biodivers. Data J.* **2019**, *7*, e47537. [CrossRef]

26. Wickel, J.; Jamon, A.; Pinault, M.; Durville, P.; Chabanet, P. Species composition and structure of marine fish communities of Mayotte Island (south-western Indian Ocean). *Cybium* **2014**, *38*, 179–203.

27. Kulbicki, M.; Williams, J.T. Checklist of the shorefishes of Ouvea Atoll New Caledonia. *Atoll Res. Bull.* **1997**, *444*, 1–26. [CrossRef]

28. Barjau-Gonzalez, E.; Rodriguez-Romero, J.; Galvan-Magana, F.; Lopez-Martinez, J. Changes in the taxonomic diversity of the reef fish community of San Jos, Island, Gulf of California, Mexico. *Biodivers. Conserv.* **2012**, *21*, 3543–3554. [CrossRef]

29. Shan, X.J.; Jin, X.S.; Yuan, W. Taxonomic diversity of fish assemblages in the Changjiang Estuary and its adjacent waters. *Acta Oceanol. Sin.* **2010**, *29*, 70–80. [CrossRef]

30. Paz-Rios, C.E.; Sosa-Lopez, A.; Torres-Rojas, Y.E.; del Rio-Rodriguez, R.E. Long-term multiscale analysis of temporal variability in the fish community in Terminos Lagoon. *Estuar. Coast. Shelf Sci.* **2022**, *277*, 108066. [CrossRef]

31. Kide, S.O.; Mante, C.; Demarcq, H.; Merigot, B. Groundfish assemblages diversity in upwelling ecosystems: Insights from the Mauritanian Exclusive Economic Zone. *Biodivers. Conserv.* **2021**, *30*, 2279–2304. [CrossRef]

32. Arai, T. Diversity and conservation of coral reef fishes in the Malaysian South China Sea. *Rev. Fish Biol. Fish.* **2015**, *25*, 85–101. [CrossRef]

33. Campbell, N.; Neat, F.; Burns, F.; Kunzlik, P. Species richness, taxonomic diversity, and taxonomic distinctness of the deep-water demersal fish community on the Northeast Atlantic continental slope (ICES Subdivision VIa). *ICES J. Mar. Sci.* **2011**, *68*, 365–376. [CrossRef]

34. Valentine, J.F.; Heck, K.L. Perspective review of the impacts of overfishing on coral reef food web linkages. *Coral Reefs* **2005**, *24*, 209–213. [CrossRef]

35. Mora, C. *Ecology of Fishes on Coral Reefs*; Cambridge University Press: Cambridge, UK, 2015.

36. Lindahl, U.; Ohman, M.C.; Schelten, C.K. The 1997/1998 mass mortality of corals: Effects on fish communities on a Tanzanian coral reef. *Mar. Pollut. Bull.* **2001**, *42*, 127–131. [CrossRef]

37. Cheal, A.J.; Wilson, S.K.; Emslie, M.J.; Dolman, A.M.; Sweatman, H. Responses of reef fish communities to coral declines on the Great Barrier Reef. *Mar. Ecol. Prog. Ser.* **2008**, *372*, 211–223. [CrossRef]

38. Alvarez-Filip, L.; Gill, J.A.; Dulvy, N.K. Complex reef architecture supports more small-bodied fishes and longer food chains on Caribbean reefs. *Ecosphere* **2011**, *2*, 1–17. [CrossRef]

39. Milne, R.; Bauch, C.T.; Anand, M. Local Overfishing Patterns Have Regional Effects on Health of Coral, and Economic Transitions Can Promote Its Recovery. *Bull. Math. Biol.* **2022**, *84*, 46. [CrossRef]

40. Shin, Y.J.; Rochet, M.J.; Jennings, S.; Field, J.G.; Gislason, H. Using size-based indicators to evaluate the ecosystem effects of fishing. *ICES J. Mar. Sci.* **2005**, *62*, 384–396. [CrossRef]

41. Bellwood, D.R.; Hughes, T.P.; Folke, C.; Nystrom, M. Confronting the coral reef crisis. *Nature* **2004**, *429*, 827–833. [CrossRef]

42. Brandl, S.J.; Goatley, C.H.R.; Bellwood, D.R.; Tornabene, L. The hidden half: Ecology and evolution of cryptobenthic fishes on coral reefs. *Biol. Rev.* **2018**, *93*, 1846–1873. [CrossRef]

43. Ansell, A.; Gibson, R.; Barnes, M.; Press, U. The ecological implications of small body size among coral-reef fishes. *Oceanogr. Mar. Biol. Annu. Rev.* **1998**, *36*, 373–411.

44. Li, Z.L.; Shan, X.J.; Jin, X.S.; Dai, F.Q. Long-term variations in body length and age at maturity of the small yellow croaker (Larimichthys polyactis Bleeker, 1877) in the Bohai Sea and the Yellow Sea, China. *Fish. Res.* **2011**, *110*, 67–74. [CrossRef]

45. Coker, D.J.; Wilson, S.K.; Pratchett, M.S. Importance of live coral habitat for reef fishes. *Rev. Fish Biol. Fish.* **2014**, *24*, 89–126. [CrossRef]

46. Emslie, M.J.; Cheal, A.J.; Sweatman, H.; Delean, S. Recovery from disturbance of coral and reef fish communities on the Great Barrier Reef, Australia. *Mar. Ecol. Prog. Ser.* **2008**, *371*, 177–190. [CrossRef]

47. LI, Y.C.; Wu, Z.J.; Liang, J.L.; Chen, S.Q.; Zhao, J.M. Analysis on the outbreak period and cause of *Acanthaster planci* in Xisha Islands in recent 15 years. *Chin. Sci. Bull.* **2019**, *64*, 3478–3484.

48. Coleman, F.C.; Williams, S.L. Overexploiting marine ecosystem engineers: Potential consequences for biodiversity. *Trends Ecol. Evol.* **2002**, *17*, 40–44. [CrossRef]

49. Du, J.G.; Chen, B.; Lu, Z.B.; Song, P.Q.; Xu, Z.C.; Yu, W.W.; Song, X.K. Changes of fish diversity and trophic levels in Quanzhou Bay. *Biodivers. Sci.* **2010**, *18*, 420–427.

50. Planque, B.; Fromentin, J.M.; Cury, P.; Drinkwater, K.F.; Jennings, S.; Perry, R.I.; Kifani, S. How does fishing alter marine populations and ecosystems sensitivity to climate? *J. Mar. Syst.* **2010**, *79*, 403–417. [CrossRef]

51. Fenner, D. Fishing down the largest coral reef fish species. *Mar. Pollut. Bull.* **2014**, *84*, 9–16. [CrossRef] [PubMed]

52. Volkoff, H.; Rnnestad, I. Effects of temperature on feeding and digestive processes in fish. *Temperature* **2020**, *7*, 307–320. [CrossRef] [PubMed]

53. Rummer, J.L.; Munday, P.L. Climate change and the evolution of reef fishes: Past and future. *Fish Fish.* **2017**, *18*, 22–39. [CrossRef]

54. Zhang, X.Y.; Li, Y.C.; Du, J.G.; Qiu, S.T.; Xie, B.; Chen, W.L.; Wang, J.J.; Hu, W.J.; Wu, Z.J.; Chen, B. Effects of ocean warming and fishing on the coral reef ecosystem: A case study of Xisha Islands, South China Sea. *Front. Mar. Sci.* **2022**, *9*, 1046106. [CrossRef]
55. Zuo, X.L.; Su, F.Z.; Wu, W.Z.; Chen, Z.K.; Shi, W. Spatial and temporal variability of thermal stress to China's coral reefs in South China Sea. *Chin. Geogr. Sci.* **2015**, *25*, 159–173. [CrossRef]
56. Halford, A.; Cheal, A.J.; Ryan, D.; Williams, D.M. Resilience to large-scale disturbance in coral and fish assemblages on the Great Barrier Reef. *Ecology* **2004**, *85*, 1892–1905. [CrossRef]
57. Holbrook, S.J.; Schmitt, R.J.; Brooks, A.J. Resistance and resilience of a coral reef fish community to changes in coral cover. *Mar. Ecol. Prog. Ser.* **2008**, *371*, 263–271. [CrossRef]
58. Eddy, T.D.; Lam, V.W.Y.; Reygondeau, G.; Cisneros-Montemayor, A.M.; Greer, K.; Palomares, M.L.D.; Bruno, J.F.; Ota, Y.; Cheung, W.W.L. Global decline in capacity of coral reefs to provide ecosystem services. *One Earth* **2021**, *4*, 1278–1285. [CrossRef]

 *biology*

Article

Effect of Environmental Variables on African Penguin Vocal Activity: Implications for Acoustic Censusing

Franziska Hacker [1,*], Francesca Terranova [2], Gavin Sean Petersen [3], Emma Tourtigues [1], Olivier Friard [2], Marco Gamba [2], Katrin Ludynia [3,4], Tess Gridley [5], Lorien Pichegru [6], Nicolas Mathevon [1,7], David Reby [1,7,†] and Livio Favaro [2,8,*,†]

[1] ENES Bioacoustics Research Team, University of Saint-Etienne, 42100 Saint-Etienne, France
[2] Department of Life Sciences and Systems Biology, University of Turin, 10124 Turin, Italy
[3] Southern African Foundation for the Conservation of Coastal Birds (SANCCOB), Cape Town 7441, South Africa
[4] Department of Biodiversity and Conservation Biology, University of the Western Cape, Robert Sobukwe Road, Bellville 7535, South Africa
[5] Statistics in Ecology, Environment and Conservation, Department of Statistical Sciences, University of Cape Town, Rondebosch, Cape Town 7701, South Africa
[6] Institute for Coastal and Marine Research, Nelson Mandela Metropolitan University, Port Elisabeth 6031, South Africa
[7] Institut Universitaire de France, Ministry of Higher Education, Research and Innovation, 1 rue Descartes, CEDEX 05, 75231 Paris, France
[8] CAPE Department, Stazione Zoologica Anton Dohrn, 80121 Naples, Italy
* Correspondence: ffeist98@gmail.com (F.H.); livio.favaro@unito.it (L.F.)
† Co-senior authors.

Citation: Hacker, F.; Terranova, F.; Petersen, G.S.; Tourtigues, E.; Friard, O.; Gamba, M.; Ludynia, K.; Gridley, T.; Pichegru, L.; Mathevon, N.; et al. Effect of Environmental Variables on African Penguin Vocal Activity: Implications for Acoustic Censusing. *Biology* **2023**, *12*, 1191. https://doi.org/10.3390/biology12091191

Academic Editors: Cong Zeng and Deliang Li

Received: 9 July 2023
Revised: 26 August 2023
Accepted: 29 August 2023
Published: 31 August 2023

Simple Summary: Most seabird species are in need of effective conservation, with 43% being near to globally threatened. Passive acoustic monitoring could serve as a cost-effective, noninvasive population monitoring tool essential for informing future conservation efforts. As such, we set out to investigate whether passive acoustic monitoring could successfully predict the African penguin density at a remote colony in Betty's Bay, South Africa. We first automated the detection and counting of penguins' vocalisations in our recordings to facilitate the handling of large datasets. Then, we investigated whether temperature, humidity, and wind speed affected the calling rate of penguins, which would be essential for an accurate census. Finally, taking into account the variations with weather conditions, we showed that passive acoustic monitoring could successfully predict the number of callers within a 10.5 m radius around our devices, indicating that it can be used for cost-effective, noninvasive censuses of African penguin colonies.

Abstract: Global biodiversity is in rapid decline, and many seabird species have disproportionally poorer conservation statuses than terrestrial birds. A good understanding of population dynamics is necessary for successful conservation efforts, making noninvasive, cost-effective monitoring tools essential. Here, we set out to investigate whether passive acoustic monitoring (PAM) could be used to estimate the number of animals within a set area of an African penguin (*Spheniscus demersus*) colony in South Africa. We were able to automate the detection of ecstatic display songs (EDSs) in our recordings, thus facilitating the handling of large datasets. This allowed us to show that calling rate increased with wind speed and humidity but decreased with temperature, and to highlight apparent abundance variations between nesting habitat types. We then showed that the number of EDSs in our recordings positively correlated with the number of callers counted during visual observations, indicating that the density could be estimated based on calling rate. Our observations suggest that increasing temperatures may adversely impact penguin calling behaviour, with potential negative consequences for population dynamics, suggesting the importance of effective conservation measures. Crucially, this study shows that PAM could be successfully used to monitor this endangered species' populations with minimal disturbance.

Keywords: passive acoustic monitoring; remote monitoring; remote census; *Spheniscus demersus*; vocalisations

1. Introduction

Global biodiversity is in rapid decline [1,2], which is evident across a range of habitats, including agricultural lands [3], freshwater ecosystems [4], the Arctic [5], and Antarctica [6,7], but especially apparent in marine and coastal regions [8,9]. Seabirds, making up around 3.5% of all avian species, have been classified as the most threatened avian group, characterised by significantly poorer conservation statuses than other birds [10]. Population declines over the last 20 years have been recorded in approximately half of all seabird species, and those of most concern are penguins (Sphenisciformes) and albatrosses and petrels (Procellariiformes), together representing around 43% of all seabird species [10]. Since many seabird species are critical indicators for the health of their respective marine ecosystems [11,12], their continued decline has drastic implications for not only their conservation but also the broader status of our oceans and coastal habitats. This is supported by the fact that key threats to seabirds are anthropogenic activities known to negatively affect marine ecosystems on larger scales, such as overfishing [13], pollution [14], habitat degradation [15], incidental mortality after by-catch in fisheries gear [16], and human disturbances at colonies, such as tourism [10,17,18].

Focusing on the conservation of umbrella species often allows for protecting the status of their habitat and that of other species living within the same space. Selecting just a single species can be beneficial for the conservation and protection of large areas [19], with studies reporting positive effects on the conservation of intact and restoration of degraded forests [20], and on aquatic biodiversity [21]. Thus, given the importance of seabirds as indicator species, conserving seabird umbrella species could help prevent both the decline of their respective populations and that of their habitat and other marine species within it.

Penguins (Spheniscidae) are a family of seabirds inhabiting most of the southern parts of our planet. Out of the 19 penguin species recognised today [22], two are classified as near threatened, four as vulnerable, and five as endangered [23]. Recently, three penguin species have been internationally voted as highest priority for conservation efforts [24]; one of which, the African penguin (*Spheniscus demersus*), is characterised by a largely depleted population [23], with many island colonies subject to drastic declines and collapses, and only two mainland colonies across South Africa [25,26]. Both individual survival and colony breeding success are strongly impacted by habitat loss owing to resource competition with industrial fisheries, the expansion of anthropogenic activities, and marine noise pollution [26–29]. Moreover, the negative consequences of oil spills alone may be sufficient to lead to the extinction of the African penguin [27,30,31]. Thus, well-founded conservation actions are urgently needed to counteract their decline and aid populations' recovery.

The efficacy of such conservation actions must be assessed to ensure they are successful in increasing breeding success or breeding numbers. In that regard, improving our understanding of population trends, threats, life history, distribution, and ecology are priority research areas that can inform and support conservation efforts [10,32]. Considering the African penguin's sensitivity to human disturbances and the recommendations within the African Penguin Management Plan [33], noninvasive monitoring tools are essential to minimise the adverse effects of further studies of this species' population trends and breeding ecology on site. Observational studies or monitoring via visual remote sensing tools, such as camera traps or drones, can provide essential insights into colony dynamics and breeding success but often still require human presence in sensitive areas and can be hindered by factors such as the need for good weather conditions and the limited spatiotemporal resolution [34]. Because African penguins rely extensively on acoustic communication for intraspecific communication, a powerful, noninvasive and cost-effective way of monitoring penguin colonies could be passive acoustic monitoring (PAM), which would keep disturbances to a minimum [35] while effectively assessing the number of birds within an area. Specifically, the distinctiveness of the ecstatic display song (EDS), which consists of a sequence of short syllables (type A) followed by a long syllable (type B) and an audible inspiration (type C), and the frequency at which it is produced could make it a good target for detection in recordings [36]. Further, EDSs are important in territory defence and mate choice in African penguins, with especially high calling rates during the beginning of a breeding season [37].

To date, PAM has been used to investigate the relationship between acoustic activity and colony density in a variety of bird species (e.g., eastern wood pewee (*Contopus virens*) [38], Forster's terns (*Sterna fosteri*) [35], bell miner (*Manorina melanophrys*) [32], short-tailed shearwaters (*Ardenna tenuirostris*) [39], Magellanic (*S. magellanicus*) and southern rockhopper (*Eudyptes chrysocome*) penguins [34]). A review of multiple PAM surveying attempts showed that 79% of studies investigating a relationship between the number of vocalisations and bird density or abundance obtained counts that agreed with those obtained from human surveyors [40]. Additionally, in African penguins, acoustic indices such as the acoustic entropy index (H) have already been shown to be valuable tools for predicting the number of both EDSs and mutual display songs (MDSs) in soundscape recordings of colonies, reflecting overall vocal activity [41]. However, not all studies report the successful application of PAM to estimate population densities [42,43] and some suggest combining them with other survey methods for accurate results [40]. Therefore, further investigations of its applicability for, and accuracy in, the remote monitoring of African penguin colonies are needed to effectively support the development of successful conservation efforts [33].

To investigate whether PAM could be a viable tool to estimate the density within a colony, we deployed a series of static acoustic monitoring devices within the Stony Point penguin colony (Betty's Bay, South Africa), a key mainland colony with around 1500 breeding pairs as of July 2022 (~7% of the total African penguin population) [41]. To facilitate data handling and provide a modern, time-efficient strategy for counting penguin vocalisations in audio recordings, we first aimed to automate EDS detection. Then, we investigated to what extent weather variables affected calling rate, which can guide the ideal timing of remote acoustic breeding censuses and improve their accuracy. Lastly, taking those variations into account, we estimated the animal density within an area by examining the relationship between total EDS numbers in our recordings and visually observed penguins and callers to trial a novel, noninvasive counting tool useful for this species' conservation. As such, our study aimed to provide the first steps for the development of a time-efficient remote monitoring tool for the endangered African penguin while simultaneously providing insights into the broader use of PAM for the management and conservation of sensitive, cryptic species.

2. Materials and Methods

2.1. Acoustic Recordings and Weather Data

We deployed three acoustic sensors (AudioMoth (AM); Open Acoustic Devices, 2022) in IPX7 waterproof cases (48 kHz sampling rate, no filters applied) throughout the African penguin colony at Stony Point (34°37′14.21″ S, 18°89′32.65″ E) in Betty's Bay, South Africa, and recorded the colony continuously from March to July 2022. The daily recording schedule, from 4:00–9:00 and 16:30–21:30 South African Standard Time (SAST), was based on previous investigations of peak acoustic activity at the same colony [41] and adjusted to local sunrise and sunset times. This schedule resulted in a total of $n = 6627$ 30 min recordings (198,810 total mins) collected over 113 consecutive days. The recorders were placed at locations with slightly different flora, shown on a map in Figure 1. One device was set up in an area of dense dune spinach (*Tetragonia decumbens*) bushes covering the ground completely (AM1), one in a grassy area dispersed with *Baccharis halimifolia* bushes (AM2), and one at the southern end of the colony, where most penguin nests are sand burrows covered by some dune spinach (AM3). AM1 and AM2 were spaced approximately 330 m apart and AM2 was around 130 m from AM3. This layout would allow for an investigation of how vegetation influenced penguin density and calling rate. The acoustic sensors were tied to wooden poles at a height of around 20 cm and oriented to face downwards to minimise directionality that could be introduced by the device casing and batteries. Devices were collected for data download once per week to minimise disturbance.

Figure 1. Map and satellite image of the penguin colony and pictures of the three respective recording sites. (**A**) A map of the location of the penguin colony within Africa and South Africa and satellite image of the entire Stony Point penguin colony, with red circles indicating the locations of the acoustic sensors. The righthand panels (**B**–**D**) show pictures of our three recording sites AM1, AM2, and AM3, respectively.

A 3879 Diastella (March–June) and a Bresser WIFI Colour Weather Station (June and July) set up according to the manufacturer's instructions measured temperature, humidity, wind direction and speed, barometric pressure, and precipitation rate at the colony through a rain gauge, wind cups and vane, and a thermohygrometer. Additionally, the temperature measured using the AM devices at the beginning of each recording session was extracted using an adapted Python script (Open Acoustic Devices, 2020; Supplementary Information).

2.2. Visual Counting of Penguins

Visual observations were conducted twice throughout the breeding season: in early May and late June/early July, representing the middle and end of one breeding cycle, respectively. Here, the middle of a breeding cycle is defined as the period during which most breeding pairs have chicks fully covered in their juvenile down feathers, whereas the end of the cycle concerns the period during which most chicks have or are about to fledge and leave the nests as "blues" with waterproof plumage. A 10.5 m radius around each AM device, measured with a rangefinder, was marked with wooden poles to allow for the accurate counting of penguins within a set radius. This radius was limited by the layout of the colony, which includes some areas as narrow as 11–12 m between the boardwalk and private property. Across a period of 10 consecutive days, an experimenter (F.H.) counted penguins twice daily, at sunrise and sunset, noting all visible penguins every five minutes and all heard EDS vocalisations and identifying all callers within the radius around the AM device for half an hour per site, resulting in a total of $n = 48$ observation periods. Since visual sexing of African penguins is unreliable [44], both males and females were counted. Observation times were shifted with sunrise and sunset in a way that the first two sites were surveyed in the half hour before and after sunrise, while the third site was surveyed either in the half hour before or after sunset. The schedule was rotated to ensure that every site was surveyed at each of the four possible timeslots (before/after sunrise/sunset) at least twice during each 10-day observation period. Additionally, we counted the number of occupied nests within the radius around the AM devices and calculated the resulting maximum penguin densities. The obtained number of callers within our defined radius and the overall EDS numbers in our recordings would then allow us to investigate whether the number of breeding pairs around the acoustic sensors could be estimated from our recordings.

2.3. Data Analysis

The collected audio files were uploaded onto the OCCAM SuperComputer [45] at the Competence Centre for Scientific Computing of the University of Turin. To automate the detection of the number of EDSs inside the audio recordings, we used the monitoR [46] package in R v.4.2.3 [47]. A typical EDS is usually composed of many short syllables (A) followed by a long one (B) and an audible inspiration (C) [48], as visualised in Figure 2. The large variations in the number of A syllables in a song [37] and the everchanging characteristics of C syllables (e.g., frequency and frequency modulation, duration, etc.) meant that focussing on the B-syllable for automatic detections would be the most reliable means of acquiring accurate EDS numbers.

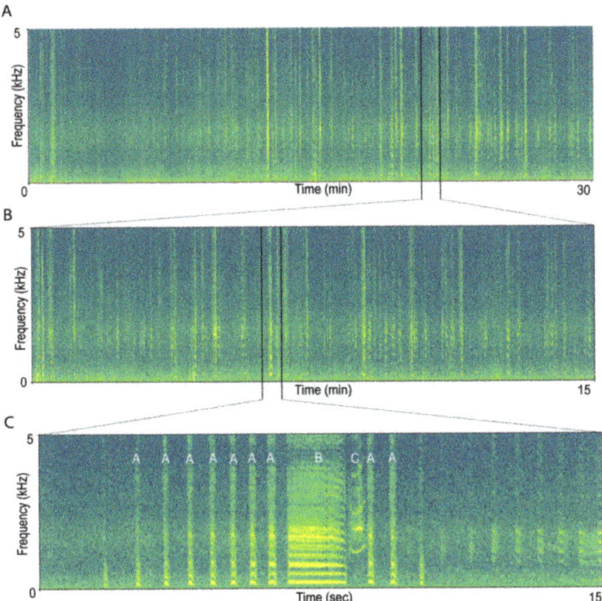

Figure 2. An example recording containing EDS vocalisations. Panel (**A**) shows a 30 min recording from one of our remote acoustic recorders taken at the Stony Point penguin colony, Betty's Bay, South Africa. Panel (**B**) represents the latter 15 min of the same recording, with single EDSs becoming visible, and panel (**C**) visualises a single, typical EDS vocalisation from the same recording, with A, B, and C indicating the repeated short syllables, the long detection target syllable, and the inspiration syllable, respectively. The figure was created using Raven Pro v. 1.5.0 (Spectrogram window size: 1024, Hann window, overlap 50%).

As such, using five different B-syllable recordings, binary point-matching templates were created through the monitoR automatic template creation tool [49]. Since different templates could detect the same syllable, we merged all detections of different templates found in the same minute (±1 s). Different threshold scores, above which any detections would be counted as EDSs, were tested (20, 21, 22, 25, 27, and 30) to identify the most accurate one based on visual spectrogram inspections and randomly generated detection curves. Figure S1 in Supplementary Information (SI) visualises one of the five created templates and detection curves of different scores. A threshold of 21 was found to be the most accurate, and as such, detections with scores ≥ 21 were classified as EDSs. To calculate the detector's average sensitivity (true positive rate), all true positive (TP) and false negative (FN) detections in $n = 678$ (around 10% of our final dataset) audio files were manually counted using spectrographic inspection. Then, the true positive rate was calculated as TP/(TP + FN) and found to be 65.79%. Additionally, the accuracy of automatic detections was crosschecked by two independent observers (F.H. and F.T.) conducting manual spectrographic counts of EDSs with visually assessed good or very good signal-to-noise ratio (SNR) [50] on a subset of audio files ($n = 199$) and validated using Spearman's rank correlation test.

2.4. Statistics–Modelling

A generalised linear mixed model (GLMM) approach was used to investigate the impact of weather conditions on calling rate and the correlation between the average number of identified callers and the number of recorded EDSs using the glmmTMB package [51] in R [47], comparing their respective full and null models. The first model analysed the effects of temperature, wind speed, humidity, and location on the number of EDSs recorded per 30 min while controlling for the recording date and time. Then, we created the second GLMM to investigate whether the number of automatically detected EDSs in our recordings correlated with the average number of penguins and callers counted within the radius around the respective AM device while taking into account call rate variations with weather conditions and controlling for location, time, and date of recording. In both, correlation among all predictors was assessed using variance inflation factor (VIF) analysis and significance was investigated using a chi-square test. Location was controlled for in the second GLMM based on the first model's results highlighting a significant effect between the recorders. This showed whether PAM could be a reliable census tool for the defined area of detection despite the influences of weather on calling rate. Graphs were made using the sjPlot package [52] in R.

3. Results

3.1. Automating EDS Counting

We first set out to automate EDS detection in the recordings of our remote acoustic sensors, for which automatic detections were compared with manual spectrographic counts. Spearman's rank correlation showed a significant correlation between manual and automatic counts, indicating that automatic counts could reflect the relative number of calls present in a 30 min recording ($n = 199$, R $= 0.61$, $p < 0.001$).

3.2. Influence of Environmental Variables on Calling Rate

EDS numbers were significantly correlated with all investigated weather variables and location (full vs. null: $x^2 = 27{,}836.21$, df $= 5$, $p < 0.001$). Both higher humidity (estimate $= 0.003$, se $= 0.001$, $p < 0.001$) and wind speeds (estimate $= 0.013$, se $= 0.001$, $p < 0.001$) led to an increase in EDS production. In contrast, higher temperatures led to a decrease in calling rate (estimate $= -0.028$, se $= 0.001$, $p < 0.001$). Lastly, AM1 recorded significantly less EDSs than AM2 (estimate $= 0.619$, se $= 0.004$, $p < 0.001$), but more than AM3 (estimate $= -0.031$, se $= 0.005$, $p < 0.001$). This was confirmed by our visual observations and the count of active nests. AM2, characterised by a mixed vegetation of grass, bushes, and artificial burrows, contained 27 nests and had the highest penguin density (0.156 penguins/m²); AM3, located at the southern end of the colony and containing mostly sand burrows and small dune spinach bushes, had 20 nests in its radius and the lowest density (0.116 penguins/m²); and AM1, which was solely surrounded by dune spinach, had a total of 22 nests and the second lowest density (0.127 penguins/m²). Thus, while the environment found at the colony did not allow for multiple replicates of each vegetation type, the high number of observation sessions supported that habitat type may influence the overall penguin density or calling rate; we therefore controlled for location in the subsequent GLMM.

3.3. Relationship between EDS Counts and Penguin Abundance

Despite the variations in EDS numbers with environmental conditions, our model showed that there was a significant correlation between the number of detected EDSs and the number of callers (the number of penguins seen calling) within the 10.5 m radius (full vs. null: $x^2 = 24.127$, df = 2, $p < 0.001$). Specifically, the number of EDSs detected in our recordings was significantly positively correlated with the number of callers visually observed within the 10.5 m radius around our devices (estimate = 0.369, se = 0.075, $p < 0.001$), as shown in Figure 3, but not with the overall average number of penguins present within a radius (estimate = -0.036, se = 0.022, $p = 0.102$), potentially because transiting penguins, those crossing through the radius without having a nest therein, were included in our counts. Thus, future studies should aim to effectively distinguish between resident and transiting penguins to assess whether there is a correlation between resident penguins and EDSs.

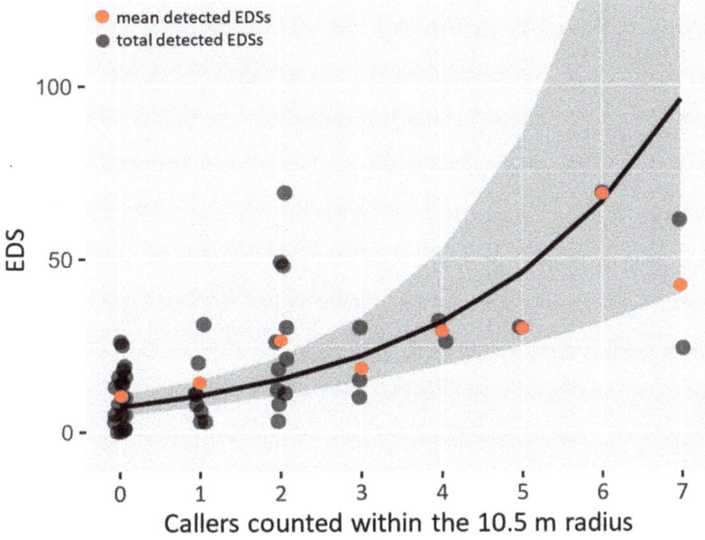

Figure 3. Correlation between the number of visually observed callers and the automatically counted ecstatic display song numbers. An overlay of the model prediction (grey line) along with the 95% confidence interval (shaded grey area). Red dots indicate the mean number of ecstatic display songs (EDS) observed for the respective number of calling individuals within the 10.5 m radius around our recording devices.

4. Discussion

4.1. Effectiveness of PAM

Passive acoustic monitoring has been shown to accurately reflect bird population estimates in several species [32,34,35,38–40]. Similarly, despite some studies reporting less accurate results [42,43], we revealed here its potential for accurate remote acoustic censuses of an African penguin colony. Despite the identified effects of environmental variables on the number of recorded EDS calls, we showed that there was a positive correlation between the number of EDSs detected in our recordings and the number of callers observed within our radii, indicating that PAM based on EDS detections could be an effective monitoring tool for African penguin colonies.

However, our visual observations showed that the number of active callers within our plots (0–7 penguins) was always lower than the number of active nests within that radius, which lowers the risk of overestimation. Further, within a site, the average number of counted penguins across our observational period (AM1: 7; AM2: 14; AM3: 11) was about half the number of active nests within the site (AM1: 22; AM2: 27; AM3: 20), and even less were identified as active callers (AM1: 0.5; AM2: 0.5; AM3: 2), which often called multiple times within a half-hour survey. However, these low medians and highly variable data may have been caused by the occurrence of days of low vocal activity likely resulting from the fact that our observations took place toward the end of the breeding season. Therefore, to ensure accuracy of a remote censusing tool, we stress the importance of conducting acoustic and visual surveys at the beginning of the breeding season when penguins are most vocally active.

Lastly, the sensitivity of our detector (65.79%) indicates that another counting approach may be needed, since reducing false negatives in monitoR is difficult, with the inclusion of more templates leading to less accurate results.

4.2. Impacts on Detection Effectiveness

We further showed that environmental conditions—temperature, wind speed, humidity, and location—affected the detection of EDSs in our recordings, which should be considered when developing this census tool further. Variations between locations and with varying weather conditions may have resulted from changes in vocal activity, penguin density or penguin presence at the colony. Alternatively, differences in sound propagation and detection between our AMs may have played a role.

First, we showed that all chosen weather variables—temperature, humidity, and wind speed—affected calling rates and should thus be considered on census days, similar to how forecasts are used in planning songbird transect surveys [53,54]. Higher wind speeds may have directly increased the calling rate of penguins by, e.g., lowering the perceived temperature, leading to higher vocal activity, or changed sound propagation in a way that altered song detection at the receiver, our AMs, thus causing variations in overall detection of EDSs at different wind speeds. Alternatively, the increased number of recorded EDSs at higher wind speeds could have resulted from an increased number of false positive detections in our automated detection. However, manual inspections of a portion of the dataset showed that overestimations of EDSs did not always occur on days with higher wind speeds. Similarly, higher humidity may be indicative of rainy days with relatively lower or lower perceived temperatures on which penguins may call longer or more frequently, which could explain the increased occurrence of EDSs.

Notably, decreasing the calling rate at higher temperatures could have important implications for both mate selection and territorial disputes [36]. First, during hot weather, fewer males may be prone to call, potentially leading to fewer new pairs formed at the beginning of a breeding season, possibly resulting in more nonbreeding adults. Investigations of whether the numbers of nonbreeding adults at colonies during a breeding season are higher after a hotter start to the season may confirm this hypothesis. Second, since vocal contests are usually the first step in territorial conflicts and often serve to assess rivals, they frequently prevent the escalation of fights [55]. Thus, higher temperatures could lead to fewer EDSs produced in territorial defence, which may increase physical territorial fights and injuries, likely negatively affecting penguin survival. Alternatively, the lower number of EDSs could reflect a higher absence of breeding individuals from the colony, spending more time at sea. This could adversely affect chick survival rates, especially in young chicks. As such, the decrease in EDSs at higher temperatures could have negative consequences for the survival rate of African penguins given the current state of our climate and the rising global temperatures [56]. Increased temperatures have already been shown to lead to a shift in the timing of the 2022 breeding season and a decrease in egg survival rate, most recently affected by a mass-abandonment following a hot spell in January 2022 at the Stony Point colony, as indicated by the colony's research manager (Van Eeden, L., Pers. Comm.). Therefore, effective conservation strategies for the African penguin are necessary.

Furthermore, the variation in detected EDSs across locations suggest that penguin density varies between locations of differing vegetation types, with some areas of the colony supporting a higher local density than others. Alternatively, differences in EDS numbers across locations may be reflective of variations in sound propagation in areas with different vegetation types or changes in calling rates rather than penguin density. However, given that the number of identified active nests and detected EDSs was highest at AM2, the location with the comparatively densest vegetation, it is suggested that habitat suitability is the likely explanation for the observed variations in density rather than changes in calling rate or sound propagation. Notably, despite habitat variations, PAM was successful in identifying the relationship between EDSs and present callers, highlighting its usefulness for the monitoring of species with cryptic nesting behaviour. Furthermore, our results can provide information on suitable vegetation that can support larger numbers of penguins, which can be important for improving existing colonies through habitat rehabilitation actions and the planning of the establishment of artificial colonies [57], which is especially essential given the expansion of urban areas into nature, decreasing the area available to breeding colonies.

4.3. Improving Detections for Successful PAM

Others have suggested that an improved understanding of the focal species' vocal behaviour significantly improves the development of acoustic detections and passive acoustic surveys [58]. Accuracy can be further improved by investigating the following important factors: the ideal census time, specifically distinguishing between breeding and nonbreeding seasons, the female-to-male singing ratio, and an expansion of the target vocalisations and sampling radius.

Since previous research has suggested that EDSs have functions of both territory defence and mate attraction [48], differences in the overall number of produced EDSs are likely to appear between different stages of the breeding season. Thus, future investigations should attempt to compare accuracy across all stages, such as the beginning, middle, and end, to identify the ideal time of year or breeding season period during which the number of EDSs most accurately reflects the number of active breeding pairs.

Furthermore, previous research on the female-to-male singing ratio reports that females produce EDSs, especially in territory defence [59], but further investigations of this ratio are necessary to accurately estimate breeding pairs. On one hand, since visual sexing is not possible in African penguins, assuming all callers to be a male may lead to an overestimation of breeding pairs. On the other hand, depending on the female-to-male singing ratio and the census timing, the fact that only one parent is usually present at a nest during the chick-rearing period and both sexes produce EDSs may be beneficial for acoustic censuses, since breeding pairs could still be counted correctly.

Additionally, the detection radius may be increased to improve the overall detection of African penguins further. The fact that only a small portion of EDSs came from callers within our radii may have resulted from including a relatively narrow radius compared with the detection range of our AM devices. Similarly, since not all present penguins were observed to produce EDSs, sampling could be expanded to include a larger portion of the African penguin vocal repertoire, which consists of six identified calls and songs [36]. This suggestion is supported by recent efforts to estimate population density in Magellanic and southern rockhopper penguins using PAM, which included sounds other than calls, such as huffs and sneezes, and obtained an accurate estimate [34].

Lastly, some have suggested that PAM is most efficient when combined with other noninvasive census strategies [40], such as thermal sensing. The usefulness of thermal sensing has recently been investigated in a study of American woodcock (*Scolopax minor*) detection probability [60], which pointed out that, in forests, its accuracy declined with vegetation density. However, given the absence of a forest-like environment with dense trees found at our study sites, thermal sensing may be useful to support remote acoustic censuses. This will be especially important for establishing whether the absence of EDSs in recordings correlates to penguin absence or silence.

5. Conclusions

We showed that PAM has the potential to become a useful, low-cost monitoring tool for sensitive seabird species such as the African penguin despite the influences of environmental variables on the number of detected EDSs in AM recordings. The detection of African penguin EDS vocalisations was successfully automated, as confirmed by a comparison between manually counted and automatically detected EDSs in our recordings. We also found that humidity, wind speed, and temperature significantly affected EDS production, with an increase at higher humidity and wind speeds and a decrease at higher temperatures. Lastly, taking these weather effects into account, we showed a significant correlation between the number of detected EDSs and the number of callers within a 10.5 m radius around our AM devices.

To further improve the accuracy of remote PAM censuses of this species and their efficiency, the above outlined limitations should be considered in the further development of this tool. Specifically, future research aiming to improve it further should focus on determining the ideal census time within a breeding season, accurate assessment of the female-to-male singing ratio to more accurately predict breeding pair numbers based on EDS vocalisations, and an expansion of the target vocalisations and sampling radius.

Supplementary Materials: The following supporting information can be downloaded at: https://www.mdpi.com/article/10.3390/biology12091191/s1. The adapted Python script for AudioMoth parameter extraction; Figure S1: Example of B-syllable recording and corresponding monitoR template; Figure S2: Example of a detected EDS in an AudioMoth recording.

Author Contributions: Conceptualisation, L.F. and D.R.; project administration, L.P., N.M., D.R. and L.F.; funding acquisition, L.P., N.M., D.R. and L.F.; methodology, F.H., M.G., L.F. and D.R.; software, F.H., F.T. and O.F., validation, M.G., L.F. and D.R.; formal analysis, F.H., F.T. and M.G.; investigation, F.H., E.T. and G.S.P.; resources, G.S.P., O.F., N.M., D.R. and L.F.; data curation, F.H., F.T. and O.F.; writing—original draft preparation, F.H.; writing—review and editing, F.H., F.T., M.G., L.F. and D.R.; visualisation, F.H. and F.T.; supervision, K.L., T.G., D.R. and L.F. All authors have read and agreed to the published version of the manuscript.

Funding: This research was partially funded by the National Geographic Society, grant number NGS-67401R-20.

Institutional Review Board Statement: Ethical approval was obtained from the Nelson Mandela University, Port Elizabeth, South Africa (ethics reference number A19-SCI-ZOO-016), and the research was conducted under a permit for the purposes of a scientific investigation or practical experiment obtained from the South African Department for Forestry, Fisheries, and the Environment (permit reference number: RES2022-40) as well as a CapeNature permit to enter a nature reserve for scientific purposes (permit number CN32-87-19920).

Informed Consent Statement: Not applicable.

Data Availability Statement: The data presented in this study are available upon request from the corresponding author. The data are not publicly available, as recordings may contain voices of staff and visitors of the colony.

Acknowledgments: We wish to thank Lizanne Van Eeden, Ziyaad Erasmus, Austin Byron Willemse, and Angelo Hufke, the rangers at the Stony Point penguin colony, who aided us massively with our field work. Additionally, we are grateful to the BirdLife South Africa team, Alistair McInnes, Pierre Pistorius, Tegan Carpenter-Kling, Eleanor Weideman, and Pierre Retief, for their support. Further thanks go to the Institut Universitaire de France and to Labex CeLyA for supporting DR and NM.

Conflicts of Interest: The authors declare no conflict of interest. The funders had no role in the design of the study; in the collection, analyses, or interpretation of data; in the writing of the manuscript; or in the decision to publish the results.

References

1. Visconti, P.; Bakkenes, M.; Baisero, D.; Brooks, T.; Butchart, S.H.M.; Joppa, L.; Alkemade, R.; Di Marco, M.; Santini, L.; Hoffmann, M.; et al. Projecting Global Biodiversity Indicators under Future Development Scenarios. *Conserv. Lett.* **2016**, *9*, 5–13. [CrossRef]
2. Waldron, A.; Miller, D.C.; Redding, D.; Mooers, A.; Kuhn, T.S.; Nibbelink, N.; Roberts, J.T.; Tobias, J.A.; Gittleman, J.L. Reductions in Global Biodiversity Loss Predicted from Conservation Spending. *Nature* **2017**, *551*, 364–367. [CrossRef] [PubMed]
3. Brühl, C.A.; Zaller, J.G. Biodiversity Decline as a Consequence of an Inappropriate Environmental Risk Assessment of Pesticides. *Front. Environ. Sci.* **2019**, *7*, 177. [CrossRef]
4. Reid, A.J.; Carlson, A.K.; Creed, I.F.; Eliason, E.J.; Gell, P.A.; Johnson, P.T.J.; Kidd, K.A.; MacCormack, T.J.; Olden, J.D.; Ormerod, S.J.; et al. Emerging Threats and Persistent Conservation Challenges for Freshwater Biodiversity. *Biol. Rev.* **2019**, *94*, 849–873. [CrossRef]
5. Humphries, G.R.W.; Huettmann, F. Putting Models to a Good Use: A Rapid Assessment of Arctic Seabird Biodiversity Indicates Potential Conflicts with Shipping Lanes and Human Activity. *Divers. Distrib.* **2014**, *20*, 478–490. [CrossRef]
6. Chown, S.L.; Clarke, A.; Fraser, C.I.; Cary, S.C.; Moon, K.L.; McGeoch, M.A. The Changing form of Antarctic Biodiversity. *Nature* **2015**, *522*, 431–438. [CrossRef]
7. Lee, J.R.; Waterman, M.J.; Shaw, J.D.; Bergstrom, D.M.; Lynch, H.J.; Wall, D.H.; Robinson, S.A. Islands in the Ice: Potential Impacts of Habitat Transformation on Antarctic Biodiversity. *Glob. Chang. Biol.* **2022**, *28*, 5865–5880. [CrossRef]
8. Roman, L.; Hardesty, B.D.; Hindell, M.A.; Wilcox, C. A Quantitative Analysis Linking Seabird Mortality and Marine Debris Ingestion. *Sci. Rep.* **2019**, *9*, 3202. [CrossRef]
9. Woodworth, B.K.; Fuller, R.A.; Hemson, G.; McDougall, A.; Congdon, B.C.; Low, M. Trends in Seabird Breeding Populations across the Great Barrier Reef. *Conserv. Biol.* **2021**, *35*, 846–858. [CrossRef]
10. Croxall, J.P.; Butchart, S.H.M.; Lascelles, B.; Stattersfield, A.J.; Sullivan, B.; Symes, A.; Taylor, P. Seabird Conservation Status, Threats and Priority Actions: A Global Assessment. *Bird Conserv. Int.* **2012**, *22*, 1–34. [CrossRef]
11. Parsons, M.; Mitchell, I.; Butler, A.; Ratcliffe, N.; Frederiksen, M.; Foster, S.; Reid, J.B. Seabirds as Indicators of the Marine Environment. *ICES J. Mar. Sci.* **2008**, *65*, 1520–1526. [CrossRef]
12. Mallory, M.L.; Gilchrist, H.G.; Braune, B.M.; Gaston, A.J. Marine Birds as Indicators of Arctic Marine Ecosystem Health: Linking the Northern Ecosystem Initiative to Long-Term Studies. *Environ. Monit. Assess.* **2006**, *113*, 31–48. [CrossRef] [PubMed]
13. Link, J.S.; Watson, R.A.; Pranovi, F.; Libralato, S. Comparative Production of Fisheries Yields and Ecosystem Overfishing in African Large Marine Ecosystems. *Environ. Dev.* **2020**, *36*, 100529. [CrossRef]
14. Häder, D.-P.; Banaszak, A.T.; Villafañe, V.E.; Narvarte, M.A.; González, R.A.; Helbling, E.W. Anthropogenic Pollution of Aquatic Ecosystems: Emerging Problems with Global Implications. *Sci. Total Environ.* **2020**, *713*, 136586. [CrossRef] [PubMed]
15. Laurance, W.F. Habitat Destruction: Death by a Thousand Cuts. In *Conservation Biology for All*; OUP Oxford: Oxford, UK, 2010; ISBN 978-0-19-157425-2.
16. Karpouzi, V.S.; Watson, R.; Pauly, D. Modelling and Mapping Resource Overlap between Seabirds and Fisheries on a Global Scale: A Preliminary Assessment. *Mar. Ecol. Prog. Ser.* **2007**, *343*, 87–99. [CrossRef]
17. Schratzberger, M.; Somerfield, P.J. Effects of Widespread Human Disturbances in the Marine Environment Suggest a New Agenda for Meiofauna Research Is Needed. *Sci. Total Environ.* **2020**, *728*, 138435. [CrossRef]
18. Courrat, A.; Lobry, J.; Nicolas, D.; Laffargue, P.; Amara, R.; Lepage, M.; Girardin, M.; Le Pape, O. Anthropogenic Disturbance on Nursery Function of Estuarine Areas for Marine Species. *Estuar. Coast. Shelf Sci.* **2009**, *81*, 179–190. [CrossRef]
19. Thornton, D.; Zeller, K.; Rondinini, C.; Boitani, L.; Crooks, K.; Burdett, C.; Rabinowitz, A.; Quigley, H. Assessing the Umbrella Value of a Range-Wide Conservation Network for Jaguars (*Panthera Onca*). *Ecol. Appl.* **2016**, *26*, 1112–1124. [CrossRef]
20. Mekonnen, A.; Fashing, P.J.; Chapman, C.A.; Venkataraman, V.V.; Stenseth, N.C. The Value of Flagship and Umbrella Species for Restoration and Sustainable Development: Bale Monkeys and Bamboo Forest in Ethiopia. *J. Nat. Conserv.* **2022**, *65*, 126117. [CrossRef]
21. Kalinkat, G.; Cabral, J.S.; Darwall, W.; Ficetola, G.F.; Fisher, J.L.; Giling, D.P.; Gosselin, M.-P.; Grossart, H.-P.; Jähnig, S.C.; Jeschke, J.M.; et al. Flagship Umbrella Species Needed for the Conservation of Overlooked Aquatic Biodiversity: Freshwater Flagship Umbrella Species. *Conserv. Biol.* **2017**, *31*, 481–485. [CrossRef]
22. Harrison, P.; Perrow, M.; Larsson, H. *Seabirds. The New Identification Guide*; Lynx Edicions: Barcelona, Spain, 2021.
23. BirdLife International IUCN Red List of Threatened Species: *Spheniscus demersus*. In *IUCN Red List of Threatened Species*; 2019; Available online: https://dx.doi.org/10.2305/IUCN.UK.2020-3.RLTS.T22697810A157423361.en (accessed on 3 August 2023).
24. Boersma, P.D.; Borboroglu, P.G.; Gownaris, N.J.; Bost, C.A.; Chiaradia, A.; Ellis, S.; Schneider, T.; Seddon, P.J.; Simeone, A.; Trathan, P.N.; et al. Applying Science to Pressing Conservation Needs for Penguins. *Conserv. Biol.* **2020**, *34*, 103–112. [CrossRef] [PubMed]
25. Crawford, R.J.M.; Dyer, B.M.; Brown, P.C. Absence of Breeding by African Penguins at Four Former Colonies. *S. Afr. J. Mar. Sci.* **1995**, *15*, 269–272. [CrossRef]
26. Crawford, R.; Altwegg, R.; Barham, B.; Barham, P.; Durant, J.; Dyer, B.; Geldenhuys, D.; Makhado, A.; Pichegru, L.; Ryan, P.; et al. Collapse of South Africa's Penguins in the Early 21st Century. *Afr. J. Mar. Sci.* **2011**, *33*, 139–156. [CrossRef]
27. Favaro, L.; Pichegru, L. Penguins: Behavioural Ecology and Vocal Communication. In *Encyclopedia of Animal Cognition and Behavior*; Vonk, J., Shackelford, T., Eds.; Springer International Publishing: Cham, Switzerland, 2018; pp. 1–9, ISBN 978-3-319-47829-6.

28. Sydeman, W.J.; Hunt, G.L.; Pikitch, E.K.; Parrish, J.K.; Piatt, J.F.; Boersma, P.D.; Kaufman, L.; Anderson, D.W.; Thompson, S.A.; Sherley, R.B. South Africa's Experimental Fisheries Closures and Recovery of the Endangered African Penguin. *ICES J. Mar. Sci.* **2021**, *78*, 3538–3543. [CrossRef]

29. Pichegru, L.; Vibert, L.; Thiebault, A.; Charrier, I.; Stander, N.; Ludynia, K.; Lewis, M.; Carpenter-Kling, T.; McInnes, A. Maritime Traffic Trends around the Southern Tip of Africa—Did Marine Noise Pollution Contribute to the Local Penguins' Collapse? *Sci. Total Environ.* **2022**, *849*, 157878. [CrossRef] [PubMed]

30. Borboroglu, P.G.; Boersma, P.D. *Penguins: Natural History and Conservation*; University of Washington Press: Seattle, WC, USA, 2015; ISBN 978-0-295-99906-7.

31. Sherley, R.B.; Crawford, R.J.M.; Blocq, A.D.; Dyer, B.M.; Geldenhuys, D.; Hagen, C.; Kemper, J.; Makhado, A.B.; Pichegru, L.; Tom, D.; et al. The Conservation Status and Population Decline of the African Penguin Deconstructed in Space and Time. *Ecol. Evol.* **2020**, *10*, 8506–8516. [CrossRef]

32. Lambert, K.T.A.; McDonald, P.G. A Low-Cost, yet Simple and Highly Repeatable System for Acoustically Surveying Cryptic Species. *Austral Ecol.* **2014**, *39*, 779–785. [CrossRef]

33. DEA. Biodiversity Management Plan for the African Penguin Spheniscus demersus. *Gov. Gaz.* **2013**, 72. Available online: https://www.dffe.gov.za/sites/default/files/docs/biodiversitymanagementplan_africanpenguin.pdf (accessed on 3 August 2023).

34. Francomano, D. Soundscape Dynamics in the Social-Ecological Systems of Tierra del Fuego. Ph.D. Thesis, Purdue University, West Lafayette, IN, USA, 2020; p. 252.

35. Borker, A.L.; Mckown, M.W.; Ackerman, J.T.; Eagles-Smith, C.A.; Tershy, B.R.; Croll, D.A. Vocal Activity as a Low Cost and Scalable Index of Seabird Colony Size. *Conserv. Biol.* **2014**, *28*, 1100–1108. [CrossRef]

36. Favaro, L.; Ozella, L.; Pessani, D. The Vocal Repertoire of the African Penguin (*Spheniscus demersus*): Structure and Function of Calls. *PLoS ONE* **2014**, *9*, e103460. [CrossRef]

37. Favaro, L.; Gamba, M.; Cresta, E.; Fumagalli, E.; Bandoli, F.; Pilenga, C.; Isaja, V.; Mathevon, N.; Reby, D. Do Penguins' Vocal Sequences Conform to Linguistic Laws? *Biol. Lett.* **2020**, *16*, 20190589. [CrossRef]

38. Doser, J.W.; Finley, A.O.; Weed, A.S.; Zipkin, E.F. Integrating Automated Acoustic Vocalization Data and Point Count Surveys for Estimation of Bird Abundance. *Methods Ecol. Evol.* **2021**, *12*, 1040–1049. [CrossRef]

39. Brownlie, K.C.; Monash, R.; Geeson, J.J.; Fort, J.; Bustamante, P.; Arnould, J.P.Y. Developing a Passive Acoustic Monitoring Technique for Australia's Most Numerous Seabird, the Short-Tailed Shearwater (*Ardenna Tenuirostris*). *Emu Austral Ornithol.* **2020**, *120*, 123–134. [CrossRef]

40. Pérez-Granados, C.; Traba, J. Estimating Bird Density Using Passive Acoustic Monitoring: A Review of Methods and Suggestions for Further Research. *Ibis* **2021**, *163*, 765–783. [CrossRef]

41. Favaro, L.; Cresta, E.; Friard, O.; Ludynia, K.; Mathevon, N.; Pichegru, L.; Reby, D.; Gamba, M. Passive Acoustic Monitoring of the Endangered African Penguin (*Spheniscus demersus*) Using Autonomous Recording Units and Ecoacoustic Indices. *Ibis* **2021**, *163*, 1472–1480. [CrossRef]

42. Oppel, S.; Hervías, S.; Oliveira, N.; Pipa, T.; Silva, C.; Geraldes, P.; Goh, M.; Immler, E.; McKown, M. Estimating Population Size of a Nocturnal Burrow-Nesting Seabird Using Acoustic Monitoring and Habitat Mapping. *Nat. Conserv.* **2014**, *7*, 1–13. [CrossRef]

43. Arneill, G.E.; Critchley, E.J.; Wischnewski, S.; Jessopp, M.J.; Quinn, J.L. Acoustic Activity across a Seabird Colony Reflects Patterns of Within-colony Flight Rather than Nest Density. *Ibis* **2020**, *162*, 416–428. [CrossRef]

44. Campbell, K.J.; Farah, D.; Collins, S.; Parsons, N.J. Sex Determination of African Penguins Spheniscus demersus Using Bill Measurements: Method Comparisons and Implications for Use. *Ostrich* **2016**, *87*, 47–55. [CrossRef]

45. Aldinucci, M.; Bagnasco, S.; Lusso, S.; Pasteris, P.; Rabellino, S.; Vallero, S. OCCAM: A Flexible, Multi-Purpose and Extendable HPC Cluster. *J. Phys. Conf. Ser.* **2017**, *898*, 82039. [CrossRef]

46. Katz, J.; Hafner, S.D.; Donovan, T. Tools for Automated Acoustic Monitoring within the R Package MonitoR. *Bioacoustics* **2016**, *25*, 197–210. [CrossRef]

47. R Core Team. *R: A Language and Environment for Statistical Computing*; R Core Team: Vienna, Austria, 2022.

48. Favaro, L.; Gamba, M.; Alfieri, C.; Pessani, D.; McElligott, A.G. Vocal Individuality Cues in the African Penguin (*Spheniscus demersus*): A Source-Filter Theory Approach. *Sci. Rep.* **2015**, *5*, 17255. [CrossRef]

49. Katz, J.; Hafner, S.D. Making Bin Templates. Available online: https://jonkatz2.github.io/monitoR/assets/makingTemplates/makingBinTemplates.html (accessed on 24 December 2022).

50. Kriesell, H.J.; Elwen, S.H.; Nastasi, A.; Gridley, T. Identification and Characteristics of Signature Whistles in Wild Bottlenose Dolphins (*Tursiops truncatus*) from Namibia. *PLoS ONE* **2014**, *9*, e106317. [CrossRef]

51. Brooks, M.E.; Kristensen, K.; van Benthem, K.J.; Magnusson, A.; Berg, C.W.; Nielsen, A.; Skaug, H.J.; Machler, M.; Bolker, B.M. GlmmTMB Balances Speed and Flexibility among Packages for Zero-Inflated Generalized Linear Mixed Modeling. *R J.* **2017**, *9*, 378–400. [CrossRef]

52. Lüdecke, D. Data Visualization for Statistics in Social Science. Available online: https://strengejacke.github.io/sjPlot/ (accessed on 24 December 2022).

53. Anderson, A.S.; Marques, T.A.; Shoo, L.P.; Williams, S.E. Detectability in Audio-Visual Surveys of Tropical Rainforest Birds: The Influence of Species, Weather and Habitat Characteristics. *PLoS ONE* **2015**, *10*, e0128464. [CrossRef]

54. Santos, M.; Travassos, P.; Repas, M.; Cabral, J.A. Modelling the Performance of Bird Surveys in Non-Standard Weather Conditions: General Applications with Special Reference to Mountain Ecosystems. *Ecol. Indic.* **2009**, *9*, 41–51. [CrossRef]
55. Radford, A.N.; du Plessis, M.A. Territorial Vocal Rallying in the Green Woodhoopoe: Factors Affecting Contest Length and Outcome. *Anim. Behav.* **2004**, *68*, 803–810. [CrossRef]
56. Hondula, D.M.; Balling, R.C.; Vanos, J.K.; Georgescu, M. Rising Temperatures, Human Health, and the Role of Adaptation. *Curr. Clim. Change Rep.* **2015**, *1*, 144–154. [CrossRef]
57. Reporter, W. Penguins Breeding at New Colony: A World-First for African Penguins. *Witness*, 21 November 2022.
58. Wood, C.M.; Peery, M.Z. What Does 'Occupancy' Mean in Passive Acoustic Surveys? *Ibis* **2022**, *164*, 1295–1300. [CrossRef]
59. Eggleton, P.; Siegfried, W.R. Displays of the Jackass Penguin. *Ostrich* **1979**, *50*, 139–167. [CrossRef]
60. Gray, L.F.; McNeil, D.J.; Larkin, J.T.; Parker, H.A.; Shaffer, D.; Larkin, J.L. Quantifying Detection Probability of American Woodcock (*Scolopax minor*) on Transects Sampled with Thermal Cameras. *Wildl. Soc. Bull.* **2022**, e1417. [CrossRef]

biology

Article

Comparative Mitogenome Analyses Uncover Mitogenome Features and Phylogenetic Implications of the Reef Fish Family Holocentridae (Holocentriformes)

Qin Tang [1], Yong Liu [2,3,4,5], Chun-Hou Li [2,3,4,5], Jin-Fa Zhao [2,3,4,5] and Teng Wang [2,3,4,5,*]

[1] College of Fisheries, Huazhong Agricultural University, Wuhan 430070, China; tangqin@mail.hzau.edu.cn
[2] Key Laboratory of South China Sea Fishery Resources Exploitation and Utilization, Ministry of Agriculture and Rural Affairs, South China Sea Fisheries Research Institute, Chinese Academy of Fishery Sciences, Guangzhou 510300, China; liuyong@scsfri.ac.cn (Y.L.); chunhou@scsfri.ac.cn (C.-H.L.); zhaojf2019@163.com (J.-F.Z.)
[3] Scientific Observation and Research Station of Xisha Island Reef Fishery Ecosystem of Hainan Province, Key Laboratory of Efficient Utilization and Processing of Marine Fishery Resources of Hainan Province, Sanya Tropical Fisheries Research Institute, Sanya 572018, China
[4] Guangdong Provincial Key Laboratory of Fishery Ecology Environment, Guangzhou 510300, China
[5] Observation and Research Station of Pearl River Estuary Ecosystem, Guangzhou 510300, China
* Correspondence: wangteng@scsfri.ac.cn

Simple Summary: Mitochondria play a critical role in the energy metabolism of coral reef fish, providing ATP to fuel cellular processes. A mitogenome study has been employed to investigate the genetic diversity, population structure, and evolutionary relationships among coral reef fish taxa. Species of the Holocentridae family play important ecological roles in coral reef communities. Two subfamilies of this family, Holocentrinae and Myripristinae, exhibit similarities in morphology and distribution, with minor differences in habitation and feeding behavior. Here, we present full mitochondrial genome sequences of eight holocentrid species and report the results of a comparative analysis with six previously published species. The results indicate that these mitogenome structures are relatively conserved, except for the high variability in control regions. The whole genomes, except for *nad6*, exhibited positive AT-skews and negative GC-skews. Furthermore, we compared the two subfamilies to explore the reasons behind their varying inhabitation and behavior. Phylogenetic analysis indicated all species formed two subfamilies, the Holocentrinae and Myripristinae, with each subfamily comprising two genera. Positive selection analysis revealed that all protein-coding genes (PCGs) were subjected to purifying selection. The data obtained from our study could serve as a valuable resource for future investigations on the evolution and conservation of holocentrid fish.

Abstract: To understand the molecular mechanisms and adaptive strategies of holocentrid fish, we sequenced the mitogenome of eight species within the family Holocentridae and compared them with six other holocentrid species. The mitogenomes were found to be 16,507–16,639 bp in length and to encode 37 typical mitochondrial genes, including 13 PCGs, two ribosomal RNAs, and 22 transfer RNA genes. Structurally, the gene arrangement, base composition, codon usage, tRNA size, and putative secondary structures were comparable between species. Of the 13 PCGs, *nad6* was the most specific gene that exhibited negative AT-skews and positive GC-skews. Most of the genes begin with the standard codon ATG, except *cox1*, which begins with the codon GTG. By examining their phylogeny, *Sargocentron* and *Neoniphon* were verified to be closely related and to belong to the same subfamily Holocentrinae, while *Myripristis* and *Ostichthys* belong to the other subfamily Myripristinae. The subfamilies were clearly distinguished by high-confidence-supported clades, which provide evidence to explain the differences in morphology and feeding habits between the two subfamilies. Selection pressure analysis indicated that all PCGs were subject to purifying selection. Overall, our study provides valuable insight into the habiting behavior, evolution, and ecological roles of these important marine fish.

Citation: Tang, Q.; Liu, Y.; Li, C.-H.; Zhao, J.-F.; Wang, T. Comparative Mitogenome Analyses Uncover Mitogenome Features and Phylogenetic Implications of the Reef Fish Family Holocentridae (Holocentriformes). *Biology* **2023**, *12*, 1273. https://doi.org/10.3390/biology12101273

Academic Editor: M. Gonzalo Claros

Received: 8 September 2023
Accepted: 20 September 2023
Published: 22 September 2023

Keywords: Holocentridae; mitogenome; codon usage; gene rearrangement; phylogeny; selection pressure

1. Introduction

Holocentridae, a family of ray-finned fish, is also known as a nocturnal coral reef fish family, with the subfamily Holocentrinae typically known as squirrelfish, while Myripristinae members are known as soldierfish [1–3]. The family Holocentridae is primarily distributed in the tropical parts of the Atlantic, Indian, and Pacific Oceans. Typically, they inhabit waters up to a depth of 100 m, although some species of the *Ostichthys* genus (soldierfish) from the subfamily Myripristinae have been detected much deeper [4]. Holocentridae fish possess large eyes and are primarily active at night, suggesting they are nocturnal in terms of their activity patterns [5]. The colors of the majority of Holocentridae are either red or silver [6]. Members of the Holocentrinae subfamily (squirrelfish) possess venomous spines near the gill opening, which can inflict painful wounds [7]. Regarding feeding habits, squirrelfish mainly feed on small fish and benthic invertebrates, while soldierfish typically feed on zooplankton [8]. Unlike adults, the larvae of Holocentridae are pelagic and can be found far out at sea [9]. Currently, according to the statistics of Fishes of the Word, a total of 83 species belonging to 8 genera are recognized, with *Sargocentron* and *Myripristis* being the most numerous genera, which contain 33 and 28 species, respectively [10]. Based on our investigation, only four genera of Holocentridae have been studied at the mitochondrial genome level so far, including *Sargocentron*, *Neoniphon*, *Myripristis*, and *Ostichthys*, with *Sargocentron* and *Neoniphon* belonging to the Holocentrinae subfamily and *Myripristis* and *Ostichthys* falling under the Myripristinae subfamily.

The Holocentridae family contributes to the ecological diversity of coral reefs. Prior to the disclosure of its complete mitochondrial genome structure, several studies concentrated on the physiology, ecology, and evolutionary aspects of this fish family. In a report by Eric et al. in 2011, Holocentrids were identified as vocal reef fishes. In their study, the authors compared sound production mechanisms across different species and found that all fish possess fast-contracting muscles and have relatively similar sound-producing mechanisms [1]. Fanny et al. (2021) studied the visual systems of Holocentridae and compared the two subfamilies, the Holocentrinae and Myripristinae, demonstrating that squirrelfish had a slightly more developed photopic visual system than soldierfish [3]. The evolution of holocentroids has also been evaluated. Andrews et al. (2023) reported a new holocentroid species from the fossil material of the early Paleocene and estimated a Danian divergence between Myripristinae and Holocentrinae from the fossil analysis via micro-computed tomography, suggesting that several holocentroid lineages crossed the Cretaceous–Palaeogene boundary [4]. These studies give us insight into the traits of holocentrid fish and their ability to adapt to the marine environment, as well as their evolutionary history and important role in the ocean ecosystem. Next, a study conducted at the mitochondrial structure level will provide proof of previous findings and reveal a systematic relationship of holocentrid species.

Mitochondria play a crucial role within eukaryotic cells, participating in various essential processes such as ATP generation through oxidative phosphorylation, cell differentiation, signaling, growth, and apoptosis [11–13]. The vertebrate mitogenome is characterized by its small size, ranging from 16 to 17 kb, and its circular double-stranded structure. A typical mitogenome contains 13 protein-coding genes (PCGs), 22 transfer RNA genes (tRNAs), 2 ribosomal RNA genes (rRNAs), and two non-coding regions, namely, the origin of L-strand replication (O_L) and the control region (CR) [14,15]. Mitochondrial DNA sequences have been extensively studied across various fields. In evolutionary biology, they have been used to investigate the evolutionary relationships and genetic variation between different species, while in biogeography, they have been used to uncover spatial distributions and migration patterns [15]. Although mitochondrial function in corals is highly conserved, it has been observed that the PCGs of some species undergo evolutionary selection in response to the metabolic demands imposed by extreme environments [16–19]. In a recent study by Ramos et al. (2023), selection tests were conducted on mitochondrial PCGs of deep-sea and shallow-water species, revealing that certain PCGs underwent adaptive evolution during their adaptation to the deep-sea environment [16]. Another extreme environment, the sub-zero habitat of the Antarctic, poses a significant challenge to the survival of fish. Thus, Antarctic icefishes have developed a unique mechanism to adapt to the inhabitants. In an earlier study conducted in 2010, O'Brien et al. observed an increase in mitochondrial density in cardiac myocytes and oxidative skeletal muscle fibers, accompanied by a proliferation of mitochondrial membranes. This expansion of membranes facilitates the efficient intracellular diffusion of oxygen [20].

Although holocentrids play an irreplaceable role in coral reef ecosystems and several studies have focused on their morphological characteristics, activity patterns, and underlying mechanisms, there has been limited research on their mitogenome characteristics and evolutionary biology. In this study, we present eight mitogenomes for the first time and report the results of a comparative analysis with six published mitogenomes. We provide comprehensive insights into the detailed features of all mitogenomes with respect to structure, gene arrangement, nucleotide composition, noncoding RNA, and codon usage. Additionally, we explore the phylogenetic relationships between species and estimate the selection pressures during their evolution. Through comparative analysis and newly generated results, we offer valuable insights into the evolutionary history of holocentrid species. Furthermore, we make meaningful contributions towards identifying and protecting these coral reef fish species.

2. Materials and Methods

2.1. Sampling, DNA Extraction, Library Construction, and Sequencing

In this study, we de novo sequenced eight holocentrid species, including *Myripristis kuntee* (Shoulderbar soldierfish), *Myripristis murdjan* (Pinecone soldierfish), *Myripristis violacea* (Lattice soldierfish), *Neoniphon opercularis* (Blackfin squirrelfish), *Sargocentron caudimaculatum* (Silverspot squirrelfish), *Sargocentron diadema* (Crown squirrelfish), *Sargocentron melanospilos* (Blackblotch squirrelfish), and *Sargocentron punctatissimum* (Speckled squirrelfish). The specimens were obtained from the Xisha Islands (15°46′~17°08′ N, 111°11′~112°54′ E), China, and deposited in the South China Sea Fisheries Research Institute, Chinese Academy of Fishery Sciences. Six published mitogenome sequences from the other six holocentrid species were downloaded from GenBank for an integrative and comparative analysis: *Neoniphon samara* (Sammara squirrelfish), NC_063501.1; *Sargocentron spiniferum* (Sabre squirrelfish), KX254549.1; *Sargocentron rubrum* (Redcoat; squirrelfish), NC_004395.1; *Myripristis vittate* (Whitetip soldierfish), NC_063496.1; *Myripristis berndti* (Blotcheye soldierfish), AP002940.1; *Ostichthys japonicus* (Japanese soldierfish), AP004431.1.

Total genomic DNA was extracted from the specimens using the E.Z.N.A.® Tissue DNA Kit (OMEGA, Beijing, China) in accordance with the manufacturer's protocols. Two distinct types of tissue were sampled: a small fragment of muscle from a specimen or a clip of the pelvic fin taken from the right side of a specimen. After DNA extraction, 1 μg of purified DNA was randomly fragmented into fragments with a length ranging from 300 to 500 bp and used for subsequent library construction. Complete genomic libraries were established using the Illumina TruSeqTM Nano DNA Sample Prep Kit (Illumina, San Diego, CA, USA) in accordance with the manufacturer's instructions. Then, libraries were sequenced using the Illumina NovaSeq 6000 platform to obtain 150 bp paired-end reads. The library construction and sequencing procedures were carried out by the Biozeron Corporation (Shanghai, China).

2.2. Sequence Assembly, Annotation, and Analyses

Prior to assembly, raw reads were filtered using Trimmomatic (v0.39) [21] to remove the low-quality reads (the reads showing a quality score below 20, Q < 20), the reads with adaptors, the reads containing a percentage of uncalled bases ("N" characters) equal to or greater than 10%, and duplicated sequences. Filtered reads were assembled into contigs using MitoZ (v2.3) [22], and potential mitochondrial contigs were extracted by aligning them against the NCBI mitogenome database. Then, GetOrganelle (v1.7.5) (https://github.com/Kinggerm/GetOrganelle, accessed on 20 March 2022) was used to assemble the mitogenomes [23]. After assembly, the starting position and orientation of the mitochondrial sequence were obtained based on a reference genome. Annotation of the mitogenomes to protein-coding genes (PCGs), tRNAs, and rRNAs was performed using MITOS [24] and Mitoannotator (v3.83) [25]. Functional annotations of PCGs were performed using sequence-similarity Blast searches with a typical cut-off E-value of 10^{-5} against several publicly available protein databases: NCBI non-redundant (Nr) protein database, Swiss-Prot, Clusters of Orthologous Groups (COGs), and Kyoto Encyclopedia of Genes and Genomes (KEGG) and Gene Ontology (GO) terms. tRNA genes were searched using tRNAscan-SE (v2.0) [26], and their secondary structures were drawn via RNAplot from the package ViennaRNA (v2.5.1) [27]. Base composition and codon distributions were analyzed using MEGA 7.0 [28], and the nucleotide composition skewness was measured using the following formulas: AT-skew = $(A - T)/(A + T)$ and GC-skew = $(G - C)/(G + C)$ [15]. Relative synonymous codon usage (RSCU) was calculated using the "cusp" of EMBOSS package (v6.6.0.0) [29]. Circular genomes were visualized using the CGView tool (http://stothard.afns.ualberta.ca/cgview_server/, accessed on 25 March 2022) online [30]. The r package "ComplexHeatmap" was used to draw heatmaps. The conserved-sequence block domains (CSBs) of control regions were determined by comparing them with public holocentrid species.

2.3. Phylogenetic Analyses

The phylogenetic relationships were reconstructed using the 13 PCGs of the 14 holocentrid fish mitogenomes; three parrotfish, namely, *Scarus frenatus* (OQ349185.1, Bridled parrotfish), *Scarus niger* (OQ349187.1, Dusky parrotfish), and *Scarus prasiognathos* (OQ349189.1, Singapore parrotfish), were used as outgroup taxa. Multiple sequence alignment was performed using MAFFT (v7.453) [31] with default parameters. The alignment results were further trimmed to eliminate the ambiguous positions using Gblocks (v0.91b) [32]. Then, trimmed sequences were concatenated into a supermatrix with FASconCAT [33]. Fasta files were converted into Nexus format using Geneious (v.2022.2.2) [34]. Phylogenetic relationships were inferred from the concatenated dataset using maximum likelihood (ML) and Bayesian inference (BI) methods. ML analyses were performed using IQ-TREE (v1.6.12) [35] with the following parameters: "-m MFP -b 1000 -bnni". By using these parameters, the best-fit substitution models (including FreeRate heterogeneity models) and partition schemes were inferred via the built-in ModelFinder [36]. "MFP" (ModelFinder Plus) allows for extended model selection followed by tree inference, while "-b 1000" ensures that 1000 bootstrap searches will be performed in order to infer the consensus trees. BI analysis was carried out using MrBayes (v3.2.7a) [37]. Before BI analysis, the best model was selected with jModeltest (v2.1.10) [38], and the model of JC was optimal for analysis with nucleotide alignment. Then, BI analysis was performed using four simultaneous Markov Chain Monte Carlo chains for 2,000,000 generations and sampled every 1000 generations, using a burn-in of 25% generations. The average standard deviation of split frequencies was set as less than 0.01. Phylogenetic trees generated from both ML and BI methods were visualized in FigTree (v1.4.4) (http://tree.bio.ed.ac.uk/software/figtree/, accessed on 10 June 2023).

2.4. Positive Selection

Positive selection refers to the evolutionary process through which genetic variants increase in frequency within a given population until they become prevalent. This phenomenon results from the advantageous traits conferred by these genetic variants, which enhance the fitness and reproductive success of the individuals carrying them. To perform positive selection analysis, first, a multiple-codon alignment was produced for each PCG from the corresponding aligned predicted protein sequences using PAL2NAL [39]. Then, positive selection analyses were performed using two codon-based maximum likelihood methods, i.e., Single Likelihood Ancestor Counting (SLAC) and Fixed Effects Likelihood (FEL), as implemented via HYPHY (v2.5.39) (MP) [40] on a Linux system. SLAC represents a substantially enhanced and refined version of the Suzuki–Gojobori counting method and is designed to assess the rate of nonsynonymous and synonymous substitutions in DNA sequences, thereby shedding light on the selective pressures operating on specific genes during the progress of evolution. FEL is also an innovative and statistically robust approach rooted in likelihood-based methods aiming to characterize the evolutionary dynamics of genetic sequences in the context of codon substitution models. In this study, the number of non-synonymous substitutions per non-synonymous site (dN) and the number of synonymous substitutions per synonymous site (dS) were estimated using both methods. And the dN/dS ratio (or ω) was taken as a judgment of the selective pressure on each codon of the PCGs. In detail, the ratio dN/dS > 1 suggests positive or diversifying selection, dN/dS < 1 suggests negative or purifying selection, and dN/dS = 1 indicates neutral evolution. The significance level of the positive selection estimated from both SLAC and FEL analyses was set as p-value < 0.05.

3. Results

3.1. General Features of Mitochondrial Genomes

The total length of the eight newly sequenced complete mitogenomes ranged from 16,507 bp in *Sargocentron punctatissimum* to 16,639 bp in *Neoniphon opercularis* (Table S1). All mitogenomes comprised 37 genes, including 13 PCGs, two rRNAs (12S rRNA and 16S rRNA, named *rrnS* and *rrnL*), one control region (CR), and 22 tRNAs (Table 1, Figure 1, Table S2, Figure S1). Taking *Myripristis kuntee* as an example, the total length of the 13 PCGs in the mitogenome is 11,439 bp, which accounts for 69.20% of the entire mitogenome. In total, 12/13 of the PCGs are encoded on the Heavy (H) strand in the positive direction, except for the gene *nad6* (NADH dehydrogenase subunit 6), which is located on the Light (L) strand in the reverse direction (Figure 1, Table 1). Fourteen of the tRNAs, namely, *trnD, trnK, trnG, trnR, trnH, trnS1, trnL1, trnT, trnF, trnV, trnL2, trnI, trnM,* and *trnW,* are located on the H-strand, while the other eight tRNAs (*trnS2, trnE, trnP, trnQ, trnA, trnN, trnC,* and *trnY*) are located on the L-strand (Figure 1, Table 1). This arrangement pattern of genes is identical among holocentrid species (Table S2) and is similar to that found in most vertebrates [41]. The comparative analysis of 14 mitochondrial genomes with respect to structure reveals that they are almost identical, and no rearrangements of genes have occurred, but the control region was the most variable region among the species in both length and nucleotide composition (Figure 2, Figure S2).

Table 1. Summary of the mitochondrial genomes of holocentrid fish. *Myripristis kuntee* was taken as an example.

Gene	Start	End	Strand	Size (bp)	Start Codon	Stop Codon	Anticodons
cox1	1	1557	+	1557	GTG	AGA	—
trnS2	1553	1623	−	71	—	—	TGA
trnD	1627	1698	+	72	—	—	GTC
cox2	1712	2402	+	691	ATG	T	—
trnK	2403	2475	+	73	—	—	TTT
atp8	2477	2644	+	168	ATG	TAA	—
atp6	2635	3318	+	684	ATG	TAA	—
cox3	3318	4103	+	786	ATG	TAA	—
trnG	4103	4173	+	71	—	—	TCC
nad3	4174	4524	+	351	ATG	TAG	—
trnR	4523	4591	+	69	—	—	TCG
nad4l	4592	4888	+	297	ATG	TAA	—
nad4	4882	6262	+	1381	ATG	T	—
trnH	6263	6331	+	69	—	—	GTG

Table 1. *Cont.*

Gene	Start	End	Strand	Size (bp)	Start Codon	Stop Codon	Anticodons
trnS1	6332	6399	+	68	–	–	GCT
trnL1	6401	6473	+	73	–	–	TAG
nad5	6474	8312	+	1839	ATG	TAA	–
nad6	8308	8829	–	522	ATG	AGG	–
trnE	8830	8898	–	69	–	–	TTC
cob	8905	10,045	+	1141	ATG	T	–
trnT	10,046	10,117	+	72	–	–	TGT
trnP	10,122	10,191	–	70	–	–	TGG
trnF	11,054	11,121	+	68	–	–	GAA
rrnS	11,122	12,070	+	949	–	–	–
trnV	12,071	12,142	+	72	–	–	TAC
rrnL	12,170	13,814	+	1645	–	–	–
trnL2	13,839	13,912	+	74	–	–	TAA
nad1	13,913	14,887	+	975	ATG	TAA	–
trnI	14,892	14,961	+	70	–	–	GAT
trnQ	14,961	15,031	–	71	–	–	TTG
trnM	15,031	15,100	+	70	–	–	CAT
nad2	15,101	16,147	+	1047	ATG	TAA	–
trnW	16,147	16,219	+	73	–	–	TCA
trnA	16,221	16,289	–	69	–	–	TGC
trnN	16,291	16,363	–	73	–	–	GTT
trnC	16,397	16,461	–	65	–	–	GCA
trnY	16,462	16,529	–	68	–	–	GTA

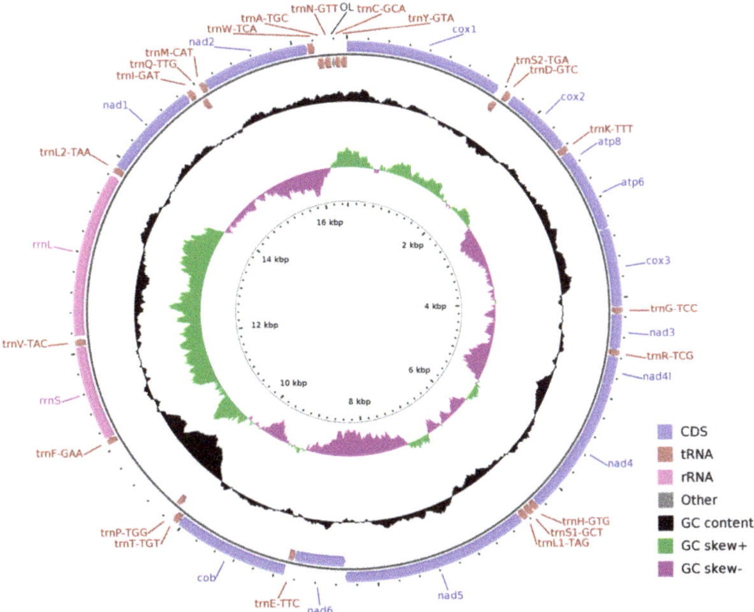

Figure 1. Map of the mitochondrial genomes of holocentrid species. *Myripristis kuntee* was taken as an example. PCGs are indicated by blue arrows, tRNA genes are indicated by brown arrows, and rRNA genes are indicated by lavender arrows. tRNAs are denoted by single-letter amino acid abbreviations followed by anticodons. Peaks on the black cycle indicate the GC content, while the outward and inward directions indicate GC content above or below average level. The purple and green cycles show the GC skew, where skew values between 0 and 1 are shown in purple and those between −1 and 0 are shown in green. Ticks in the inner cycle indicate the sequence length.

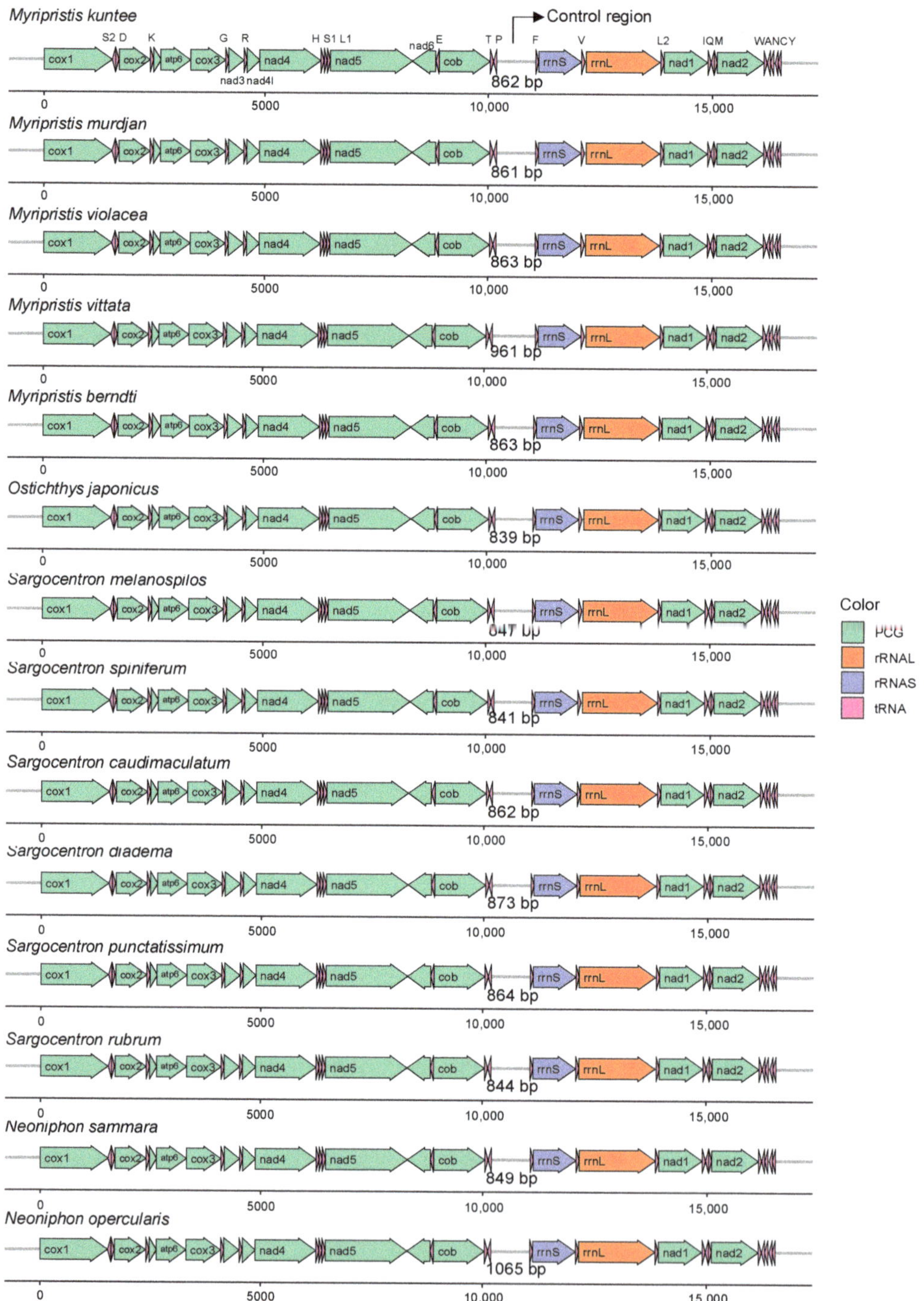

Figure 2. Gene arrangement and comparative genome analysis of 14 mitochondrial genomes of holocentrid species.

Two types of start codons and five types of stop codons were used in 14 species. Taking *M. kuntee* as an example, most of the PCGs begin with the standard codon ATG, except *cox1* (cytochrome c oxidase subunit I), which begins with the codon GTG (Table 1). Table 1 also displays the utilization of stop codons in the *M. kuntee* mitochondrial genome. Overall, seven PCGs (*atp8, atp6, cox3, nad4l, nad5, nad1,* and *nad2*) end with TAA, while *nad3* terminates with TAG, *nad6* terminates with AGG, *cox1* terminates with AGA, and the remaining three PCGs (*cox2, nad4,* and *cob*) end with an incomplete terminating codon T-- (Table 1).

3.2. Nucleotide Composition of Protein-Coding Genes of the Holocentrid Mitogenomes and the Codon Usage

The nucleotide compositions were comparable among all 14 species. For the whole mitochondrial genome, the overall A + T content ranges from 53.12% in *M. murdjan* to 56.85% in *N. sammara*, while the G + C content ranges from 43.15% in *N. sammara* to 46.88% in *M. murdjan* (Table S3). For the PCGs of the 14 mitogenomes, the average A + T content is 53.95%, which is slightly lower than the average of the whole genome, 54.70%. When focusing on each of the 13 PCGs, the lowest A + T content was found in *nad4l* (50.63 ± 1.50%), while the highest was found in *cox2* (55.33 ± 1.61%) (Table S3). Taking *M. kuntee* as an example, the A + T content of its PCGs was 52.20% (Table S3). All the holocentrid mitogenomes exhibited AT bias in the whole mitogenome, tRNAs, rRNAs, and most of the PCGs (Figure 3, Table S3). The largest AT-skew values were observed in rRNAs, all of which were positive, while the smallest and most negative values were found in the PCG nad6 (Figure 3, Table S3). For most of the PCGs, the AT-skew was higher than the GC-skew, except for nad6, which exhibited an unusual AT-skew and GC-skew (Figure 3). The negative AT-skew and positive GC-skew observed indicated that nad6 displayed an excess of T over A and G over C. Moreover, the average AT-skew and GC-skew values of all 13 PCGs were negative (Figure 3, Table S3).

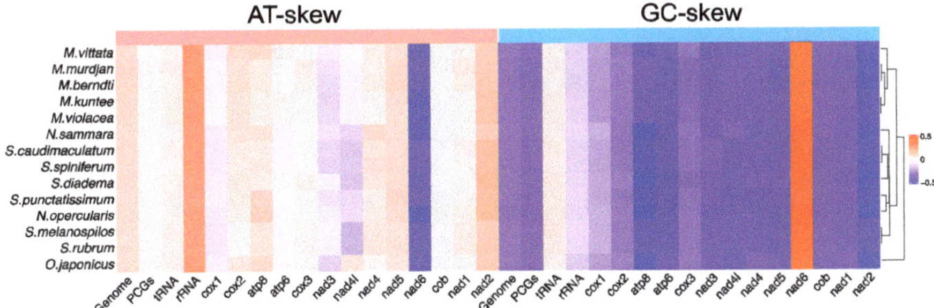

Figure 3. The base skew of various datasets among 14 mitogenomes, with hierarchical clustering of holocentrid species (*y*-axis) based on AT skew and GC skew.

The relative synonymous codon usage (RSCU) values for the PCGs are summarized in Figure 4 and Table S4. Excluding stop codons, there are 3813 codons in the mitogenome of *M. kuntee*. The codons encoding Arg, Leu, and Ser are the most frequent, while those encoding Trp and Met are scarce (Figure 4A). The heatmap, which was generated based on RSCU values, illustrates the resemblance of codon usage patterns among 14 mitogenomes (Figure 4B). Among the codons coding Ala, GCC (RSCU = 1.75) is the most frequently used. Also, it is the most frequently used codon among the 61 codons that encode amino acids.

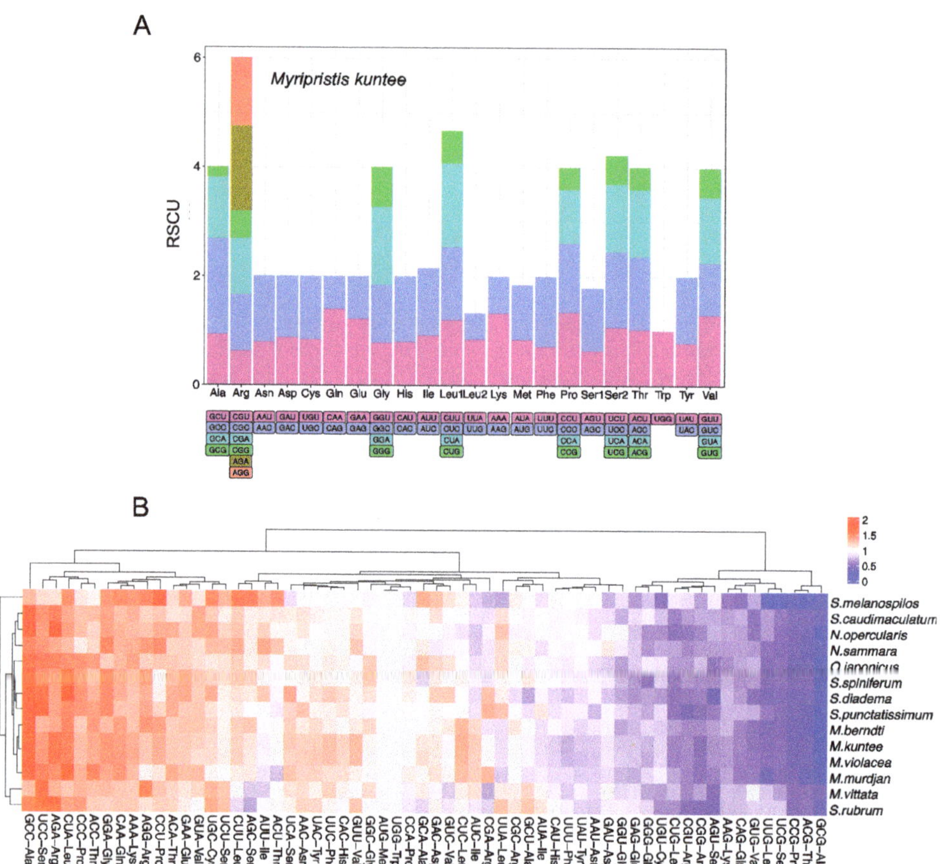

Figure 4. (**A**) Relative synonymous codon usage (RSCU) in the holocentrid mitogenome, for which *Myripristis kuntee* was taken as an example. Codon families are indicated below the *x*-axis. (**B**) Heatmap based on RSCU of 14 mitogenomes.

3.3. Transfer RNA and Ribosomal RNA

All 22 typical tRNAs of the vertebrate mitochondrial genome were found in the mitogenomes (Figure 5A). Taking *M. kuntee* as an example, the tRNA size ranged from 65 to 74 bp. Most tRNAs could be folded into the canonical clover leaf secondary structure. The secondary structure of tRNAs generally contains four domains and a short variable loop: the amino acid acceptor (AA) stem, the dihydrouridine arm (D stem and loop, D), the thymidine arm (T stem and loop, T), the anticodon arm (AC stem and loop, AC), and the variable (V) loop (Figure 5A). However, as determined from the comparison of four representative species from the genera *Myripristis*, *Neoniphon*, *Sargocentron*, and *Ostichthys*, *trnC*-GCA (Cys) is supposed to lose the D loop in *M. kuntee* and *O. japonicus* (Figure 5B). The tRNA *trnS1*-GCT (Ser) is characterized by a special V loop and a large ring structure in the D loop in three of the four typical holocentrid species, but *O. japonicus* does not have these two structures (Figure 5B). Moreover, all species lack a D stem in *trnS1*-GCT (Ser), thus leading to failure in the formation of the typical clover leaf structure (Figure 5B). Regarding another special tRNA, *trnH*-GTG (His), when focusing on the T stem, *N. opercularis* and *S. caudimaculatum* possess a small ring due to the high GC content (Figure 5B).

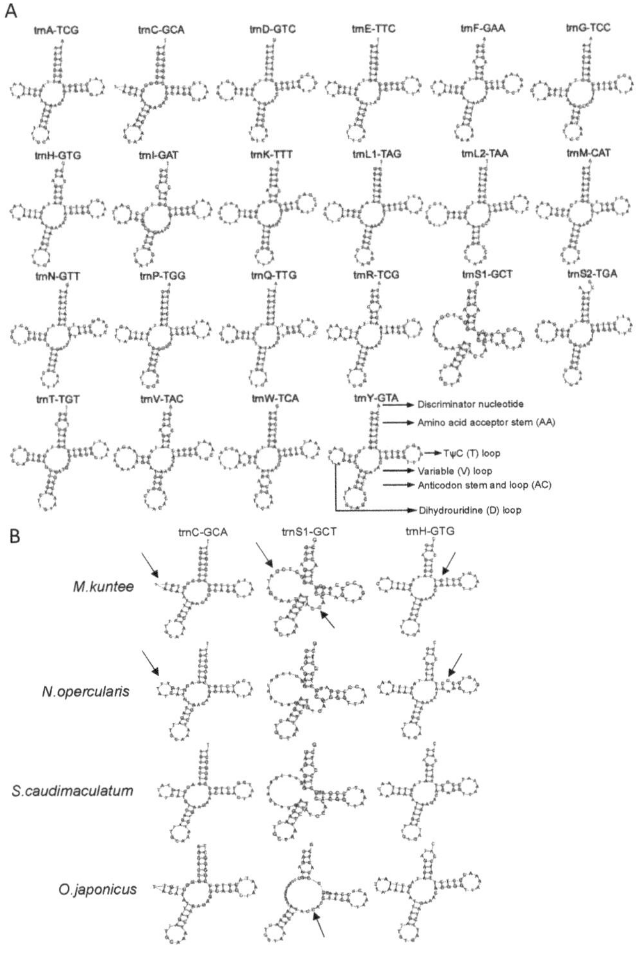

Figure 5. (**A**) Putative secondary structure of tRNAs in holocentrid mitogenomes. (**B**) Differential secondary structure of *trnC*-GCA (Ala), *trnS1*-GCT (Ala), and *trnH*-GTG (Val) of four holocentrid species. Arrows were used to highlight the different structural features of the three tRNAs.

Two ribosomal RNAs (12S rRNA and 16S rRNA, or *rrnS* and *rrnL*) are located on the H-strand in all holocentrid mitogenomes. These two genes were separated by *trnV*-TAC (Val), a feature often found in the mitochondrial genomes of vertebrates [41–43]. Taking *M. kuntee* as an example, the lengths of its 12S rRNA and 16S rRNA genes were 949 bp and 1645 bp, respectively. The total A + T content of these two rRNA genes was 53.93%, higher than their G + C content. Moreover, this mitogenome had a positive AT-skew (0.27) and a negative GC-skew (−0.12) (Table S3). Other species showed similar results to *M. kuntee*.

3.4. Overlaps and Control Regions

When focusing on the 12 protein-coding genes on the heavy strand in the positive direction, a total of three overlaps from genes were detected in the mitogenomes of holocentrids (Table 1, Table S2). Taking *Myripristis kuntee* as an example, the longest overlap was found between *atp8* and *atp6*, with a highly conserved 10 bp motif of "AGCTTCTTCG", while the second longest overlap was found between *nad4l* and *nad4*, with a 7 bp sequence, "ATGCTAA". Apart from that, a 5 bp overlapped sequence, "CCTAA", was observed between *nad5* and *nad6*. Also on the same strand, the control region, located between *trnP* and *trnF*, with a range from 838 bp in *Sargocentron diadema* to 960 bp in *Neoniphon opercularis*, was the most variable region among species. Notably, this variable region accounted for the predominant portion of length discrepancies observed within the mitogenomes of holocentrids. Five conserved sequence blocks (CSB), CSB-I, CSB-II, CSB-III CSB-IV, and CSB-V, were detected (Figure 6) from the alignment of control regions. The base composition was extremely unique to each CSB, with CSB-I being T- and A-rich, CSB-II being AT- and C-rich, CSB-III being T- and C-rich, CSB-IV being C-rich, and CSB-V being A- and C-rich (Table 2).

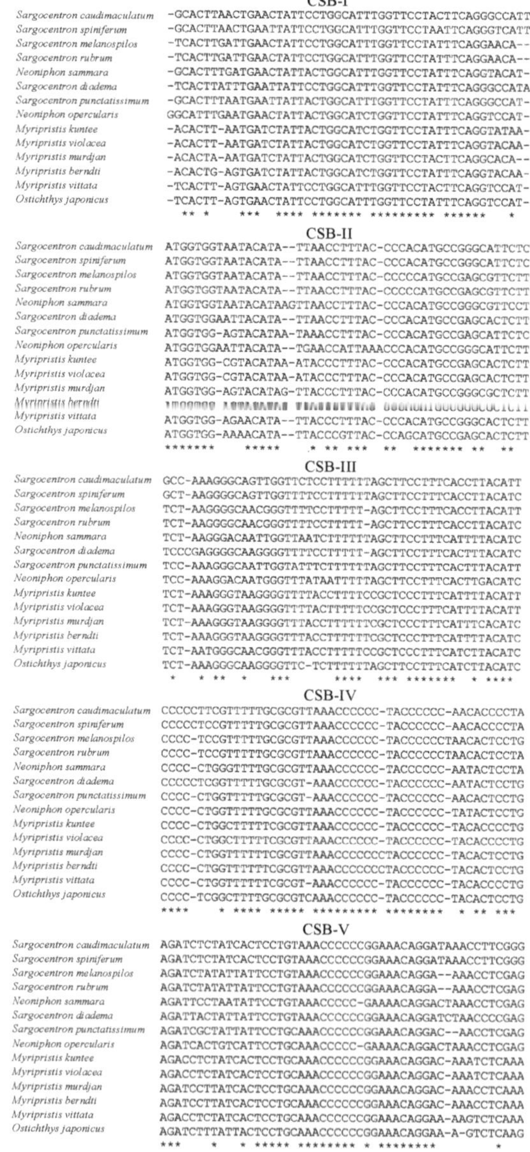

Figure 6. Conserved sequence blocks (CSBs) of the control region in the holocentrid mitogenomes. The asterisks are used to indicate the conserved sites.

Table 2. The base composition of the CSBs of the control region of parrotfish mitogenomes.

Base Composition (%)	CSB-I	CSB-II	CSB-III	CSB-IV	CSB-V
A	27.66	25.53	18.37	12.77	36.73
T	38.30	25.53	42.86	23.40	14.29
G	14.89	17.02	16.33	10.64	12.24
C	19.15	31.91	22.45	53.19	36.73

On the light strand, the special non-coding region O_L (the origin of light strand replication), with a length ranging from 28 to 37 bp among species, is known to regulate the encoding of the *nad6* gene and eight tRNAs. Structurally, O_L is situated within the cluster of five tRNA genes (WANCY), and its secondary structure exhibits a stable stem-loop configuration, characterized by a tight structure with seven G-C pairs in the genera *Neoniphon*, *Sargocentron*, and *Ostichthys* and eight G-C pairs in the genus *Myripristis* (Figure 7). The G-C base pairs forming the stem exhibited a high level of conservation, maintaining their stability across different instances. In contrast, the composition of bases within the loop region displayed variability, with the presence of T being notably limited.

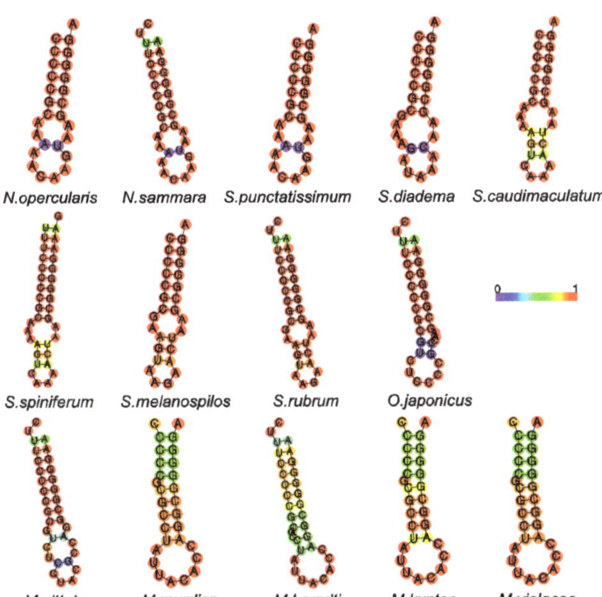

Figure 7. Putative secondary structure of the origin of L strand replication (O_L) in 14 holocentrid species. The color-scaled bar represents base pairing probabilities calculated using CentroidFold software (http://rtools.cbrc.jp/centroidfold/, accessed on 1 September 2023).

3.5. Phylogenetic Analysis

To explore the evolutionary patterns of the 14 holocentrid species, phylogenetic trees were constructed using both ML and BI methods based on 13 PCGs, and three parrotfish from our previous study were used as outgroups [15]. The topological structures of the phylogenetic trees obtained using the two methods were congruent, except that the BI tree had higher support values on the clade of *Neoniphon* (Figure 8). Both trees delimited two prominent clades: clade A and clade B. Clade A consists of species from the genera *Ostichthys* and *Myripristis*, while clade B consists of the other two genera, *Neoniphon* and *Sargocentron*. In general, different species of the same genus clustered into the same clade. In clade B, the genera *Neoniphon* and *Sargocentron* clustered together, with BI posterior probabilities (PP) equal to 1 and an ML bootstrap (BP) equal to 100, which implies a close relationship phylogenetically. However, the fact that the *Neoniphon* clade has become independent suggests it is a new population or a close relative of the genus *Sargocentron*. *O.japonicus* in clade A clustered together with species of *Myripristis*, suggesting its morphological similarity with that genus. Interestingly, two subfamilies of Holocentridae, Holocentrinae and Myripristinae, were clearly distinguished with high confidence, with a PP equal to 1 and a BP equal to 100, suggesting the independence of the squirrelfish and the soldierfish groups and potential differences in morphology and feeding habits between these two subfamilies.

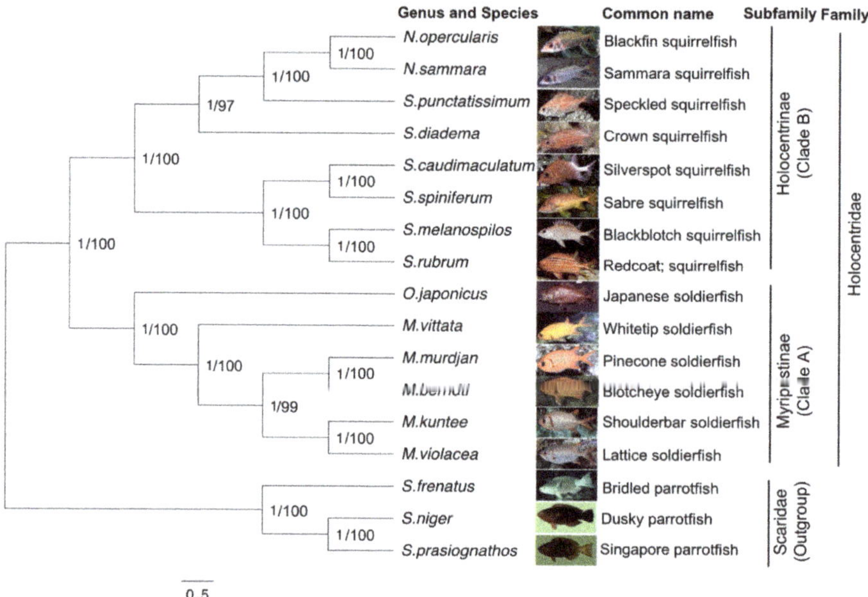

Figure 8. Phylogenetic trees were constructed using the nucleotide sequences of the 13 PCGs of the holocentrid mitogenomes, and three parrotfish mitogenomes were taken as the outgroup. Both the BI method and the ML method were used. The numbers beside the nodes are posterior probabilities (BI) and bootstraps (ML), respectively.

3.6. Non-Synonymous, Synonymous Substitutions, and Positive Selection

To better understand the role of selective pressure, the dN and dS values of the PCGs were calculated using two codon-based maximum likelihood methods, SLAC and FEL. A total of 3381 amino acid sites was calculated using both methods (Figure 9). As determined from the SLAC result, 378 sites are prone to undergoing positive/diversifying selection (dN − dS > 0), but the p-values were not significant (>0.05). Of the rest 3003 sites, 1784 are under negative/purifying selection, with dN − dS < 0 and a p-value < 0.05 (Figure 9A), while the others have a p-value > 0.05. From the FEL analysis, 2459 sites were detected with dN/dS < 1 (ω < 1) (Table S5). *nad5* and *atp8* presented the highest and lowest numbers of amino acids under purifying selection, respectively (Figure 9B), while *nad4* presented the highest percentage of amino acids under purifying selection (Figure 9C). Collectively, all PCGs were subject to purifying selection, with most dN − dS values lower than 0 or dN/dS values lower than 1 (ω < 1), taking a p-value < 0.05 as a threshold.

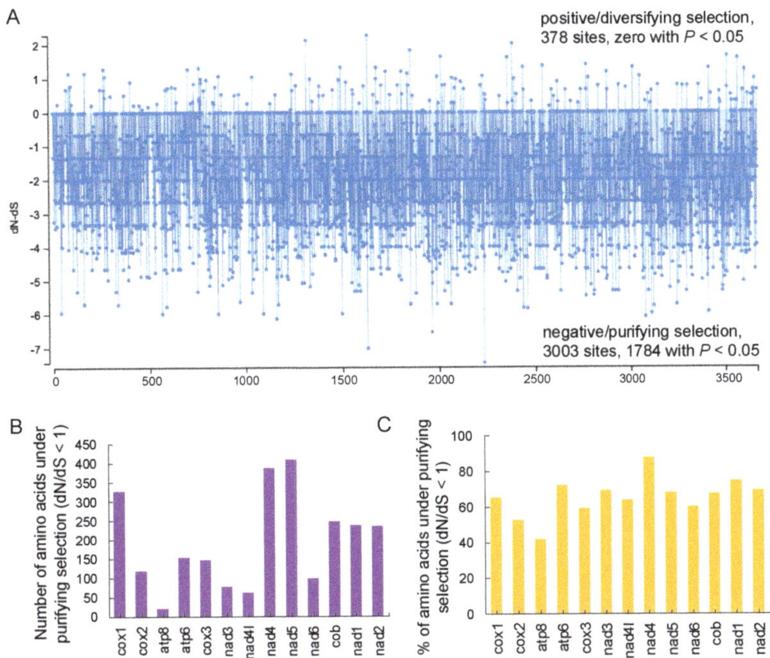

Figure 9. Positive selection of the 13 PCGs in the mitogenomes. (**A**) Detailed site-by-site dN − dS results from the SLAC analysis. (**B**) Number of amino acids under purifying selection (dN/dS < 1 or ω < 1) determined via the FEL method. (**C**) % of amino acids under purifying selection determined using the FEL method.

4. Discussion

In this study, we found that the overall codon usage among the 14 holocentrid species is similar. But when focusing on the PCGs within each species, most of the genes begin with the standard codon ATG, except *cox1*, which begins with the codon GTG. Previous research reported that ATG was the most prevalent in the mitochondrial genome of vertebrates and was exclusively used in the *cox3* gene, while GTG was primarily utilized in the *cox1* gene in over 95% of species [14]. In fish, such as the fathead minnow (*Pimephales promelas*) and parrotfishes, ATG acts as the start codon for all PCGs except *cox1*, which uses GTG as the start codon [15,44]. Unlike bony fish, in certain marine animals, like sea cucumbers, the start codon GTG is frequently employed in the genes responsible for encoding NADH dehydrogenase subunits, including genes like *nad1*, *nad4l*, and *nad5* [45]. Stop codon usage of the mitogenomes was also found to be similar among species, with more than half of the PCGs (7/13) using TAA as a stop codon; three PCGs terminating with TAG, AGG, and AGA, respectively; and the other PCGs ending with an incomplete terminating codon, i.e., T. The usage of start and stop codons in this study is comparable to findings for other fish. Satoh et al. (2016) conducted a codon usage analysis on 250 fish and discovered the utilization of nine types of start codons and seven types of stop codons [14]. The most frequently used start codons were ATG and GTG, whereas TAA, TAG, AGA, and AGG were all used as complete stop codons. Additionally, three types of incomplete stop codons (TA-, T--, and AG-) were also employed.

The RSCU analysis revealed that the codons were more prone to using A and T than C and G. The codons encoding Arg, Leu, and Ser were highly abundant, whereas those encoding Trp appeared infrequently (Figure 4A). These results are similar to those on the codon usage in the Characidae family [46] but different from this usage for invertebrate species such as *Lysmata vittate* [47]. Moreover, GCC-Ala (RSCU = 1.75) is the most frequently used codon in the mitogenome of *M. kuntee*. Usually, the RSCU intuitively reflects the preference for codon usage [48]. The observed bias of A and T nucleotides in holocentrid species likely contributed to a corresponding bias in the usage of codons. Previous studies reported that a notable characteristic of the mitochondrial genome in other teleost species is the A and T bias, resulting in a consequential bias in the encoded amino acids [14,15].

The comprehensive phylogenetic tree generated using the mitogenomes of holocentrid fish indicates the evolutionary position of the two subfamilies. The subfamilies distinctly diverged, which was supported with high confidence, providing evidence explaining the differences in morphology and feeding habits between the two subfamilies. According to previous studies, fish from these two subfamilies share some similarities but also demonstrate unique characteristics in their habitats and lifestyles. During the larval stage, both subfamilies live in the upper pelagic ocean and feed on zooplankton [49]. As they transition into juvenile life, the majority of holocentrids migrate to shallow tropical coral reef habitats [10], where the subfamily Holocentrinae adopts a nocturnal lifestyle and feeds on benthic crustaceans, while the subfamily Myripristinae feeds on zooplankton in the water column [50]. Our previous study on parrotfish indicated that ecological differences in habitats affect the formation of morphology and feeding habits and might act as the primary driving force in species diversification [15]. And the visual system of the two subfamilies differed after settlement, with Myripristinae showing a more pronounced adaptation for scotopic vision than Holocentrinae [2]. Moreover, high-confidence clades support the notion of two genera in each clade. Within Holocentrinae, a previous analysis indicated that there was strong support for the genera *Neoniphon* and *Sargocentron* being paraphyletic [51]. In more specific terms, the genome characteristics of the mitochondrion indicate that the species *N. opercularis* and *N. samara* are phylogenetically closer to *S. punctatissimum* and *S. diadema* than to other species within the genus *Sargocentron*. This could suggest the existence of an evolutionary correlation between the two genera. However, the cited author proposed that *Neoniphon* and *Sargocentron* probably underwent a complex evolution but did not derive from a common ancestry as determined via Bayesian ancestral state reconstruction methods [51].

Positive selection is an evolutionary process in which advantageous genetic variations (mutations) increase in frequency within a population. However, under natural conditions, the primary form of selection is purifying selection, which continually removes harmful mutations that occur in each generation [52,53]. In this study, the purifying selection for PCGs was prominent, thus ensuring that deleterious mutations cannot take over the population of Holocentridae. Previous studies on other reef fish have also shown that PCGs of the mitochondrial genome undergo purification selection [15], indicating fewer amino acid variations during evolution. The *nad5* is a core subunit of NADH dehydrogenase, which is located on the mitochondrial membrane and is involved in the function of the respiratory chain, while *atp8* synthesizes ATP and serves as the primary energy source for mitochondrial oxidative phosphorylation. The properties of purification exhibited by *nad5* and *atp8* make them potential markers for identifying holocentrids.

5. Conclusions

In this study, a mitogenome study was employed to investigate the genetic diversity, population structure, and evolutionary relationships of the reef fish family Holocentridae. Eight holocentrid mitogenomes were sequenced, and a comparative mitogenome analysis with six published holocentrid fish was performed. Our characteristic analysis indicated that a typical holocentrid mitogenome is 16,239 bp in length and encodes 37 genes. For all the species, the mitogenome structures were relatively conserved. The whole genomes of the mitochondria exhibited positive AT-skews and negative GC-skews. Among the 13 PCGs, *nad6* was the most specific gene that exhibited negative AT-skews and positive GC-skews. GCC-Ala is the most frequently used codon in the mitogenome. Most of the genes, except *cox1*, begin with the standard codon ATG. Our phylogenetic analysis supports the notion that the genera *Sargocentron* and *Neoniphon* belong to the subfamily Holocentrinae, while the genera *Myripristis* and *Ostichthys* belong to the subfamily Myripristinae. The clades provide evidence to explain the differences in morphology and feeding habits between different subfamilies. The conducted positive selection analysis indicates that all the PCGs were under purifying selection. *nad5* and *atp8* are potential markers for identifying holocentrids. Our study contributes to an in-depth understanding of the biological characteristics and evolutionary relationships of holocentrid fish and enriches the mitochondrial genome resources of coral reef fish.

Supplementary Materials: The following supporting information can be downloaded at https://www.mdpi.com/article/10.3390/biology12101273/s1. Figure S1: Map of the mitochondrial genome of 13 holocentrid species (except *Myripristis kuntee*); Figure S2: The alignment of control regions of 14 holocentrid species. Table S1: Total length and GC content of the eight newly sequenced complete mitogenomes of holocentrid fish; Table S2: Gene summary of the eight mitochondrial genomes of holocentrid fish; Table S3: AT-skew and GC-skew of the complete genome, tRNA, rRNA, and PCGs; Table S4: The relative synonymous codon usage (RSCU) of the PCGs among 14 holocentrid species; Table S5: Estimation of selection pressure at each amino acid site via Fixed Effects Likelihood (FEL) analysis.

Author Contributions: Conceptualization, T.W. and Q.T.; methodology, Y.L., C.-H.L., J.-F.Z., Q.T. and T.W.; formal analysis, Q.T. and T.W.; resources, T.W.; data curation, Q.T. and T.W.; writing—original draft preparation, Q.T.; writing—review and editing, T.W.; supervision, T.W.; project administration, T.W.; funding acquisition, T.W. and Q.T. All authors have read and agreed to the published version of the manuscript.

Funding: This work was supported by the Hainan Provincial Natural Science Foundation (323MS124, 322MS153), the Fundamental Research Funds for the Central Universities in China (2662022SCQD002), the Fundamental and Applied Fundamental Research Major Program of Guangdong Province (2019B030302004-05), the Hainan Provincial Natural Science Foundation (322CXTD530), the Science and Technology Planning Project of Guangdong Province (2019B121201001), the Central Public-interest Scientific Institution Basal Research Fund, CAFS (2020TD16), the Financial Fund of the Ministry of Agriculture and Rural Affairs, P. R. of China (NFZX2021). The funders had no role in the study's design, data collection and analysis, the decision to publish, or the preparation of the manuscript.

Institutional Review Board Statement: Not applicable.

Informed Consent Statement: Not applicable.

Data Availability Statement: The newly generated mitogenome sequences from the current study were deposited in the NCBI nucleotide database under the accession numbers OR148894-OR148901.

Acknowledgments: The authors would like to thank the National Key Laboratory of Crop Genetic Improvement at Huazhong Agricultural University for providing of the computing platform. All bioinformatic analyses in this paper were run on the platform.

Conflicts of Interest: The authors declare that they have no competing interest. The funders had no role in the design of the study; in the collection, analyses, or interpretation of data; in the writing of the manuscript, or in the decision to publish the results.

References

1. Parmentier, E.; Vandewalle, P.; Brie, C.; Dinraths, L.; Lecchini, D. Comparative study on sound production in different Holocentridae species. *Front. Zool.* **2011**, *8*, 12. [CrossRef] [PubMed]
2. Fogg, L.G.; Cortesi, F.; Lecchini, D.; Gache, C.; Marshall, N.J.; de Busserolles, F. Development of dim-light vision in the nocturnal reef fish family Holocentridae. I: Retinal gene expression. *J. Exp. Biol.* **2022**, *225*, jeb244513. [CrossRef] [PubMed]
3. de Busserolles, F.; Cortesi, F.; Fogg, L.; Stieb, S.M.; Luehrmann, M.; Marshall, N.J. The visual ecology of Holocentridae, a nocturnal coral reef fish family with a deep-sea-like multibank retina. *J. Exp. Biol.* **2021**, *224*, jeb233098. [CrossRef]
4. Andrews, J.V.; Schein, J.P.; Friedman, M. An earliest Paleocene squirrelfish (Teleostei: Beryciformes: Holocentroidea) and its bearing on the timescale of holocentroid evolution. *J. Syst. Palaeontol.* **2023**, *21*, 2168571. [CrossRef]
5. Schmitz, L.; Wainwright, P.C. Nocturnality constrains morphological and functional diversity in the eyes of reef fishes. *BMC Evol. Biol.* **2011**, *11*, 338. [CrossRef] [PubMed]

6. Quimpo, T.J.R.; Cabaitan, P.C.; Olavides, R.D.D.; Dumalagan, E.E., Jr.; Munar, J.; Siringan, F.P. Spatial variability in reef-fish assemblages in shallow and upper mesophotic coral ecosystems in the Philippines. *J. Fish. Biol.* **2019**, *94*, 17–28. [CrossRef]
7. Randall, J.E. *Reef and Shore Fishes of the South Pacific*; University of Hawaii Press: Honolulu, HI, USA, 2003.
8. Page, H.M.; Brooks, A.J.; Kulbicki, M.; Galzin, R.; Miller, R.J.; Reed, D.C.; Schmitt, R.J.; Holbrook, S.J.; Koenigs, C. Stable isotopes reveal trophic relationships and diet of consumers in temperate kelp forest and coral reef ecosystems. *Oceanography* **2013**, *26*, 181–189. [CrossRef]
9. Paxton, J.R.; Eschmeyer, W.N. *Encyclopedia of Fishes*, 2nd ed.; (Natural World); Academic Press: San Diego, CA, USA, 1998.
10. Nelson, J.S.; Grande, T.C.; Wilson, M.V.H. *Fishes of the World*, 5th ed.; John Wiley & Sons: Hoboken, NJ, USA, 2016.
11. Embley, T.M.; Martin, W. Eukaryotic evolution, changes and challenges. *Nature* **2006**, *440*, 623–630. [CrossRef]
12. Martijn, J.; Vosseberg, J.; Guy, L.; Offre, P.; Ettema, T.J.G. Deep mitochondrial origin outside the sampled alphaproteobacteria. *Nature* **2018**, *557*, 101–105. [CrossRef]
13. Yang, M.; Gong, L.; Sui, J.; Li, X. The complete mitochondrial genome of *Calyptogena marissinica* (Heterodonta: Veneroida: Vesicomyidae): Insight into the deep-sea adaptive evolution of vesicomyids. *PLoS ONE* **2019**, *14*, e0217952. [CrossRef]
14. Satoh, T.P.; Miya, M.; Mabuchi, K.; Nishida, M. Structure and variation of the mitochondrial genome of fishes. *BMC Genom.* **2016**, *17*, 719. [CrossRef]
15. Gao, J.; Li, C.; Yu, D.; Wang, T.; Lin, L.; Xiao, Y.; Wu, P.; Liu, Y. Comparative mitogenome analyses uncover mitogenome features and phylogenetic implications of the Parrotfishes (Perciformes: Scaridae). *Biology* **2023**, *12*, 410. [CrossRef]
16. Ramos, N.I.; DeLeo, D.M.; Horowitz, J.; McFadden, C.S.; Quattrini, A.M. Selection in coral mitogenomes, with insights into adaptations in the deep sea. *Sci. Rep.* **2023**, *13*, 6016. [CrossRef]
17. Shang, Y.; Wang, X.; Liu, G.; Wu, X.; Wei, Q.; Sun, G.; Mei, X.; Dong, Y.; Sha, W.; Zhang, H. Adaptability and evolution of Gobiidae: A genetic exploration. *Animals* **2022**, *12*, 1741. [CrossRef] [PubMed]
18. Rothschild, L.J.; Mancinelli, R.L. Life in extreme environments. *Nature* **2001**, *409*, 1092–1101. [CrossRef] [PubMed]
19. Danovaro, R.; Snelgrove, P.V.; Tyler, P. Challenging the paradigms of deep-sea ecology. *Trends Ecol. Evol.* **2014**, *29*, 465–475. [CrossRef] [PubMed]
20. O'Brien, K.M.; Mueller, I.A. The unique mitochondrial form and function of Antarctic channichthyid icefishes. *Integr. Comp. Biol.* **2010**, *50*, 993–1008. [CrossRef]
21. Bolger, A.M.; Lohse, M.; Usadel, B. Trimmomatic: A flexible trimmer for Illumina sequence data. *Bioinformatics* **2014**, *30*, 2114–2120. [CrossRef]
22. Meng, G.; Li, Y.; Yang, C.; Liu, S. MitoZ: A toolkit for animal mitochondrial genome assembly, annotation and visualization. *Nucleic Acids Res.* **2019**, *47*, e63. [CrossRef]
23. Jin, J.J.; Yu, W.B.; Yang, J.B.; Song, Y.; dePamphilis, C.W.; Yi, T.S.; Li, D.Z. GetOrganelle: A fast and versatile toolkit for accurate de novo assembly of organelle genomes. *Genome Biol.* **2020**, *21*, 241. [CrossRef]
24. Bernt, M.; Donath, A.; Juhling, F.; Externbrink, F.; Florentz, C.; Fritzsch, G.; Putz, J.; Middendorf, M.; Stadler, P.F. MITOS: Improved de novo metazoan mitochondrial genome annotation. *Mol. Phylogenet Evol.* **2013**, *69*, 313–319. [CrossRef] [PubMed]
25. Iwasaki, W.; Fukunaga, T.; Isagozawa, R.; Yamada, K.; Maeda, Y.; Satoh, T.P.; Sado, T.; Mabuchi, K.; Takeshima, H.; Miya, M.; et al. MitoFish and MitoAnnotator: A mitochondrial genome database of fish with an accurate and automatic annotation pipeline. *Mol. Biol. Evol.* **2013**, *30*, 2531–2540. [CrossRef] [PubMed]
26. Chan, P.P.; Lin, B.Y.; Mak, A.J.; Lowe, T.M. tRNAscan-SE 2.0: Improved detection and functional classification of transfer RNA genes. *Nucleic Acids Res.* **2021**, *49*, 9077–9096. [CrossRef]
27. Lorenz, R.; Bernhart, S.H.; Honer Zu Siederdissen, C.; Tafer, H.; Flamm, C.; Stadler, P.F.; Hofacker, I.L. ViennaRNA Package 2.0. *Algorithms Mol. Biol.* **2011**, *6*, 26. [CrossRef] [PubMed]
28. Kumar, S.; Stecher, G.; Tamura, K. MEGA7: Molecular evolutionary genetics analysis Version 7.0 for bigger datasets. *Mol. Biol. Evol.* **2016**, *33*, 1870–1874. [CrossRef]
29. Rice, P.; Longden, I.; Bleasby, A. EMBOSS: The European molecular biology open software suite. *Trends Genet.* **2000**, *16*, 276–277. [CrossRef]
30. Grant, J.R.; Stothard, P. The CGView Server: A comparative genomics tool for circular genomes. *Nucleic Acids Res.* **2008**, *36*, W181–W184. [CrossRef]
31. Katoh, K.; Misawa, K.; Kuma, K.; Miyata, T. MAFFT: A novel method for rapid multiple sequence alignment based on fast Fourier transform. *Nucleic Acids Res.* **2002**, *30*, 3059–3066. [CrossRef]
32. Castresana, J. Selection of conserved blocks from multiple alignments for their use in phylogenetic analysis. *Mol. Biol. Evol.* **2000**, *17*, 540–552. [CrossRef]
33. Kuck, P.; Longo, G.C. FASconCAT-G: Extensive functions for multiple sequence alignment preparations concerning phylogenetic studies. *Front. Zool.* **2014**, *11*, 81. [CrossRef]
34. Kearse, M.; Moir, R.; Wilson, A.; Stones-Havas, S.; Cheung, M.; Sturrock, S.; Buxton, S.; Cooper, A.; Markowitz, S.; Duran, C.; et al. Geneious Basic: An integrated and extendable desktop software platform for the organization and analysis of sequence data. *Bioinformatics* **2012**, *28*, 1647–1649. [CrossRef] [PubMed]
35. Nguyen, L.T.; Schmidt, H.A.; von Haeseler, A.; Minh, B.Q. IQ-TREE: A fast and effective stochastic algorithm for estimating maximum-likelihood phylogenies. *Mol. Biol. Evol.* **2015**, *32*, 268–274. [CrossRef] [PubMed]

36. Kalyaanamoorthy, S.; Minh, B.Q.; Wong, T.K.F.; von Haeseler, A.; Jermiin, L.S. ModelFinder: Fast model selection for accurate phylogenetic estimates. *Nat. Methods* **2017**, *14*, 587–589. [CrossRef] [PubMed]
37. Ronquist, F.; Teslenko, M.; van der Mark, P.; Ayres, D.L.; Darling, A.; Hohna, S.; Larget, B.; Liu, L.; Suchard, M.A.; Huelsenbeck, J.P. MrBayes 3.2: Efficient Bayesian phylogenetic inference and model choice across a large model space. *Syst. Biol.* **2012**, *61*, 539–542. [CrossRef]
38. Posada, D. jModelTest: Phylogenetic model averaging. *Mol. Biol. Evol.* **2008**, *25*, 1253–1256. [CrossRef]
39. Suyama, M.; Torrents, D.; Bork, P. PAL2NAL: Robust conversion of protein sequence alignments into the corresponding codon alignments. *Nucleic Acids Res.* **2006**, *34*, W609–W612. [CrossRef]
40. Pond, S.L.K.; Poon, A.F.Y.; Velazquez, R.; Weaver, S.; Hepler, N.L.; Murrell, B.; Shank, S.D.; Magalis, B.R.; Bouvier, D.; Nekrutenko, A.; et al. HyPhy 2.5-A customizable platform for evolutionary hypothesis testing using phylogenies. *Mol. Biol. Evol.* **2020**, *37*, 295–299. [CrossRef]
41. Cheng, Y.; Wang, R.; Xu, T. The mitochondrial genome of the spinyhead croaker *Collichthys lucida*: Genome organization and phylogenetic consideration. *Mar. Genom.* **2011**, *4*, 17–23. [CrossRef]
42. Jin, X.; Wang, R.; Xu, T.; Shi, G. Complete mitochondrial genome of *Oxuderces dentatus* (Perciformes, Gobioidei). *Mitochondrial DNA* **2012**, *23*, 142–144. [CrossRef]
43. Chen, L.; Lin, Y.; Xiao, Q.; Lin, Y.; Du, Y.; Lin, C.; Ward-Fear, G.; Hu, C.; Qu, Y.; Li, H. Characterization of the complete mitochondrial genome of the many-lined sun skink (*Eutropis multifasciata*) and comparison with other Scincomorpha species. *Genomics* **2021**, *113*, 2526–2536. [CrossRef]
44. Liu, S.; Zhang, K.; Xiao, L. The complete mitochondrial genome of *Pimephales promelas* (Cypriniformes: Cyprinidae). *Mitochondrial DNA A DNA Mapp. Seq. Anal.* **2016**, *27*, 3711–3712. [CrossRef] [PubMed]
45. Li, Z.; Ma, B.; Li, X.; Lv, Y.; Jiang, X.; Ren, C.; Hu, C.; Luo, P. The complete mitochondrial genome of *Stichopus naso* (Aspidochirotida: Stichopodidae: *Stichopus*) and Its Phylogenetic Position. *Genes* **2022**, *13*, 825. [CrossRef]
46. Liu, H.; Sun, C.; Zhu, Y.; Li, Y.; Wei, Y.; Ruan, H. Mitochondrial genomes of four American characins and phylogenetic relationships within the family Characidae (Teleostei: Characiformes). *Gene* **2020**, *762*, 145041. [CrossRef] [PubMed]
47. Zhu, L.; Zhu, Z.; Zhu, L.; Wang, D.; Wang, J.; Lin, Q. The complete mitogenome of *Lysmata vittata* (Crustacea: Decapoda: Hippolytidae) with implication of phylogenomics and population genetics. *PLoS ONE* **2021**, *16*, e0255547. [CrossRef] [PubMed]
48. Gun, L.; Yumiao, R.; Haixian, P.; Liang, Z. Comprehensive analysis and comparison on the codon usage pattern of whole *Mycobacterium tuberculosis* coding genome from different area. *Biomed. Res. Int.* **2018**, *2018*, 3574976. [CrossRef]
49. Sampey, A.; McKinnon, A.D.; Meekan, M.G.; McCormick, M.I. Glimpse into guts: Overview of the feeding of larvae of tropical shorefishes. *Mar. Ecol. Prog. Ser.* **2007**, *339*, 243–257. [CrossRef]
50. Greenfield, D.W.; Randall, J.E.; Psomadakis, P.N. A review of the soldierfish genus *Ostichthys* (Beryciformes: Holocentridae), with descriptions of two new species from Myanmar. *J. Ocean. Sci. Found.* **2017**, *26*, 1–33.
51. Dornburg, A.; Moore, J.A.; Webster, R.; Warren, D.L.; Brandley, M.C.; Iglesias, T.L.; Wainwright, P.C.; Near, T.J. Molecular phylogenetics of squirrelfishes and soldierfishes (Teleostei: Beryciformes: Holocentridae): Reconciling more than 100 years of taxonomic confusion. *Mol. Phylogenet Evol.* **2012**, *65*, 727–738. [CrossRef]
52. Phung, T.N.; Huber, C.D.; Lohmueller, K.E. Determining the effect of natural selection on linked neutral divergence across species. *PLoS Genet.* **2016**, *12*, e1006199. [CrossRef]
53. Elyashiv, E.; Bullaughey, K.; Sattath, S.; Rinott, Y.; Przeworski, M.; Sella, G. Shifts in the intensity of purifying selection: An analysis of genome-wide polymorphism data from two closely related yeast species. *Genome Res.* **2010**, *20*, 1558–1573. [CrossRef]

Article

Insights into the Mitochondrial Genetic Makeup and Miocene Colonization of Primitive Flatfishes (Pleuronectiformes: Psettodidae) in the East Atlantic and Indo-West Pacific Ocean

Shantanu Kundu [1,2,†], Flandrianto Sih Palimirmo [2,3,†], Hye-Eun Kang [4], Ah Ran Kim [5], Soo Rin Lee [5], Fantong Zealous Gietbong [6], Se Hyun Song [7] and Hyun-Woo Kim [1,2,5,*]

[1] Institute of Fisheries Science, Pukyong National University, Busan 48513, Republic of Korea
[2] Department of Marine Biology, Pukyong National University, Busan 48513, Republic of Korea
[3] Research Center for Conservation of Marine and Inland Water Resources, National Research and Innovation Agency, Cibinong 16911, Indonesia
[4] Institute of Marine Life Science, Pukyong National University, Busan 48513, Republic of Korea
[5] Marine Integrated Biomedical Technology Center, National Key Research Institutes in Universities, Pukyong National University, Busan 48513, Republic of Korea
[6] The Ministry of Livestock, Fisheries and Animal Industries (MINEPIA), Yaounde 00237, Cameroon
[7] Fisheries Resources Management Division, National Institute of Fisheries Science, Busan 46083, Republic of Korea
[*] Correspondence: kimhw@pknu.ac.kr; Tel.: +82-51-629-5926; Fax: +82-51-629-5930
[†] These authors contributed equally to this work.

Simple Summary: The present research enriches our comprehension of the mitogenomic genetic features, genetic diversity, evolutionary past, and conservation prerequisites of *Psettodes* flatfishes on a global scale. This study focuses on the matrilineal evolutionary path of these primitive groups, with a specific emphasis on the complete mitogenome of the *Psettodes belcheri* and casting light on its genetic composition, structural traits, and evolutionary chronicle. Exploring genetic variations and phylogenetic relationships uncovers the intricate evolutionary links between *Psettodes* species and their broader context within the Pleuronectiformes species. The complex interplay of hydrographic conditions, ocean currents, and ecological factors emerges as pivotal in shaping the evolutionary landscape of these flatfishes. Given the potential consequences for conservation, this study highlights the necessity for a holistic comprehension of marine environments and the ramifications of climate change and human interventions on flatfish species.

Citation: Kundu, S.; Palimirmo, F.S.; Kang, H.-E.; Kim, A.R.; Lee, S.R.; Gietbong, F.Z.; Song, S.H.; Kim, H.-W. Insights into the Mitochondrial Genetic Makeup and Miocene Colonization of Primitive Flatfishes (Pleuronectiformes: Psettodidae) in the East Atlantic and Indo-West Pacific Ocean. *Biology* **2023**, *12*, 1317. https://doi.org/10.3390/biology12101317

Academic Editors: Cong Zeng and Deliang Li

Received: 21 August 2023
Revised: 25 September 2023
Accepted: 6 October 2023
Published: 9 October 2023

Abstract: The mitogenomic evolution of the *Psettodes* flatfishes is still poorly known from their range distribution in eastern Atlantic and Indo-West Pacific Oceans. The study delves into the matrilineal evolutionary pathway of these primitive flatfishes, with a specific focus on the complete mitogenome of the *Psettodes belcheri* species, as determined through next-generation sequencing. The mitogenome in question spans a length of 16,747 base pairs and comprises a total of 37 genes, including 13 protein-coding genes, 2 ribosomal RNA genes, 22 transfer RNA genes, and a control region. Notably, the mitogenome of *P. belcheri* exhibits a bias towards AT base pairs, with a composition of 54.15%, mirroring a similar bias observed in its close relative, *Psettodes erumei*, which showcases percentages of 53.07% and 53.61%. Most of the protein-coding genes commence with an ATG initiation codon, except for Cytochrome c oxidase I (COI), which initiates with a GTG codon. Additionally, four protein-coding genes commence with a TAA termination codon, while seven others exhibit incomplete termination codons. Furthermore, two protein-coding genes, namely NAD1 and NAD6, terminate with AGG and TAG stop codons, respectively. In the mitogenome of *P. belcheri*, the majority of transfer RNAs demonstrate the classical cloverleaf secondary structures, except for tRNA-serine, which lacks a DHU stem. Comparative analysis of conserved blocks within the control regions of two Psettodidae species unveiled that the CSB-II block extended to a length of 51 base pairs, surpassing the other blocks and encompassing highly variable sites. A comprehensive phylogenetic analysis using mitochondrial genomes (13 concatenated PCGs) categorized various Pleuronectiformes species, highlighting the basal position of the Psettodidae family and showed monophyletic clustering of

Psettodes species. The approximate divergence time (35−10 MYA) between *P. belcheri* and *P. erumei* was estimated, providing insights into their separation and colonization during the early Miocene. The TimeTree analysis also estimated the divergence of two suborders, Psettodoidei and Pleuronectoidei, during the late Paleocene to early Eocene (56.87 MYA). The distribution patterns of *Psettodes* flatfishes were influenced by ocean currents and environmental conditions, contributing to their ecological speciation. In the face of climate change and anthropogenic activities, the conservation implications of *Psettodes* flatfishes are emphasized, underscoring the need for regulated harvesting and adaptive management strategies to ensure their survival in changing marine ecosystems. Overall, this study contributes to understanding the evolutionary history, genetic diversity, and conservation needs of *Psettodes* flatfishes globally. However, the multifaceted exploration of mitogenome and larger-scale genomic data of *Psettodes* flatfish will provide invaluable insights into their genetic characterization, evolutionary history, environmental adaptation, and conservation in the eastern Atlantic and Indo-West Pacific Oceans.

Keywords: marine fish; ancient lineages; mitogenome; phylogeny; evolution; oceanography

1. Introduction

Flatfishes, also known as flounders, soles, halibuts, turbots, plaices, and tonguefishes, are classified as Pleuronectiformes. This order is split into two suborders: Psettodoidei, which includes the single-family Psettodidae, and Pleuronectoidei, which includes 13 families [1]. Flatfishes live mostly in tropical and subtropical marine habitats, preferring shallow parts of the continental shelf with soft sandy bottoms [2]. Their evolutionary journey, particularly marked by cranial asymmetry and unique scale structures, has facilitated their successful adaptation and dominance in benthic aquatic habitats [3–5]. The genus *Psettodes* is a "primitive" flatfish group within the monotypic Psettodidae family, with three recognized species: *Psettodes belcheri*, *Psettodes bennettii*, and *Psettodes erumei* [6]. While *P. erumei*, commonly referred to as the Indian halibut, ranges expansively across the Red Sea and the Indo-West Pacific Ocean, the other two species are confined to the eastern Atlantic. Intriguingly, the ranges of *P. belcheri* and *P. bennettii* exhibit partial overlap, spanning from the west Sahara to the Liberian coast.

The economic significance of flatfishes has led to their frequent capture through demersal trawling in marine ecosystems [7]. The Indian halibut (*P. erumei*) is heavily harvested within the tropical fishing zone as defined by the United Nations Food and Agriculture Organization (FAO) [8]. Despite this heavy exploitation, the IUCN Red List classifies information on the status of these three *Psettodes* species as 'Data Deficient.' [9]. Furthermore, the extraordinary transformation from bilateral pelagic larval symmetry to adult flattened symmetry limits flatfishes to demersal zones on the sea bottom, owing to their unusual temperature-mediated spawning behavior [10]. While most flatfish species have either a right-sided (dextral) or a left-sided (sinistral) mouth, several species, including *Psettodes*, display varying degrees of dextral to sinistral polymorphism [11]. This sexual dimorphism in external morphology often poses challenges in species identification. Moreover, this characteristic complexity in external appearance frequently hinders precise species differentiation among flatfishes [12].

The integration of molecular data has become integral in the exploration of global flatfish species, encompassing diverse aspects such as species identification [13], systematic classification [14], phylogenetic relationships [15], and population structure estimation [16]. Genetic information has also found utility in identifying valuable commercial species and managing aquaculture stocks [17,18]. The acquisition of comprehensive genomic resources is pivotal for advancing flatfish research and its manifold applications [19]. While the origins of flatfishes have long been debated, recent molecular insights lend support to the concept of a 'lower-percoid' origin, a perspective gaining prominence in the field of teleost ichthyology [20–22]. The taxonomic placement of *Psettodes* has stirred argument, especially in the context of analyses involving nuclear and mitogenomic data. This genus consistently diverges from other flatfishes (Pleuronectiformes), instead aligning with other carangimorphs [23–26]. A comprehensive phylogenetic analysis utilizing 1000 ultraconserved DNA element loci across 45 carangimorphs subsequently solidified the monophyly of flatfishes [27]. Later studies, involving complete mitogenome-based phylogeny and molecular clock analyses, offered insights into the enigmatic placement and evolutionary divergence of the *Psettodes* genus within the context of other carangimorphs and pleuronectoids [28,29]. However, these investigations were limited to a single *Psettodes* taxon (*P. erumei*) from the Indo-Pacific region, highlighting the need for expanded inquiries involving other congeners from the Atlantic Ocean.

An integrated strategy that looks beyond superficial investigation is required for an accurate appraisal of speciation within maritime ecosystems. Advances in technology, spanning maritime engineering to genomics, are facilitating the investigation of marine speciation [30,31]. To enhance our grasp of *Psettodes'* maternal evolutionary path, this study seeks to construct the complete mitogenome of *P. belcheri*, elucidate its genomic attributes, and establish its phylogenetic relationships. This endeavor contributes to understanding the maternal evolution of these ancient lineages in marine environments, strengthening the mitogenomic repository of flatfishes, and expanding our scientific insight into the Psettodidae family on a global scale. While marine habitats harbor a diverse array of life forms, our understanding of speciation in marine ecosystems is comparatively limited compared with freshwater systems, necessitating urgent exploration of the mechanisms driving speciation and adaptation [32]. Given the unique distribution of Psettodidae flatfishes across the eastern Atlantic, Red Sea, and Indo-West Pacific Ocean, this study also aims to estimate the divergence time between two Psettodidae species (*P. belcheri* and *P. erumei*) and investigate potential evolutionary scenarios within the marine environment.

2. Materials and Methods

2.1. Sampling and Species Identification

A solitary specimen of the spottail spiny turbot, scientifically known as *P. belcheri*, was acquired from the estuaries of the Kineke River (latitude 2.938611° N, longitude 9.911667° E) located in Kribi, Cameroon (Figure 1). The identification process was meticulously carried out in accordance with available taxonomic keys [33]. Upon euthanizing the specimen with MS-222 (200 mg/L), muscle tissue was aseptically collected from the ventral thoracic region. The voucher specimen was duly archived in 10% formaldehyde at the Fisheries and Animal Industries department (MINEPIA) in Yaoundé, Cameroon. Simultaneously, a tissue sample was preserved at the Department of Marine Biology, Pukyong National University in Busan, South Korea. The Institutional Animal Care and Use Committee of the host institute granted approval (Approval Code: PKNUIACUC-2022-72, dated 16 December 2022) for the utilization of deceased fish muscle tissue in molecular investigations. To enhance our comprehension of their geographical distribution, global maps of Pleuronectiformes, inclusive of the range distributions of *P. belcheri* and its congeners (*P. erumei* and *P. bennettii*), were obtained in .shp file format from the IUCN database (https://www.iucnredlist.org/, accessed on 15 August 2023) (Figure 1).

2.2. DNA Extraction, Mitogenome Sequencing, and Assembly

The AccuPrep® Genomic DNA extraction kit, manufactured by Bioneer in Daejeon, Republic of Korea, was utilized to extract genomic DNA, following the established standard protocol. The quality and quantity of the resulting genomic DNA were meticulously evaluated employing a NanoDrop spectrophotometer (Thermo Fisher Scientific D1000, WA, USA). To obtain the comprehensive mitogenome of *P. belcheri*, sequencing procedures were executed using the NovaSeq platform, provided by Illumina and accessible at Macrogen (https://dna.macrogen.com/, accessed on 15 August 2023) in Daejeon, Republic of Korea. The sequencing libraries were prepared following the manufacturer's instructions for the TruSeq Nano DNA High-Throughput Library Prep Kit (Illumina, Inc., San Diego, CA, USA). In short, 100 ng of genomic DNA underwent fragmentation using adaptive focused acoustic technology (Covaris, Woburn, MA, USA), resulting in blunt-ended dsDNA molecules with 5′-phosphorylation. After the end-repair step, DNA fragments were size selected using a bead-based method. These fragments were then modified with the addition of a single 'A' base and ligated with TruSeq DNA UD Indexing adapters. Subsequently, the products were purified and enriched through PCR to create the final DNA library. Library quantification was performed using qPCR, following the qPCR Quantification Protocol Guide (KAPA Library Quantification kits for Illumina Sequencing platforms), and quality assessment was carried out using Agilent Technologies 4200 TapeStation D1000 screentape (Agilent Technologies, Santa Clara, CA, USA). Finally, paired-end (2 × 150 bp) sequencing was performed by Macrogen using the NovaSeq platform (Illumina, Inc., San Diego, CA, USA). Over 20 million raw reads underwent processing using the Cutadapt tool (http://code.google.com/p/cutadapt/, accessed on 15 August 2023) to trim adapters and remove low-quality bases, with a Phred quality score (Q score) cutoff of 20. The Geneious Prime version 2023.0.1 was used to assemble the targeted genome from the high-quality paired-end NGS reads. This assembly was accomplished by employing reference mapping with the mitogenome of a closely related species as a reference, and we employed default mapping algorithms. Mitogenome assembly was accomplished by scrutinizing the alignment of overlapping regions via MEGA X [34]. The boundaries and orientations of individual genes were validated using the MITOS v806 (http://mitos.bioinf.uni-leipzig.de, accessed on 15 August 2023) and MitoAnnotator (http://mitofish.aori.u-tokyo.ac.jp/annotation/input/, accessed on 15 August 2023) web servers [35,36]. To further corroborate protein-coding genes (PCGs), the translated putative amino acid sequences were scrutinized using the Open Reading Frame Finder web tool (https://www.ncbi.nlm.nih.gov/orffinder/, accessed on 15 August 2023), based on the vertebrate mitochondrial genetic code. The resultant mitogenome of *P. belcheri* was duly submitted to the global GenBank database.

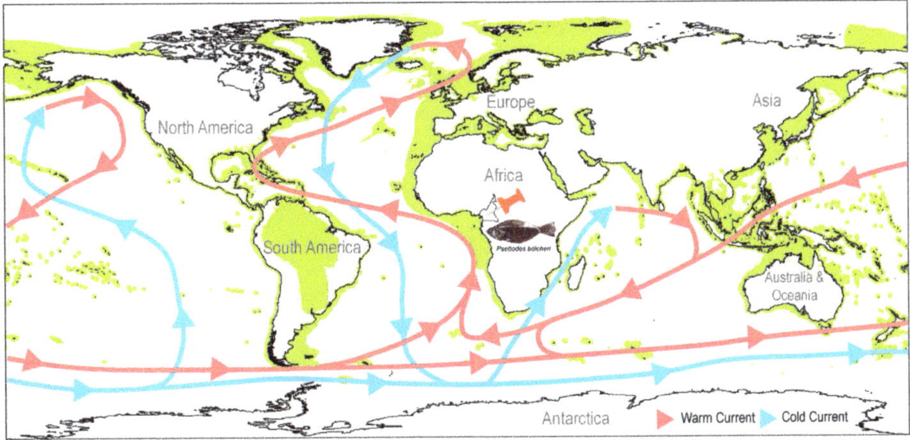

Figure 1. Global distribution of Pleuronectiformes along with the ocean surface currents (warm and cold currents are marked by the red and blue arrow, respectively). Collection locality of *Psettodes belcheri* is marked by a red pin from Cameroon, Africa. The .shp files were acquired from the IUCN database (accessed on 15 August 2023). The species photographs were taken by the sixth author (F.Z.G) and edited manually in Adobe Photoshop CS 8.0.

2.3. Mitogenomic Characterization

A spherical representation of the generated mitogenome was crafted using MitoAnnotator (http://mitofish.aori.u-tokyo.ac.jp/annotation/input/, accessed on 15 August 2023). A comprehensive comparative analysis was carried out to assess the mitogenomic architecture and variations within our generated sequence in relation to two pre-existing mitogenomes from a single congener, *P. erumei* (FJ606835, sourced from China, and AP006835, sourced from Japan). The calculation of intergenic spacers, which separate adjacent genes, and overlapping regions was performed manually. To determine the nucleotide compositions of protein-coding genes (PCGs), ribosomal RNA (rRNA), transfer RNA (tRNA), and the control region (CR), we employed MEGA X [34]. Similarly, base composition skews, as previously detailed, were computed using the following formulas: AT-skew = [A − T]/[A + T], GC-skew = [G − C]/[G + C] [37]. The verification of initiation and termination codons for each PCG, as well as adherence to the vertebrate mitochondrial genetic code, was carried out using MEGA X. Additionally, the boundaries of rRNA and tRNA genes were confirmed through the use of the tRNAscan-SE Search Server 2.0 in conjunction with ARWEN 1.2 [38,39]. Structural domains within the control region were delineated through CLUSTAL X alignments [40], and tandem repeats were explored utilizing the online Tandem Repeats Finder web tool (https://tandem.bu.edu/trf/trf.html, accessed on 15 August 2023) [41].

2.4. Genetic Distance, Phylogenetic Analyses, and TimeTree Estimation

Genetic distances were computed using the Kimura 2-parameter (K2P) method within MEGA X. Due to the unavailability of the complete mitogenome of *P. bennettii*, intra-species and inter-species distances were determined using the widely employed mitochondrial COI gene. To elucidate the matrilineal phylogenetic connections, 14 mitogenomes from 13 Pleuronectiformes species were sourced from the GenBank database (accessed on 15 August 2023) [28,42–49] (Table S1). Dataset preparation adhered to methodologies outlined in two recent Pleuronectiformes studies [26,50]. The spotfin flounder, *Cyclopsetta fimbriata*, was categorized within the recently established family Cyclopsettidae (=Paralichthyidae II). Additionally, two Cynoglossidae species were included: *Cynoglossus gracilis* (=Cynoglossidae I) and *Symphurus orientalis* (=Cynoglossidae II). The mitogenome of *Lates calcarifer* (DQ010541), from the Centropomidae family, was incorporated as an outgroup. Concatenation of all 13 PCGs was executed using the iTaxoTools 0.1 tool to construct the dataset for phylogenetic analysis [51]. In order to prevent inadvertent gaps within the dataset alignment, we consciously excluded the non-coding rRNA genes and control regions from the current phylogenetic analysis. Model selection yielded the 'GTR + G + I' model as the most suitable, determined by the lowest Bayesian Information Criterion (BIC) score using PartitionFinder 2 through CIPRES Science Gateway v3.3 and JModelTest v2 [52–54]. Employing Mr. Bayes 3.1.2, a Bayesian (BA) tree was constructed, employing nst = 6, along with one cold and three hot Metropolis-coupled Markov Chain Monte Carlo (MCMC) chains. The analysis spanned 10,000,000 generations with tree sampling at every 100th generation and 25% of samples discarded as burn in [55]. Visualization of the BA tree was accomplished using the iTOL v4 web server (https://itol.embl.de/login.cgi, accessed on 15 August 2023) [56]. Additionally, the divergence time estimation was performed using the RelTime method following standard protocol as implemented in MEGA X [57]. This approach was intended to reduce the large computational time involved in Bayesian methods [58,59]. After loading the sequences data, the constructed maximum-likelihood topology (.nwk format) was used as a baseline tree. After specifying the outgroup taxa, the TimeTree computation incorporated two calibration constraints through the calibration editor: the divergence from the sister lineage of Citharidae (55.54 MYA) and Achiridae (49.73 MYA), as established in a previous study [28].

3. Results and Discussion

3.1. Mitogenomic Structure and Organization

In this study, the mitogenome of *P. belcheri* (16,739 bp) was characterized and assigned the GenBank accession number OR231239. The circular mitogenome of *P. belcheri* comprised 13 protein-coding genes (PCGs), 22 transfer RNA genes (tRNAs), 2 ribosomal RNA genes (rRNAs), and a non-coding AT-rich control region (CR). Among these, 28 genes (12 PCGs, 2 rRNAs, and 14 tRNAs) were positioned on the heavy strand, while NAD6 and 8 tRNAs (trnQ, trnA, trnN, trnC, trnY, trnS2, trnE, and trnP) were situated on the light strand (Table 1, Figure 2).

Table 1. List of annotated mitochondrial genes of *Psettodes belcheri*.

Genes	Start	End	Strand	Size (bp)	Intergenic Nucleotide	Anti-Codon	Start Codon	Stop Codon
tRNA-Phe	1	69	H	69	0	TTC	.	.
12S rRNA	70	1030	H	961	0	.	.	.
tRNA-Val	1031	1102	H	72	26	GTA	.	.
16S rRNA	1129	2830	H	1702	0	.	.	.
tRNA-Leu	2831	2903	H	73	0	TTA	.	.
ND1	2904	3878	H	975	4	.	ATG	AGG
tRNA-Ile	3883	3952	H	70	1	ATC	.	.
tRNA-Gln	3954	4024	L	71	−1	CAA	.	.
tRNA-Met	4024	4093	H	70	0	ATG	.	.
ND2	4094	5138	H	1045	0	.	ATG	T--
tRNA-Trp	5139	5211	H	73	2	TGA	.	.
tRNA-Ala	5214	5282	L	69	1	GCA	.	.
tRNA-Asn	5284	5356	L	73	38	AAC	.	.
tRNA-Cys	5395	5460	L	66	0	TGC	.	.
tRNA-Tyr	5461	5530	L	70	1	TAC	.	.
COI	5532	7082	H	1551	0	.	GTG	TAA
tRNA-Ser	7083	7153	L	71	8	TCA	.	.
tRNA-Asp	7162	7230	H	69	8	GAC	.	.
COII	7239	7929	H	691	0	.	ATG	T--
tRNA-Lys	7930	8004	H	75	1	AAA	.	.
ATP8	8006	8170	H	165	−7	.	ATG	TAA
ATP6	8164	8844	H	681	2	.	ATG	TA-
COIII	8847	9629	H	783	2	.	ATG	TA-
tRNA-Gly	9632	9702	H	71	0	GGA	.	.
ND3	9703	10,050	H	348	1	.	ATG	T--
tRNA-Arg	10,052	10,120	H	69	0	CGA	.	.
ND4L	10,121	10,414	H	294	−4	.	ATG	TAA
ND4	10,411	11,791	H	1381	0	.	ATG	T--
tRNA-His	11,792	11,859	H	68	0	CAC	.	.
tRNA-Ser	11,860	11,927	H	68	6	AGC	.	.
tRNA-Leu	11,934	12,006	H	73	0	CTA	.	.
ND5	12,007	13,845	H	1839	−1	.	ATG	TAA
ND6	13,845	14,363	L	519	0	.	ATG	TAG
tRNA-Glu	14,364	14,432	L	69	5	GAA	.	.
Cytb	14,438	15,578	H	1141	0	.	ATG	T--
tRNA-Thr	15,579	15,652	H	74	−1	ACA	.	.
tRNA-Pro	15,652	15,724	L	73	0	CCA	.	.
Control region	15,725	16,747	H	1023	.	.	.	.

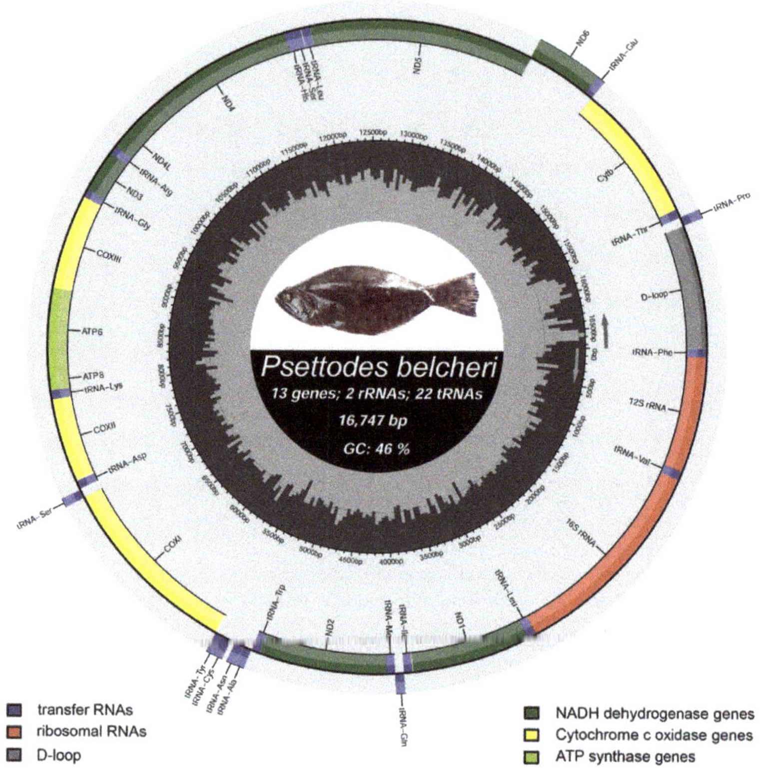

Figure 2. The mitochondrial genome of *Psettodes belcheri* drawn by the MitoAnnotator online server. PCGs, rRNAs, tRNAs, and CR are indicated by different color arcs. The species photographs were taken by the sixth author (F.Z.G) and edited manually in Adobe Photoshop CS 8.0.

A similar distribution of genes on heavy and light strands was evident in other Pleuronectiformes species, with total lengths ranging from 16,506 bp (*C. fimbriata*, family Cyclopsettidae, AP014590) to 18,706 bp (*Samariscus latus*, family Samaridae, KF494223). Variability in mitogenomic lengths was largely attributed to duplications in the control regions (CRs) [44]. The mitogenome of *P. belcheri* showcased five overlapping regions, spanning a total of 14 bp. The longest overlap (7 bp) was observed between ATP synthase 8 (atp8) and ATP synthase 6 (atp6). Similarly, both *P. erumei* mitogenomes (FJ606835 and AP006835) displayed five overlapping regions totaling 23 bp, with the longest overlap (10 bp) between atp8 and atp6, mirroring *P. belcheri*. Moreover, *P. belcheri* exhibited 15 intergenic spacer regions spanning a total of 106 bp, with the longest (38 bp) situated between trnN and trnC. Conversely, *P. erumei* mitogenomes contained 11 intergenic spacer regions spanning 77–78 bp, with the longest (37–38 bp) found between trnN and trnC (Table S2). Nucleotide composition analysis revealed the AT-biased nature of *P. belcheri* mitogenome (54.15%), encompassing 28.09% A, 16.24% G, 29.56% C, and 26.06% T. Similar AT richness was observed in the nucleotide composition of the other two *P. erumei* mitogenomes, ranging from 53.07% to 53.61% (Table 2). The AT skew and GC skew of *P. belcheri* mitogenome were recorded as 0.037 and −0.291, respectively. Comparative analysis of *P. erumei* mitogenomes demonstrated an AT skew range of 0.076 to 0.077, and a GC skew range of −0.323 to −0.328 (Table 2). This consistent nucleotide composition pattern and AT bias were also observed in other previously described fish mitogenomes [60,61]. The genetic variations identified within the *Psettodes* mitogenome may be tied to their evolutionary progression and energy metabolism, echoing findings in other fish species [62]. This research offers valuable insights into the structural attributes of *Psettodes* mitogenomes, which hold significance in deciphering the functionalities of these mitogenomes and their encoded genes.

Table 2. Nucleotide composition of the mitochondrial genomes of two *Psettodes* species.

Species Name	Size (bp)	A%	T%	G%	C%	A + T%	AT-Skew	GC-Skew
Complete mitogenome								
P. belcheri (OR231239)	16,747	28.09	26.06	16.24	29.56	54.15	0.037	−0.291
P. erumei (FJ606835)	17,315	28.83	24.78	15.71	30.68	53.61	0.076	−0.323
P. erumei (AP006835)	16,683	28.57	24.50	15.78	31.15	53.07	0.077	−0.328
PCGs								
P. belcheri (OR231239)	11,427	25.00	27.73	16.00	31.27	52.74	−0.052	−0.323
P. erumei (FJ606835)	11,427	25.37	25.75	15.62	33.25	51.13	−0.008	−0.361
P. erumei (AP006835)	11,426	25.33	25.62	15.62	33.43	50.95	−0.006	−0.363
rRNAs								
P. belcheri (OR231239)	2689	31.42	21.38	21.87	25.33	52.81	0.190	−0.073
P. erumei (FJ606835)	2680	32.43	20.56	20.86	26.16	52.99	0.224	−0.113
P. erumei (AP006835)	2680	32.43	20.45	20.90	26.23	52.87	0.227	−0.113
tRNAs								
P. belcheri (OR231239)	1556	27.83	27.31	23.14	21.72	55.14	0.009	0.032
P. erumei (FJ606835)	1553	27.17	26.85	23.82	22.15	54.02	0.006	0.036
P. erumei (AP006835)	1554	27.28	26.83	23.68	22.20	54.12	0.008	0.032
CRs								
P. belcheri (OR231239)	1015	30.25	33.40	12.51	23.84	63.65	−0.050	−0.312
P. erumei (FJ606835)	1601	38.91	33.92	10.31	16.86	72.83	0.069	−0.241
P. erumei (AP006835)	968	42.15	36.36	7.13	14.36	78.51	0.074	−0.337

3.2. Protein-Coding Genes

A cumulative length of 11,427 bp in *P. belcheri* mitogenome was occupied by a total of 13 protein-coding genes (PCGs), accounting for 68.27% of the whole sequence. Among these, the shortest PCG was ATP8, spanning 165 bp, while NAD5 represented the longest PCG with a length of 1839 bp. In both *P. erumei* mitogenomes, the total length of PCGs ranged from 11,426 bp (68.49%) to 11,427 bp (65.99%). The PCGs of *P. belcheri* were characterized by an AT bias of 52.73%, accompanied by AT skew and GC skew values of −0.052 and −0.323, respectively (Table 2). Similarly, the mitogenomes of *P. erumei* displayed an AT bias ranging from 50.95% to 51.12%, coupled with AT skew values of −0.008 to −0.006 and GC skew values of −0.363 to −0.361. Most of the PCGs commenced with an ATG (Methionine) initiation codon, except for COI, which began with a GTG (Valine) codon. A parallel pattern of initiation codons was apparent in all PCGs of the other two *P. erumei* mitogenomes. Among the PCGs, the conventional TAA termination codon was observed in four instances (COI, ATP8, NAD4L, and NAD5), while seven PCGs featured incomplete stop codons (T--/TA-). Additionally, two PCGs, NAD1 and NAD6, terminated with AGG and TAG stop codons, respectively. A corresponding distribution of stop codons was also noted in *P. erumei* mitogenomes (Tables 2 and S3). These incomplete stop codons could potentially be completed with TAA during RNA processing, as previously suggested [63]. As observed in other fish species, the identified genetic disparities might lead to the independent selection of PCGs [64]. PCGs play pivotal roles in oxidative phosphorylation, ATP synthesis, and the encoding of proteins within the electron transport pathways. Consequently, the inclusion of mitogenomes from various *Psettodes* species could facilitate the exploration of variations in gene expression and energy utilization.

3.3. Ribosomal RNA and Transfer RNA Genes

Within the *P. belcheri* mitogenome, the ribosomal RNA genes collectively spanned 2689 bp, equivalent to 16.06% of the complete mitogenome. This encompassed a small ribosomal RNA (12S rRNA) measuring 961 bp and a large ribosomal RNA (16S rRNA) with a length of 1702 bp. Comparatively, the total length of *P. erumei*'s rRNAs (2680 bp) was shorter than that of *P. belcheri* (Table 2). The ribosomal RNAs AT composition ranged from 52.81% (*P. belcheri*) to 52.99% (*P. erumei*). Comparative analysis indicated AT skew values ranging from 0.190 (*P. belcheri*) to 0.227 (*P. erumei*), while GC skew values ranged from −0.113 (*P. erumei*) to −0.073 (*P. belcheri*) in the ribosomal RNA (Table 2). These rRNA genes' structural arrangement, notably the conserved loops, offer critical insights into the catalytic chemical processes underlying protein synthesis [65]. Additionally, the *P. belcheri* mitogenome featured 22 tRNA genes, exhibiting varying lengths from 66 bp (trnC) to 75 bp (trnK), collectively constituting 1556 bp or 9.29% of the entire mitogenome. The combined tRNA length in both *P. erumei* mitogenomes was shorter (1553 bp and 1554 bp) than that of *P. belcheri* (Table 2). The tRNAs displayed an AT bias in both *P. belcheri* (55.14%) and *P. erumei* (54.02% to 54.12%). AT skew values ranged from 0.006 (*P. erumei*) to 0.009 (*P. belcheri*), while GC skew values ranged from 0.032 (*P. belcheri* and *P. erumei*, AP006835) to 0.036 (*P. erumei*, FJ606835) (Table 2). Most tRNAs were predicted to adopt the typical cloverleaf secondary structure, except for trnS1, which lacked the DHU stem, consistent with findings in other Pleuronectiformes [66,67]. These genetic attributes are pivotal for shaping transfer RNAs secondary structures and their functional roles within diverse biological systems [68]. In terms of comparative structural features, 15 tRNA genes (trnA, trnF, trnQ, trnM, trnW, trnN, trnC, trnY, trnS2, trnD, trnG, trnR, trnH, trnE, trnP) were constructed through both conventional Watson–Crick base (A=T and G≡C) pairing and wobble base pairing (G-T), while the remaining seven tRNA genes were exclusively built with Watson–Crick base pairs (Figure 3).

3.4. Features of Control Region and Gene Arrangements

The comprehensive length of *P. belcheri* control region (CR) reached 1015 bp, comprising 63.65% AT content. Conversely, the complete lengths of *P. erumei*'s CRs varied, ranging from 968 bp to 1601 bp. The AT skew fluctuated between −0.050 (*P. belcheri*) and 0.074 (*P. erumei*), while the GC skew ranged from −0.337 (*P. erumei*, AP006835) to −0.241 (*P. erumei*, FJ606835) (Table 2). *P. belcheri* mitogenome recorded more than two copies of 72 bp tandem repeats, whereas the mitogenomes of *P. erumei* hosted over eight copies (FJ606835) and more than twelve copies (AP006835) of 56 bp repeats. Four conserved blocks (CSB-D, CSB-I, CSB-II, and CSB-III) were detected in both *P. belcheri* and *P. erumei* mitogenomes, consistent with their presence in other teleost fishes [66,69]. Of these blocks, CSB-II was the longest at 51 bp, compared with CSB-D (27 bp), CSB-I (36 bp), and CSB-III (35 bp) (Figure 4A). Comparative analyses unveiled substantial nucleotide variability and parsimony informative nucleotides within CSB-II relative to the other three conserved domains. This AT-rich regulatory region holds the potential for assessing population structures and identifying inter- and intra-specific differences among *Psettodes* species via these variable nucleotides. As demonstrated in other species, such conserved domains are integral for mitochondrial genome replication and transcription [69,70]. Intriguingly, these primitive Pleuronectiformes fishes maintain a gene order within their mitochondrial genomes that aligns with that of ancestral teleosts [71] (Figure 4B). However, Pleuronectiformes mitogenomes exhibit repeated instances of control region duplications and gene rearrangements [72–76]. These mechanisms involving genomic rearrangement through double replications, random loss, dimer-mitogenomes, and non-random loss contribute to understanding the structural diversity of mitogenomes and the intricacies of mitochondrial genome evolution.

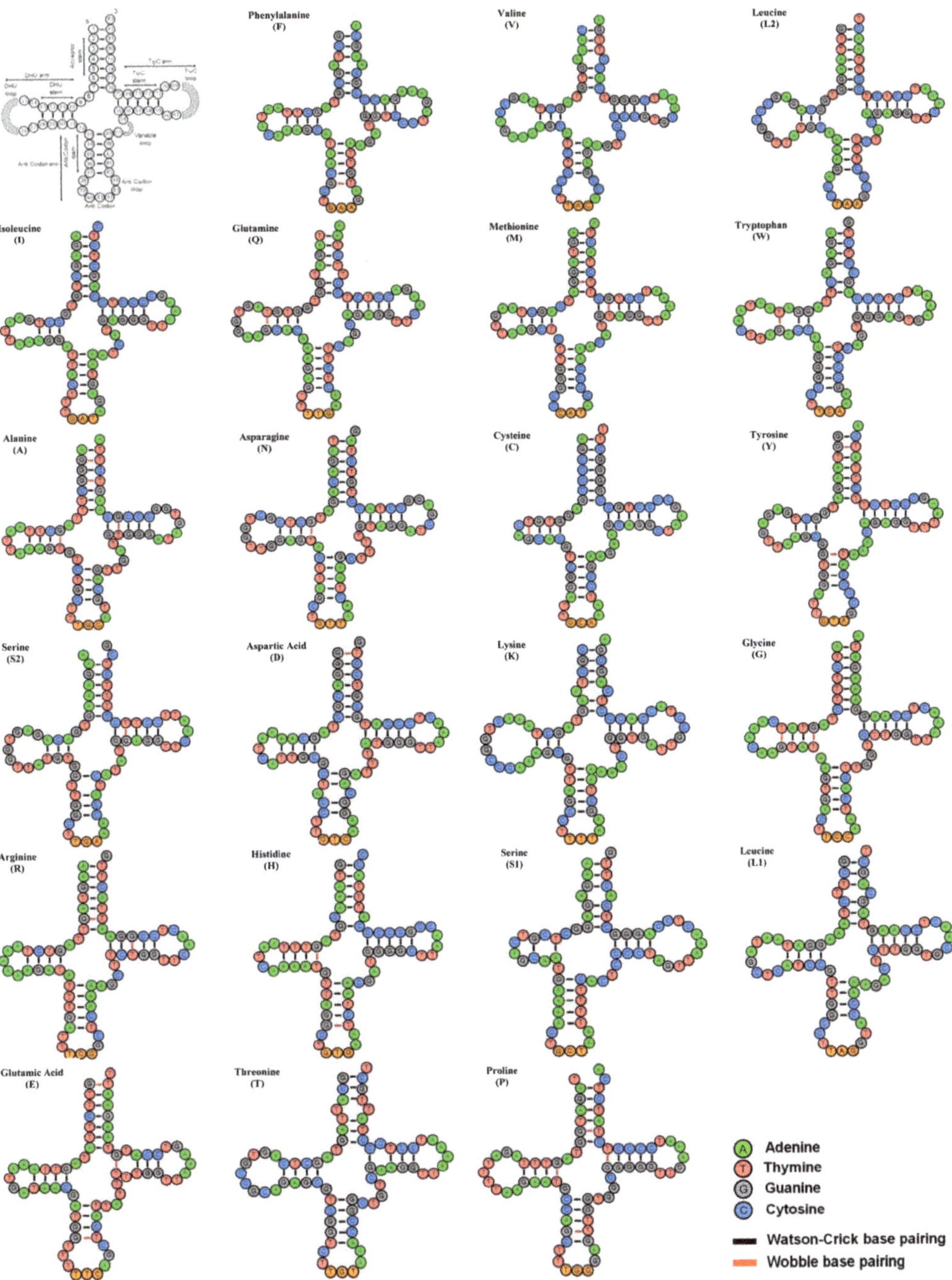

Figure 3. Secondary structures of 22 transfer RNAs (tRNAs) of *P. belcheri* display the structural variation. The tRNAs are denoted by full names and IUPAC-IUB single-letter amino acid codes. The first structure shows the nucleotide positions and details of the stem–loop of tRNAs. Watson–Crick and wobble base pairing are marked by black and red color bars, respectively.

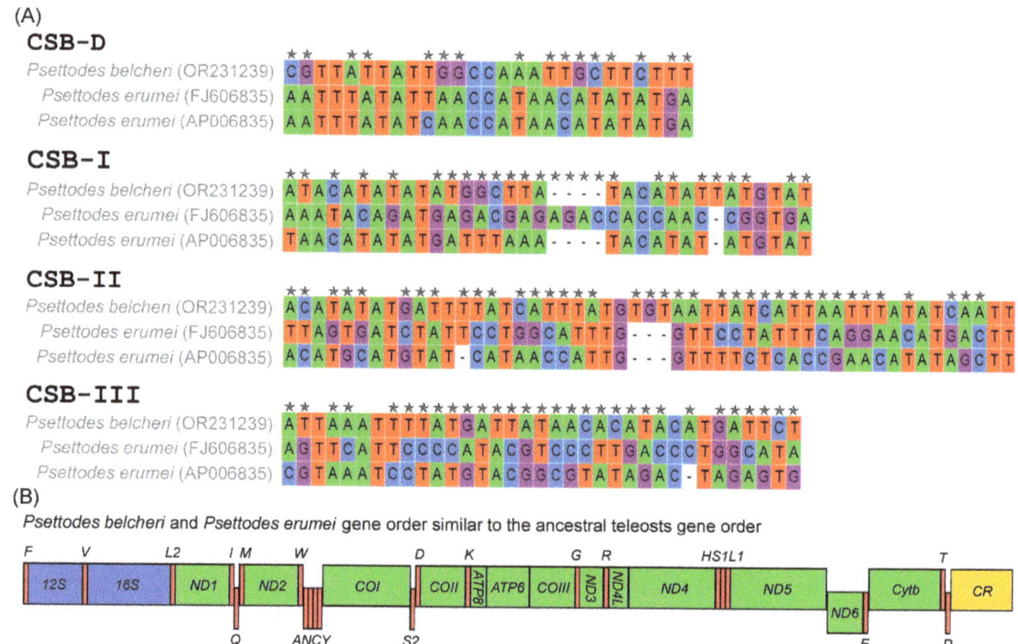

Figure 4. (**A**) Comparison of length and nucleotide composition of four conserved domains of two *Psettodes* species control regions. The variable nucleotides are marked in stars; (**B**) The gene arrangements of two *Psettodes* species mitogenomes.

3.5. Genetic Distances and Mitogenomic Phylogeny

The species under investigation, *P. belcheri*, displayed substantial inter-species genetic distances of 14.00% and 17.30% when compared with its congeners *P. bennettii* and *P. erumei*, respectively, as confirmed through the analysis of the COI gene (Figure S1). A particularly noteworthy finding is the pronounced genetic divergence of 14% observed between *P. belcheri* and *P. bennettii*, despite their shared presence in the western Atlantic. This genetic disparity, noteworthy in the context of their parapatric distribution and significant overlap, points to a significant level of reproductive isolation between these species. The examination of population genetic structures within *P. belcheri* and *P. bennettii* will provide valuable insights into their migratory patterns within the mutual habitat of the eastern Atlantic Ocean. Employing a phylogenetic analysis grounded in mitogenomes, the study effectively categorized all studied Pleuronectiformes flatfishes using a concatenation of 13 PCGs, supported by robust posterior probability (Figure 5). The resultant phylogeny derived from mitogenomes aligns harmoniously with previous evolutionary hypotheses concerning Pleuronectiformes species [26,28,29]. Importantly, representative species from the suborders Psettodoidei and Pleuronectoidei formed coherent monophyletic clusters within the present topology. The spottail spiny turbot, *P. belcheri*, clustered consistently with its congeners, most notably *P. erumei*. The family Psettodidae emerged as the basal node of Pleuronectiformes, occupying a unique position as an ancestral group among other flatfish families. Furthermore, the conducted cladistic analysis showcased a sister relationship between *C. gracilis* (=Cynoglossidae I) and *S. orientalis* (=Cynoglossidae II) (Figure 5). The application of mitochondrial genome-based and phylogenomic assessments has proven successful in elucidating higher teleostean phylogenies, encompassing flatfishes [77–79]. To establish the precise matrilineal evolution of *Psettodes* flatfishes within the monotypic family Psettodidae, the generation of the *P. bennettii* mitogenome emerges as a pivotal task. Notably, a wealth of genetic data, spanning multi-locus exon-capture data and whole genome sequencing, has recently provided fresh insights into the phylogeny and genetic evolution of flatfishes [80–85]. The integration of such extensive genetic information depicts the potential to illuminate the evolutionary landscape of primitive *Psettodes* flatfishes in the near future.

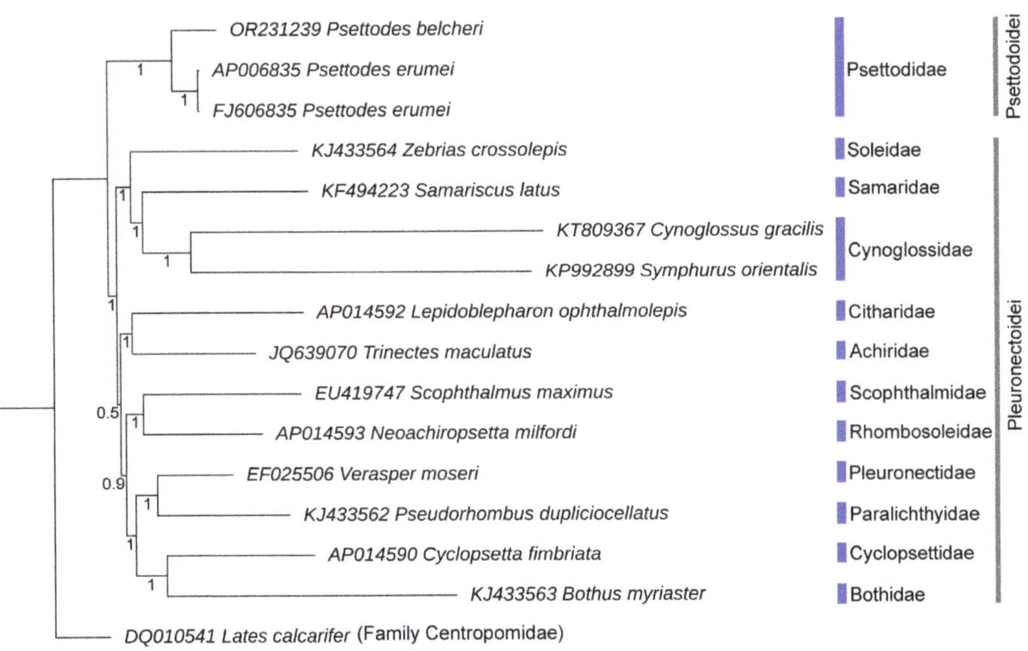

Figure 5. The Bayesian matrilineal phylogeny based on the concatenated sequences of 13 PCGs exhibits the evolutionary relationship of *Psettodes* species with other Pleuronectiformes. The Bayesian posterior probability values are displayed with each node.

3.6. Divergence Time and Diversification

The application of TimeTree analyses has illuminated the temporal aspects of the evolutionary history of *P. belcheri* and *P. erumei*. These two species have a wide divergence time during the late Eocene to early Miocene, approximately 35 to 10 million years ago (MYA) (Figure 6). Remarkably, the family Psettodidae (suborder Psettodoidei) demonstrated an early divergence of approximately 56.87 MYA from other flatfish families (suborder Pleuronectoidei) during the late Paleocene to early Eocene (Figure 6). The other Pleuronectoidei flatfish families diverged from each other between the Oligocene and early Eocene periods, with dates ranging from 29.93 MYA to 54.98 MYA (Figure 6). The distribution pattern of *Psettodes* flatfishes, specifically within the family Psettodidae, has fascinated many ichthyologists. The interannual variability in early life phenology and dispersal of flatfishes has been established as being influenced by bathymetry, changes in water salinity, oceanic temperature, and wind conditions [86,87]. Thus, the study of the evolution and diversification of marine fishes necessitates a discussion that takes into account genetic connectivity, divergent selection, as well as possible demographic and ecological opportunities [88].

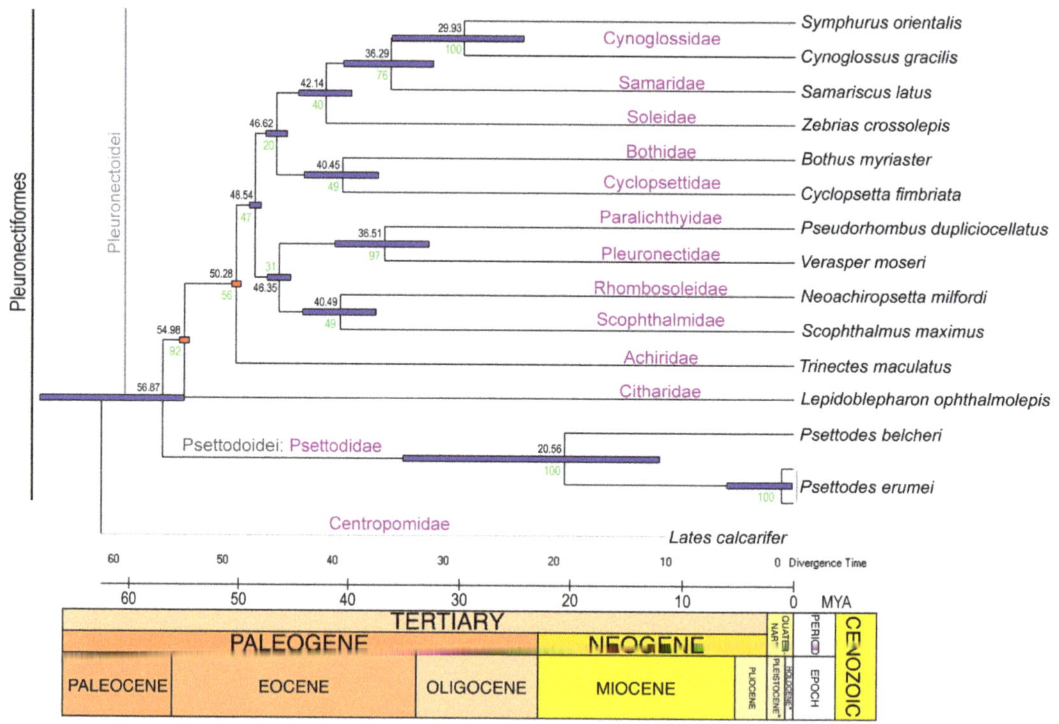

Figure 6. The maximum-likelihood-based TimeTree elucidates the approximate divergence time of the *Psettodes* species along with other *Pleuronectiformes* species. The approximate divergence times are marked displayed with each node. The red squares represent the calibration points obtained from the prior work [26].

The Miocene epoch, spanning from around 23 million to 5.3 million years ago, marked a period of substantial environmental transformation on Earth [89]. This era witnessed shifts in both climate and oceanic conditions, representing a transition between the stable warmth of the early Cenozoic and the later variable and colder conditions [90]. During the Miocene, elevated temperatures contributed to higher sea levels through the melting of polar ice and seawater expansion [91]. Furthermore, the tectonic activity in the Miocene shaped ocean basins and seafloor topography, with the collision of tectonic plates, influencing ocean currents and marine habitats. Consequently, coastlines took on different configurations than those seen today, with numerous present-day land areas submerged under shallow seas. The epoch was characterized by significant biological evolution, with the emergence of contemporary marine species lineages influencing the rich biodiversity of modern oceans [92,93]. Such diversification was prompted by factors such as varying oceanic temperatures, changing sea levels, and the availability of ecological niches.

In the context of understanding the evolutionary scenarios of *Psettodes* flatfishes, particularly following their colonization in the eastern Atlantic and Indo-West Pacific regions by the demersal lineage, a critical integration of the maximum-likelihood time-tree computation framework and marine ecological factors becomes essential. Notably, while *P. belcheri* is distributed across both the South and North Atlantic Oceans, spanning the western coast of Africa from Angola to western Sahara, *P. bennettii* is confined solely to the North Atlantic, with a distribution encompassing Gambia, Guinea, Guinea-Bissau, Mauritania, Morocco, Senegal, and Western Sahara. The sympatric speciation of these two species may be attributed to ecological selection within the pelagic environment.

Since ancient time, the hydrographic and climatic conditions in the North and South Atlantic Oceans have differed significantly due to the Coriolis effect induced by Earth's rotation (Figure 7A). In the North Atlantic, the circulation of oceanic currents results in distinct oceanic gyres, with the warm Gulf Stream current flowing northward and the cold Canary Current flowing southward [94]. The interplay of these currents, along with the North Equatorial Current, contributes to the formation of the North Atlantic gyre (Figure 7B). Conversely, the southern Atlantic Ocean features counterclockwise current circulation, dominated by the anticyclonic subtropical gyre and bounded by several major surface ocean currents (the Antarctic Circumpolar Current, the Benguela Current, the South Equatorial Current, the Brazil Current, and the Malvinas Current) (Figure 7C).

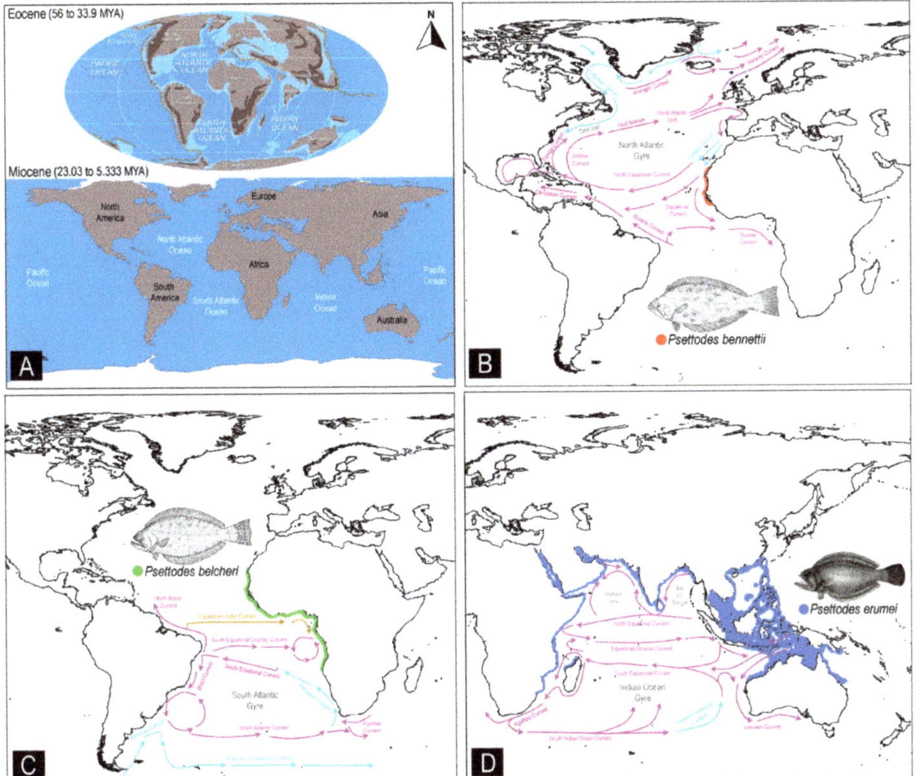

Figure 7. (**A**) Maps displaying the maritime environments during the Eocene and Miocene periods (source: Encyclopædia Britannica, Inc.). Schematic representation of the major current systems, as defined by the International Group for Marine Ecological Time Series, in the North Atlantic, South Atlantic, and Indo-West Pacific regions. This section also explores the potential diversification and colonization of primitive flatfishes: (**B**) *P. bennettii*, (**C**) *P. belcheri*, and (**D**) *P. erumei*. The illustration of the *Psettodes* species was sourced from the free media repository (Wikimedia Commons), as well as a previous study [33]. The maps were generated using the DIVA-GIS platform, utilizing IUCN range distribution data (.shp files). Additionally, the illustration of ocean currents is based on insights from a previous study [94]. Violet arrows indicate the warmer currents, while the blue arrows signify the cooler currents.

The Indian Ocean, characterized by the presence of the Arabian Sea and the Bay of Bengal, as well as warm and cold currents such as the Western Australian cold current, forms the Indian Ocean gyre. The divergence time of Psettodidae from other Pleuronectiformes during the late Paleocene to early Eocene suggests that these primitive flatfishes may have emerged in the eastern Atlantic Ocean. The currents and salinity levels of both the North and South Atlantic Oceans may have constrained the distribution of the two Psettodidae species, *P. belcheri* and *P. bennettii*, to the continental shelves of western Africa. The other congener, *P. erumei*, likely evolved during the early Miocene and colonized into the Red Sea to the Indo-West Pacific Ocean due to the formation of the Antarctic Circumpolar Current (Figure 7D). Notably, the Agulhas Current may have acted as a barrier to the distribution of *P. erumei* between the Indian Ocean and the Atlantic Ocean. Despite the differences in hydrography, the high saline outflow from the Red Sea to the Indian Ocean results in similar characteristics of these two marine environments, allowing for similar flatfish communities to thrive. However, the Indian Ocean and Western Pacific Ocean are seismically active due to the presence of tectonic plate boundaries. Ecological speciation scenarios are frequently invoked to explain such patterns, yet distinguishing between ecological adaptation and allopatric speciation remains challenging [95]. The dispersal of *P. erumei* from the Indian Ocean to the West Pacific Ocean may have been driven by the Equatorial Counter Current of the Indian Ocean gyre. Such ecological features greatly influence the evolution and adaptation of other flatfish species, including *Psettodes*, and have the potential to increase endemism in certain demersal ecosystems in marine settings. An intriguing case of rapid ecological speciation has been observed in Baltic flounder species, involving the development of a new ecological niche through a demersal spawning behavior, constituting the fastest speciation event documented for a marine vertebrate [32]. As a result, the peculiar spawning behavior attributes of *Psettodes* flatfishes, as well as ecological variables that may lead to reproductive isolation, suggest a process of ecological speciation.

3.7. Conservation Implication

Marine ecosystems remain relatively unexplored on a global scale, facing substantial gaps and challenges in terms of biodiversity assessment and the generation of genetic data [96,97]. In addition to the overarching influence of climate change, anthropogenic activities consistently pose threats to marine organisms, including flatfishes, across all oceanic regions [98–100]. Given the conservation significance of *Psettodes* flatfishes, the co-occurrence of both East Atlantic species within a portion of their range underscores the necessity for regulated harvesting within the framework of a multispecies fishery strategy. As global climate change progresses, marked alterations in environmental conditions such as elevated water temperatures, decreased oxygen levels, and modified wind patterns impact fish reproduction and recruitment on a global scale [101]. Consequently, increasing environmental variability is poised to exert negative repercussions on fish stock dynamics. Adaptations in fisheries and fishery management will likely encompass adjustments in fishing locales, target species, and the establishment of Marine Protected Areas (MPAs), highlighting the need for flexibility in both exploitation practices and management approaches. Nonetheless, predictions indicate that climate change will lead to heightened freshwater runoff in the eastern Atlantic Ocean (including basins of the Senegal, Gambia, Volta, Niger, and Congo Rivers), Indian Ocean (encompassing basins of the Zambezi, Limpopo, Indus, Ganges, Godavari, Brahmaputra, Irrawaddy, and Mekong Rivers), and western Pacific Ocean (such as the Pearl and Yangtze River Basins). This rise in runoff is anticipated to lower salinity levels, thereby constricting the distribution of numerous marine species. The reduction in salinity, driven by climate change, may exert severe selective pressure on pelagic-spawning species, potentially prompting a shift in reproductive behavior towards demersal spawning and even leading to local extinctions. Concurrently, the combined impact of eutrophication and climate change has accelerated the occurrence of hypoxia and anoxia in bottom waters where salinity conditions are conducive to pelagic spawning [102,103]. Consequently, the spawning habitat of pelagic flatfishes, currently delimited by geographic constraints, is likely to diminish in the future. This evolving scenario raises concerns about the potential local extinction of these species, underscoring the urgent need for comprehensive conservation strategies.

4. Conclusions

In conclusion, this study investigated the detailed analysis of the mitogenome of *P. belcheri*, shedding light on its genetic composition, structural organization, and evolutionary history. Furthermore, the genetic distances and phylogenetic relationships were explored, unveiling the evolutionary relationships among *Psettodes* species and their broader placement within the Pleuronectiformes order. Notably, the study identified significant genetic divergence between *P. belcheri* and *P. bennettii* in the mitochondrial COI gene, suggesting reproductive isolation despite their shared habitat. The phylogenetic relationship, divergence times, and diversification patterns were also estimated, providing insights into the emergence and distribution of *Psettodes* species in various oceanic regions. The interplay of hydrographic conditions, ocean currents, and ecological factors played crucial roles in shaping the evolutionary landscape of these flatfishes. Given the potential conservation implications, this study highlighted the need for a comprehensive understanding of marine ecosystems and the effects of climate change and anthropogenic activities on flatfish species. Conservation strategies and the establishment of MPAs were emphasized as essential components in safeguarding the diversity and sustainability of marine life, particularly species such as *Psettodes* flatfishes. In summary, the present analysis of the mitogenome and evolutionary history of *Psettodes* flatfishes offers valuable insights into the genetic attributes and possible ecological adaptations in East Atlantic and Indo-West Pacific Oceans.

Supplementary Materials: The following supporting information can be downloaded at: https://www.mdpi.com/article/10.3390/biology12101317/s1, Table S1. Dataset of Pleuronectiformes flatfishes for present phylogenetic analyses (GenBank Accessed on 15 August 2023); Table S2. Comparison of the intergenic nucleotides of two *Psettodes* species mitogenomes.; Table S3. Start and stop codons of all 13 PCGS in two *Psettodes* species mitogenomes; Figure S1. Mitochondrial COI-based Bayesian phylogeny clearly discriminate three *Psettodes* species with high posterior probabilities branch supports. The K2P genetic distances of three *Psettodes* species were overlaid on the topology.

Author Contributions: Conceptualization: S.K. and H.-W.K.; methodology: F.S.P., A.R.K. and S.R.L.; software: S.K., H.-E.K., A.R.K. and S.R.L.; validation: S.K. and H.-E.K.; formal analysis: S.K., H.-E.K. and A.R.K.; investigation: S.K., F.S.P. and F.Z.G.; resources: S.H.S. and H.-W.K.; data curation: F.S.P., S.R.L. and F.Z.G.; writing—original draft: S.K., F.S.P. and H.-E.K.; writing—review and editing: S.K., S.H.S. and H.-W.K.; visualization: S.K. and H.-W.K.; supervision: S.K. and H.-W.K.; project administration: H.-W.K.; funding acquisition: S.H.S. and H.-W.K. All authors have read and agreed to the published version of the manuscript.

Funding: This research was supported by a grant from the National Institute of Fisheries Science, Korea (Grant No. R2023002).

Institutional Review Board Statement: Not applicable.

Informed Consent Statement: Not applicable.

Data Availability Statement: The genome sequence data that support the findings of this study are openly available in GenBank of NCBI at https://www.ncbi.nlm.nih.gov, accessed on 15 August 2023, under the accession no. OR231239.

Acknowledgments: The first author (S.K.) acknowledges the Global Postdoc Program of the Pukyong National University (PKNU), Republic of Korea. The second author (F.S.P.) acknowledges the Korea International Cooperation Agency (KOICA)-PKNU scholarship for his master's program.

Conflicts of Interest: The authors declare no conflict of interest.

References

1. Fricke, R.; Eschmeyer, W.N.; Van der Laan, R. (Eds.) Eschmeyer's Catalog of Fishes: Genera, Species. Electronic Version. 2022. Available online: http://researcharchive.calacademy.org/research/ichthyology/catalog/fishcatmain.asp (accessed on 15 August 2023).
2. Nelson, J.S.; Grande, T.C.; Wilson, M.V.H. *Fishes of the World*, 5th ed.; John Wiley and Sons: Hoboken, NJ, USA, 2016; 752p.
3. Friedman, M. The evolutionary origin of flatfish asymmetry. *Nature* **2008**, *454*, 209–212. [CrossRef] [PubMed]
4. Spinner, M.; Kortmann, M.; Traini, C.; Gorb, S.N. Key role of scale morphology in flatfishes (Pleuronectiformes) in the ability to keep sand. *Sci. Rep.* **2016**, *20*, 26308. [CrossRef] [PubMed]

5. Evans, K.M.; Larouche, O.; Watson, S.J.; Farina, S.; Habegger, M.L.; Friedman, M. Integration drives rapid phenotypic evolution in flatfishes. *Proc. Natl. Acad. Sci. USA* **2021**, *118*, e2101330118. [CrossRef] [PubMed]
6. Cooper, J.A.; Chapleau, F. Monophyly and intrarelationships of the family Pleuronectidae (Pleuronectiformes), with a revised classification. *Fish. Bull.* **1998**, *96*, 686–726.
7. Eighani, M.; Paighambari, S.Y. Shrimp, bycatch and discard composition of fish caught by small-scale shrimp trawlers in the Hormuzgan coast of Iran in the Persian Gulf. *Philipp. Agric. Sci.* **2013**, *96*, 314–319.
8. Gibson, R.N.; Nash, R.D.M.; Geffen, A.J.; Van der Veer, H.W. *Flatfishes: Biology and Exploitation*, 2nd ed.; Wiley-Blackwell: West Sussex, UK, 2015; 576p.
9. IUCN. *The IUCN Red List of Threatened Species*; Version 2022-2; IUCN: Gland, Switzerland, 2022; Available online: www.iucnredlist.org (accessed on 15 August 2023).
10. Coulson, P.G.; Poad, J.A. Biological characteristics of the primitive flatfish Indian halibut (*Psettodes erumei*) from the tropical northeastern Indian Ocean, including implications of the use of incorrect aging methods on mortality estimates. *Fish. Bull.* **2021**, *119*, 168–183. [CrossRef]
11. Bergstrom, C.A.; Alba, J.; Pacheco, J.; Fritz, T.; Tamone, S.L. Polymorphism and multiple correlated characters: Do flatfish asymmetry morphs also differ in swimming performance and metabolic rate? *Ecol. Evol.* **2019**, *9*, 4772–4782. [CrossRef]
12. Tomiyama, T. Sexual dimorphism in scales of marbled flounder Pseudopleuronectes yokohamae (Pleuronectiformes: Pleuronectidae), with comments on the relevance to their spawning behaviour. *J. Fish. Biol.* **2013**, *83*, 1334–1343. [CrossRef]
13. Lee, M.Y.; Munroe, T.A. Unraveling cryptic diversity among shallow-water tonguefishes (Pleuronectiformes: Cynoglossidae: Symphurus) from the Indo-West Pacific region, with descriptions of five new species. *Zootaxa* **2021**, *5039*, 1–55. [CrossRef]
14. Betancur, R.R.; Li, C.; Munroe, T.A.; Ballesteros, J.A.; Ortí, G. Addressing gene tree discordance and non-stationarity to resolve a multi-locus phylogeny of the flatfishes (Teleostei: Pleuronectiformes). *Syst. Biol.* **2013**, *62*, 763–785. [CrossRef]
15. Bitencourt, J.A.; Affonso, P.R.A.M.; Ramos, R.T.C.; Schneider, H.; Sampaio, I. Phylogenetic relationships and the origin of New World soles (Teleostei: Pleuronectiformes: Achiridae): The role of estuarine habitats. *Mol. Phylogenet. Evol.* **2023**, *178*, 107631. [CrossRef] [PubMed]
16. Xiao, Y.; Gao, T.; Zhang, Y.; Yanagimoto, T. Demographic History and Population Structure of Blackfin Flounder (Glyptocephalus stelleri) in Japan Revealed by Mitochondrial Control Region Sequences. *Biochem. Genet.* **2010**, *48*, 402–417. [CrossRef] [PubMed]
17. Robledo, D.; Hermida, M.; Rubiolo, J.A.; Fernández, C.; Blanco, A.; Bouza, C.; Martínez, P. Integrating genomic resources of flatfish (Pleuronectiformes) to boost aquaculture production. *Comp. Biochem. Physiol. Part D Genom. Proteom.* **2017**, *21*, 41–55. [CrossRef] [PubMed]
18. Liu, K.; Zhao, S.; Yu, Z.; Zhou, Y.; Yang, J.; Zhao, R.; Yang, C.; Ma, W.; Wang, X.; Feng, M.; et al. Application of DNA barcoding in fish identification of supermarkets in Henan province, China: More and longer COI gene sequences were obtained by designing new primers. *Food Res. Int.* **2020**, *136*, 109516. [CrossRef] [PubMed]
19. Cerdà, J.; Douglas, S.; Reith, M. Genomic resources for flatfish research and their applications. *J. Fish Biol.* **2010**, *77*, 1045–1070. [CrossRef] [PubMed]
20. Chen, W.-J.; Bonillo, C.; Lecointre, G. Repeatability of clades as a criterion of reliability: A case study for molecular phylogeny of Acanthomorpha (Teleostei) with larger number of taxa. *Mol. Phylogenetics Evol.* **2003**, *26*, 262–288. [CrossRef]
21. Kawahara, R.; Miya, M.; Mabuchi, K.; Lavoue, S.; Inoue, J.G.; Satoh, T.P.; Kawaguchi, A.; Nishida, M. Interrelationships of the 11 gasterosteiform families (sticklebacks, pipefishes, and their relatives): A new perspective based on whole mitogenome sequences from 75 higher teleosts. *Mol. Phylogenetics Evol.* **2008**, *46*, 224–236. [CrossRef]
22. Wainwright, P.C.; Smith, W.L.; Price, S.A.; Tang, K.L.; Sparks, J.S.; Ferry, L.A.; Kuhn, K.L.; Eytan, R.I.; Near, T.J. The Evolution of Pharyngognathy: A Phylogenetic and Functional Appraisal of the Pharyngeal Jaw Key Innovation in Labroid Fishes and Beyond. *Syst. Biol.* **2012**, *61*, 1001–1027. [CrossRef]
23. Campbell, M.A.; Chen, W.J.; López, J.A. Are flatfishes (Pleuronectiformes) monophyletic? *Mol. Phylogenet. Evol.* **2013**, *69*, 664–673. [CrossRef]
24. Campbell, M.A.; Chen, W.J.; López, J.A. Molecular data do not provide unambiguous support for the monophyly of flatfishes (Pleuronectiformes): A reply to Betancur-R and Ortí. *Mol. Phylogenetics Evol.* **2014**, *75*, 149–153. [CrossRef]
25. Betancur, R.R.; Orti, G. Molecular evidence for the monophyly of flatfishes (Carangimorpharia: Pleuronectiformes). *Mol. Phylogenetics Evol.* **2014**, *73*, 18–22. [CrossRef] [PubMed]
26. Campbell, M.A.; Chanet, B.; Chen, J.N.; Lee, M.Y.; Chen, W.J. Origins and relationships of the Pleuronectoidei: Molecular and morphological analysis of living and fossil taxa. *Zool. Scr.* **2019**, *48*, 640–656. [CrossRef]
27. Harrington, R.C.; Faircloth, B.C.; Eytan, R.I.; Smith, W.L.; Near, T.J.; Alfaro, M.E.; Friedman, M. Phylogenomic analysis of carangimorph fishes reveals flatfish asymmetry arose in a blink of the evolutionary eye. *BMC Evol. Biol.* **2016**, *16*, 22. [CrossRef] [PubMed]
28. Campbell, M.A.; López, J.A.; Satoh, T.P.; Chen, W.-J.; Miya, M. Mitochondrial genomic investigation of flatfish monophyly. *Gene* **2014**, *551*, 176–182. [CrossRef] [PubMed]
29. Shi, W.; Chen, S.; Kong, X.; Si, L.; Gong, L.; Zhang, Y.; Yu, H. Flatfish monophyly refereed by the relationship of Psettodes in Carangimorphariae. *BMC Genom.* **2018**, *19*, 400. [CrossRef]
30. Potkamp, G.; Fransen, C.H.J.M. Speciation with gene flow in marine systems. *Contrib. Zool.* **2019**, *88*, 133–172. [CrossRef]

31. Faria, R.; Johannesson, K.; Stankowski, S. Speciation in marine environments: Diving under the surface. *J. Evol. Biol.* **2021**, *34*, 4–15. [CrossRef]
32. Momigliano, P.; Jokinen, H.; Fraimout, A.; Florin, A.B.; Norkko, A.; Merilä, J. Extraordinarily rapid speciation in a marine fish. *Proc. Natl. Acad. Sci. USA* **2017**, *114*, 6074–6079. [CrossRef]
33. Carpenter, K.E. *The Living Marine Resources of the Eastern Central Atlantic. FAO Species Identification Guide for Fishery Purposes 2016*; Food and Agriculture Organization of the United Nations: Rome, Italy, 2016.
34. Kumar, S.; Stecher, G.; Li, M.; Knyaz, C.; Tamura, K. MEGA X: Molecular Evolutionary Genetics Analysis across Computing Platforms. *Mol. Biol. Evol.* **2018**, *35*, 1547–1549. [CrossRef]
35. Bernt, M.; Donath, A.; Jühling, F.; Externbrink, F.; Florentz, C.; Fritzsch, G.; Pütz, J.; Middendorf, M.; Stadler, P.F. MITOS: Improved de novo metazoan mitochondrial genome annotation. *Mol. Phylogenetics Evol.* **2013**, *69*, 313–319. [CrossRef]
36. Iwasaki, W.; Fukunaga, T.; Isagozawa, R.; Yamada, K.; Maeda, Y.; Satoh, T.P.; Sado, T.; Mabuchi, K.; Takeshima, H.; Miya, M. MitoFish and MitoAnnotator: A Mitochondrial Genome Database of Fish with an Accurate and Automatic Annotation Pipeline. *Mol. Biol. Evol.* **2013**, *30*, 2531–2540. [CrossRef] [PubMed]
37. Perna, N.T.; Kocher, T.D. Patterns of nucleotide composition at fourfold degenerate sites of animal mitochondrial genomes. *J. Mol. Evol.* **1995**, *41*, 353–359. [CrossRef] [PubMed]
38. Chan, P.P.; Lin, B.Y.; Mak, A.J.; Lowe, T.M. tRNAscan-SE 2.0: Improved detection and functional classification of transfer RNA genes. *Nucleic Acids Res.* **2021**, *49*, 9077–9096. [CrossRef] [PubMed]
39. Laslett, D.; Canbäck, B. ARWEN: A program to detect tRNA genes in metazoan mitochondrial nucleotide sequences. *Bioinformatics* **2007**, *24*, 172–175. [CrossRef] [PubMed]
40. Thompson, J.D.; Gibson, T.J.; Plewniak, F.; Jeanmougin, F.; Higgins, D.G. The CLUSTAL_X windows interface: Flexible strategies for multiple sequence alignment aided by quality analysis tools. *Nucleic Acids Res.* **1997**, *25*, 4876–4882. [CrossRef]
41. Benson, G. Tandem repeats finder: A program to analyze DNA sequences. *Nucleic Acids Res.* **1999**, *27*, 573–580. [CrossRef]
42. Lin, G.; Lo, L.C.; Zhu, Z.Y.; Feng, F.; Chou, R.; Yue, G.H. The Complete Mitochondrial Genome Sequence and Characterization of Single-Nucleotide Polymorphisms in the Control Region of the Asian Seabass (Lates calcarifer). *Mar. Biotechnol.* **2006**, *8*, 71–79. [CrossRef]
43. He, C.; Han, J.; Ge, L.; Zhou, Z.; Gao, X.; Mu, Y.; Liu, W.; Cao, J.; Liu, Z. Sequence and organization of the complete mitochondrial genomes of spotted halibut (*Verasper variegatus*) and barfin flounder (*Verasper moseri*). *DNA Seq.* **2008**, *19*, 246–255. [CrossRef]
44. Shi, W.; Miao, X.-G.; Kong, X.-Y. A novel model of double replications and random loss accounts for rearrangements in the Mitogenome of Samariscus latus (Teleostei: Pleuronectiformes). *BMC Genom.* **2014**, *15*, 352. [CrossRef]
45. Shi, W.; Gong, L.; Wang, S.-Y.; Miao, X.-G.; Kong, X.-Y. Tandem Duplication and Random Loss for mitogenome rearrangement in Symphurus (Teleost: Pleuronectiformes). *BMC Genom.* **2015**, *16*, 355. [CrossRef]
46. Gong, L.; Si, L.-Z.; Shi, W.; Kong, X.-Y. The complete mitochondrial genome of Zebrias crossolepis (Pleuronectiformes: Soleidae). *Mitochondrial DNA A DNA Mapp. Seq. Anal.* **2016**, *27*, 1235–1236. [CrossRef] [PubMed]
47. Gong, L.; Shi, W.; Yang, M.; Li, D.; Kong, X. Novel gene arrangement in the mitochondrial genome of *Bothus myriaster* (Pleuronectiformes: Bothidae): Evidence for the Dimer-Mitogenome and Non-random Loss model. *Mitochondrial DNA A DNA Mapp. Seq. Anal.* **2016**, *27*, 3089–3092. [CrossRef] [PubMed]
48. Wei, M.; Liu, Y.; Guo, H.; Zhao, F.; Chen, S. Characterization of the complete mitochondrial genome of Cynoglossus gracilis and a comparative analysis with other Cynoglossinae fishes. *Gene* **2016**, *591*, 369–375. [CrossRef] [PubMed]
49. Si, L.-Z.; Gong, L.; Shi, W.; Yang, M.; Kong, X.-Y. The complete mitochondrial genome of Pseudorhombus dupliocellatus (Pleuronectiformes: Paralichthyidae). *Mitochondrial DNA A DNA Mapp. Seq. Anal.* **2017**, *28*, 58–59. [CrossRef] [PubMed]
50. Redin, A.D.; Kartavtsev, Y.P. The Mitogenome Structure of Righteye Flounders (Pleuronectidae): Molecular Phylogeny and Systematics of the Family in East Asia. *Diversity* **2022**, *14*, 805. [CrossRef]
51. Vences, M.; Miralles, A.; Brouillet, S.; Ducasse, J.; Fedosov, A.; Kharchev, V.; Kostadinov, I.; Kumari, S.; Patmanidis, S.; Scherz, M.D. iTaxoTools 0.1: Kickstarting a specimen-based software toolkit for taxonomists. *Megataxa* **2021**, *6*, 77–92. [CrossRef]
52. Lanfear, R.; Frandsen, P.B.; Wright, A.M.; Senfeld, T.; Calcott, B. PartitionFinder 2: New Methods for Selecting Partitioned Models of Evolution for Molecular and Morphological Phylogenetic Analyses. *Mol. Biol. Evol.* **2016**, *34*, 772–773. [CrossRef]
53. Miller, M.A.; Schwartz, T.; Pickett, B.E.; He, S.; Klem, E.B.; Scheuermann, R.H.; Passarotti, M.; Kaufman, S.; O'Leary, M.A. A RESTful API for Access to Phylogenetic Tools via the CIPRES Science Gateway. *Evol. Bioinform.* **2015**, *11*, 43–48. [CrossRef]
54. Darriba, D.; Taboada, G.L.; Doallo, R.; Posada, D. JModelTest 2: More models, new heuristics and parallel computing. *Nat. Methods* **2012**, *9*, 772. [CrossRef]
55. Ronquist, F.; Huelsenbeck, J.P. MrBayes 3: Bayesian phylogenetic inference under mixed models. *Bioinformatics* **2003**, *19*, 1572–1574. [CrossRef]
56. Letunic, I.; Bork, P. Interactive Tree Of Life (iTOL): An online tool for phylogenetic tree display and annotation. *Bioinformatics* **2007**, *23*, 127–128. [CrossRef]
57. Mello, B. Estimating TimeTrees with MEGA and the TimeTree Resource. *Mol. Biol. Evol.* **2018**, *35*, 2334–2342. [CrossRef] [PubMed]
58. Tamura, K.; Battistuzzi, F.U.; Billing-Ross, P.; Murillo, O.; Filipski, A.; Kumar, S. Estimating divergence times in large molecular phylogenies. *Proc. Natl. Acad. Sci. USA* **2012**, *109*, 19333–19338. [CrossRef] [PubMed]
59. Mello, B.; Tao, Q.; Tamura, K.; Kumar, S. Fast and Accurate Estimates of Divergence Times from Big Data. *Mol. Biol. Evol.* **2017**, *34*, 45–50. [CrossRef] [PubMed]

60. Kundu, S.; Binarao, J.D.; De Alwis, P.S.; Kim, A.R.; Lee, S.R.; Andriyono, S.; Gietbong, F.Z.; Kim, H.W. First Mitogenome of Endangered *Enteromius thysi* (Actinopterygii: Cypriniformes: Cyprinidae) from Africa: Characterization and Phylogeny. *Fishes* **2023**, *8*, 25. [CrossRef]

61. De Alwis, P.S.; Kundu, S.; Gietbong, F.Z.; Amin, M.H.F.; Lee, S.R.; Kim, H.W.; Kim, A.R. Mitochondriomics of *Clarias* Fishes (Siluriformes: Clariidae) with a New Assembly of *Clarias camerunensis*: Insights into the Genetic Characterization and Diversification. *Life* **2023**, *13*, 482. [CrossRef]

62. Wang, X.; Wang, Y.; Zhang, Y.; Yu, H.; Tong, J. Evolutionary analysis of cyprinid mitochondrial genomes: Remarkable variation and strong adaptive evolution. *Front. Genet.* **2016**, *7*, 156.

63. Ojala, D.; Montoya, J.; Attardi, G. tRNA punctuation model of RNA processing in human mitochondria. *Nature* **1981**, *290*, 470–474. [CrossRef]

64. Molina-Quirós, J.L.; Hernández-Muñoz, S.; Baeza, J.A. The complete mitochondrial genome of the roosterfish Nematistius pectoralis Gill 1862: Purifying selection in protein coding genes, organization of the control region, and insights into family-level phylogenomic relationships in the recently erected order Carangiformes. *Gene* **2022**, *845*, 146847.

65. Sato, N.S.; Hirabayashi, N.; Agmon, I.; Yonath, A.; Suzuki, T. Comprehensive genetic selection revealed essential bases in the peptidyl-transferase center. *Proc. Natl. Acad. Sci. USA* **2006**, *103*, 15386–15391. [CrossRef]

66. Satoh, T.P.; Miya, M.; Mabuchi, K.; Nishida, M. Structure and variation of the mitochondrial genome of fishes. *BMC Genom.* **2016**, *17*, 719. [CrossRef] [PubMed]

67. Shi, W.; Kong, X.Y.; Wang, Z.M.; Jiang, J.X. Utility of tRNA genes from the complete mitochondrial genome of Psetta maxima for implying a possible sister-group relationship to the Pleuronectiformes. *Zool. Stud.* **2011**, *50*, 665–681.

68. Varani, G.; McClain, W.H. The G-U wobble base pair: A fundamental building block of RNA structure crucial to RNA function in diverse biological systems. *EMBO Rep.* **2000**, *1*, 18–23. [CrossRef] [PubMed]

69. Kundu, S.; De Alwis, P.S.; Kim, A.R.; Lee, S.R.; Kang, H.-E.; Go, Y.; Gietbong, F.Z.; Wibowo, A.; Kim, H.-W. Mitogenomic Characterization of Cameroonian Endemic *Coptodon camerunensis* (Cichliformes: Cichlidae) and Matrilineal Phylogeny of Old-World Cichlids. *Genes* **2023**, *14*, 1591. [CrossRef] [PubMed]

70. Clayton, D.A. Replication of animal mitochondrial DNA. *Cell* **1982**, *28*, 693–705. [CrossRef]

71. Boore, J. Animal mitochondrial genomes. *Nucleic Acids Res.* **1999**, *27*, 1767–1700. [CrossRef]

72. Kong, X.; Dong, X.; Zhang, Y.; Shi, W.; Wang, Z.; Yu, Z. A novel rearrangement in the mitochondrial genome of tongue sole, Cynoglossus semilaevis: Control region translocation and a tRNA gene inversion. *Genome* **2009**, *52*, 975–984. [CrossRef]

73. Shi, W.; Dong, X.L.; Wang, Z.M.; Miao, X.G.; Wang, S.Y.; Kong, X.Y. Complete mitogenome sequences of four flatfishes (Pleuronectiformes) reveal a novel gene arrangement of L-strand coding genes. *BMC Evol. Biol.* **2013**, *13*, 173. [CrossRef]

74. Shi, W.; Gong, L.; Yu, H. Double control regions of some flatfish mitogenomes evolve in a concerted manner. *Int. J. Biol. Macromol.* **2020**, *142*, 11–17. [CrossRef]

75. Gong, L.; Lu, X.; Luo, H.; Zhang, Y.; Shi, W.; Liu, L.; Lü, Z.; Liu, B.; Jiang, L. Novel gene rearrangement pattern in Cynoglossus melampetalus mitochondrial genome: New gene order in genus Cynoglossus (Pleuronectiformes: Cynoglossidae). *Int. J. Biol. Macromol.* **2020**, *149*, 1232–1240. [CrossRef]

76. Wang, C.; Chen, H.; Tian, S.; Yang, C.; Chen, X. Novel Gene Rearrangement and the Complete Mitochondrial Genome of *Cynoglossus monopus*: Insights into the Envolution of the Family Cynoglossidae (Pleuronectiformes). *Int. J. Mol. Sci.* **2020**, *21*, 6895. [CrossRef] [PubMed]

77. Tinti, F.; Piccinetti, C.; Tommasini, S.; Vallisneri, M. Mitochondrial DNA Variation, Phylogenetic Relationships, and Evolution of Four Mediterranean Genera of Soles (Soleidae, Pleuronectiformes). *Mar. Biotechnol.* **2000**, *2*, 274–284. [CrossRef] [PubMed]

78. Saitoh, K.; Hayashizaki, K.; Yokoyama, Y.; Asahida, T.; Toyohara, H.; Yamashita, Y. Complete nucleotide sequence of Japanese flounder (Paralichthys olivaceus) mitochondrial genome: Structural properties and cue for resolving teleostean relationships. *J. Hered.* **2000**, *91*, 271–278. [CrossRef] [PubMed]

79. Miya, M.; Takeshima, H.; Endo, H.; Ishiguro, N.B.; Inoue, J.G.; Mukai, T.; Satoh, T.P.; Yamaguchi, M.; Kawaguchi, A.; Mabuchi, K. Major patterns of higher teleostean phylogenies: A new perspective based on 100 complete mitochondrial DNA sequences. *Mol. Phylogenet. Evol.* **2003**, *26*, 121–138. [CrossRef] [PubMed]

80. Chen, S.; Zhang, G.; Shao, C.; Huang, Q.; Liu, G.; Zhang, P.; Song, W.; An, N.; Chalopin, D.; Volff, J.N.; et al. Whole-genome sequence of a flatfish provides insights into ZW sex chromosome evolution and adaptation to a benthic lifestyle. *Nat. Genet.* **2014**, *46*, 253–260. [CrossRef]

81. Figueras, A.; Robledo, D.; Corvelo, A.; Hermida, M.; Pereiro, P.; Rubiolo, J.A.; Gómez-Garrido, J.; Carreté, L.; Bello, X.; Gut, M.; et al. Whole genome sequencing of turbot (*Scophthalmus maximus*; Pleuronectiformes): A fish adapted to demersal life. *DNA Res.* **2016**, *23*, 181–192. [CrossRef]

82. Shao, C.; Bao, B.; Xie, Z.; Chen, X.; Li, B.; Jia, X.; Yao, Q.; Ortí, G.; Li, W.; Li, X.; et al. The genome and transcriptome of Japanese flounder provide insights into flatfish asymmetry. *Nat. Genet.* **2017**, *49*, 119–124. [CrossRef]

83. Zhao, N.; Guo, H.; Jia, L.; Guo, B.; Zheng, D.; Liu, S.; Zhang, B. Genome assembly and annotation at the chromosomal level of first Pleuronectidae: Verasper variegatus provides a basis for phylogenetic study of Pleuronectiformes. *Genomics* **2021**, *113*, 717–726. [CrossRef]

84. Lü, Z.; Gong, L.; Ren, Y.; Chen, Y.; Wang, Z.; Liu, L.; Li, H.; Chen, X.; Li, Z.; Luo, H.; et al. Large-scale sequencing of flatfish genomes provides insights into the polyphyletic origin of their specialized body plan. *Nat. Genet.* **2021**, *53*, 742–751. [CrossRef]

85. Atta, C.J.; Yuan, H.; Li, C.; Arcila, D.; Betancur, R.R.; Hughes, L.C.; Ortí, G.; Tornabene, L. Exon-capture data and locus screening provide new insights into the phylogeny of flatfishes (Pleuronectoidei). *Mol. Phylogenet. Evol.* **2022**, *166*, 107315. [CrossRef]

86. Lacroix, G.; Barbut, L.; Volckaert, F.A.M. Complex effect of projected sea temperature and wind change on flatfish dispersal. *Glob. Chang. Biol.* **2018**, *24*, 85–100. [CrossRef]

87. Vaz, A.; Primo, A.L.; Crespo, D.; Pardal, M.; Martinho, F. Interannual variability in early life phenology is driven by climate and oceanic processes in two NE Atlantic flatfishes. *Sci. Rep.* **2023**, *13*, 4057. [CrossRef]

88. Reis-Santos, P.; Tanner, S.E.; Aboim, M.A.; Vasconcelos, R.P.; Laroche, J.; Charrier, G.; Pérez, M.; Presa, P.; Gillanders, B.M.; Cabral, H.N. Reconciling differences in natural tags to infer demographic and genetic connectivity in marine fish populations. *Sci. Rep.* **2018**, *8*, 10343. [CrossRef] [PubMed]

89. Ao, H.; Rohling, E.J.; Zhang, R.; Roberts, A.P.; Holbourn, A.E.; Ladant, J.B.; Dupont-Nivet, G.; Kuhnt, W.; Zhang, P.; Wu, F.; et al. Global warming-induced Asian hydrological climate transition across the Miocene–Pliocene boundary. *Nat. Commun.* **2021**, *12*, 6935. [CrossRef] [PubMed]

90. Shevenell, A.E.; Kennett, J.P.; Lea, D.W. Middle Miocene Southern Ocean Cooling and Antarctic Cryosphere Expansion. *Science* **2004**, *305*, 1766–1770. [CrossRef] [PubMed]

91. Methner, K.; Campani, M.; Fiebig, J.; Löffler, N.; Kempf, O.; Mulch, A. Middle Miocene long-term continental temperature change in and out of pace with marine climate records. *Sci. Rep.* **2020**, *10*, 7989. [CrossRef] [PubMed]

92. Herbert, T.D.; Lawrence, K.T.; Tzanova, A.; Peterson, L.C.; Caballero-Gill, R.; Kelly, C.S. Late Miocene global cooling and the rise of modern ecosystems. *Nat. Geosci.* **2016**, *9*, 843–847. [CrossRef]

93. Fenton, I.S.; Aze, T.; Farnsworth, A.; Valdes, P.; Saupe, E.E. Origination of the modern-style diversity gradient 15 million years ago. *Nature* **2023**, *614*, 708–712. [CrossRef]

94. O'Brien, T.D.; Lorenzoni, L.; Isensee, K.; Valdés, L. What are Marine Ecological Time Series telling us about the ocean. In *A Status Report IOC-UNESCO 2017*; IOC Technical Series, No. 129; IOC-UNESCO: Paris, France, 2017; 297p.

95. Teske, P.R.; Sandoval-Castillo, J.; Golla, T.R.; Emami-Khoyi, A.; Tine, M.; von der Heyden, S.; Beheregaray, L.B. Thermal selection as a driver of marine ecological speciation. *Proc. Biol. Sci.* **2019**, *286*, 20182023. [CrossRef]

96. Costello, M.J.; Coll, M.; Danovaro, R.; Halpin, P.; Ojaveer, H.; Miloslavich, P. A Census of Marine Biodiversity Knowledge, Resources, and Future Challenges. *PLoS ONE* **2010**, *5*, e12110. [CrossRef]

97. Hutchings, J.A. Collapse and recovery of marine fishes. *Nature* **2000**, *406*, 882–885. [CrossRef] [PubMed]

98. Law, K.L.; Morét-Ferguson, S.; Maximenko, N.A.; Proskurowski, G.; Peacock, E.E.; Hafner, J.; Reddy, C.M. Plastic Accumulation in the North Atlantic Subtropical Gyre. *Science* **2010**, *329*, 1185–1188. [CrossRef] [PubMed]

99. Crego-Prieto, V.; Martinez, J.L.; Roca, A.; Garcia-Vazquez, E. Interspecific Hybridization Increased in Congeneric Flatfishes after the Prestige Oil Spill. *PLoS ONE* **2012**, *7*, e34485. [CrossRef]

100. Feekings, J.; Bartolino, V.; Madsen, N.; Catchpole, T. Fishery Discards: Factors Affecting Their Variability within a Demersal Trawl Fishery. *PLoS ONE* **2012**, *7*, e36409. [CrossRef] [PubMed]

101. Pörtner, H.O.; Roberts, D.C.; Poloczanska, E.S.; Mintenbeck, K.; Tignor, M.; Alegría, A.; Craig, M.; Langsdorf, S.; Löschke, S.; Möller, V.; et al. *Climate Change 2022: Impacts, Adaptation and Vulnerability. Contribution of Working Group II to the Sixth Assessment Report of the Intergovernmental Panel on Climate Change*; Cambridge University Press: Cambridge, UK; New York, NY, USA, 2022; pp. 3–33.

102. Carstensen, J.; Andersen, J.H.; Gustafsson, B.G.; Conley, D.J. Deoxygenation of the Baltic Sea during the last century. *Proc. Natl. Acad. Sci. USA* **2014**, *111*, 5628–5633. [CrossRef] [PubMed]

103. Nissling, A.; Wallin, I. Recruitment variability in Baltic flounder (*Platichthys solemdali*)—Effects of salinity with implications for stock development facing climate change. *J. Sea Res.* **2020**, *162*, 10191. [CrossRef]

MDPI

St. Alban-Anlage 66

4052 Basel

Switzerland

www.mdpi.com

Biology Editorial Office

E-mail: biology@mdpi.com

www.mdpi.com/journal/biology